Advances in Circuits and Systems

SELECTED PAPERS ON

COMPUTER-AIDED DESIGN OF VERY LARGE SCALE INTEGRATED CIRCUITS

EDITED BY

ALBERTO L. SANGIOVANNI-VINCENTELLI

Department of Electrical Engineering and Computer Sciences
University of California, Berkeley

A SERIES PUBLISHED FOR THE IEEE CIRCUITS AND SYSTEMS SOCIETY

IEEE PRESS

The Institute of Electrical and Electronics Engineers, Inc., New York

PRINTED IN THE UNITED STATES OF AMERICA

IEEE Order Number: PC0224-6

Library of Congress Cataloging-in-Publication Data

Advances in circuits and systems.

"A series published for the IEEE Circuits and Systems Society."
Includes bibliographies.
1. Integrated circuits—Very large scale integration—Design and construction—Data processing.
2. Computer-aided design—Data processing.
I. Sangiovanni-Vincentelli, Alberto. II. IEEE Circuits and Systems Society.
TK7874.A337 1987 621.381'73 87-22561

ISBN 0-87942-227-0 (pbk.)

Contents

Part I: Introduction 1

Computer-Aided Design for VLSI Circuits, *A. R. Newton and A. L. Sangiovanni-Vincentelli* (*Computer,* April 1986) 3

Part II: University Programs: An Historical Perspective

A Structured Design Methodology and Associated Software Tools, *S. Trimberger, J. A. Rowson, C. R. Lang, and J. P. Gray* (*IEEE Transactions on Circuits and Systems,* July 1981) 25

VLSI Design Automation Activities at M.I.T., *J. Allen and P. Penfield, Jr.* (*IEEE Transactions on Circuits and Systems,* July 1981) 41

Stanford Overview in VLSI Research, *R. Dutton* (*IEEE Transactions on Circuits and Systems,* July 1981) 50

Design Aids for VLSI: The Berkeley Perspective, *A. R. Newton, D. O. Pederson, A. L. Sangiovanni-Vincentelli, and C. H. Sequin* (*IEEE Transactions on Circuits and Systems,* July 1981) 62

Design Aids for VLSI: A Perspective Revisited, *A. R. Newton, D. O. Pederson, and A. L. Sangiovanni-Vincentelli* (*IEEE Design and Test of Computers,* April 1985) 76

A Design Methodology and Computer Aids for Digital VLSI Systems, *S. W. Director, A. C. Parker, D. P. Siewiorek, and D. E. Thomas, Jr.* (*IEEE Transactions on Circuits and Systems,* July 1981) 86

Integrated CAD, CAM, and CAT of VLSI Circuits and Systems: The CMU Perspective, *S. W. Director, W. Maly, R. A. Rutenbar, J. P. Shen, D. P. Siewiorek, A. J. Strojwas, and D. E. Thomas* (*IEEE Design and Test of Computers,* June 1985) 97

Part III: The Industry Programs: USA

A VLSI Design System, *A. K. Bose, B. R. Chawla, and H. K. Gummel* (*IEEE Int. Symp. Circuit Syst.,* June 1983) 111

The VLSI Design Automation Assistant: From Algorithms to Silicon, *T. J. Kowalski, D. J. Geiger, W. H. Wolf, and W. Fichtner* (*IEEE Design and Test of Computers,* August 1985) 117

VLSI CAD Research at MCC, *T. Rhyne* (*IEEE Design and Test of Computers,* November 1984) 127

Chipbuster VLSI Design System, *W. McCalla, B. Infante, D. Brzezinski, and J. Beyers* (*IEEE Int. Conf. CAD,* 1986) 131

Part IV: The Industry Program: Japan

CAD Systems for VLSI in Japan, *T. Sudo, T. Ohtsuki, and S. Goto* (*Proceedings of the IEEE,* January 1983) . 135

An Integrated Design Automation System for VLSI Circuits, *O. Karatsu, T. Hoshino, M. Endo, H. Kitazawa, T. Adachi, and K. Ueda* (*IEEE Design and Test of Computers,* October 1985) 150

Author Index 161

Editor's Biography 163

Part I: Introduction

Moore's law predicted the doubling of the number of devices integrated on a chip every two years. This empirical law has been surprisingly accurate for a number of years and only now shows a decay due to the physical limits to integration that are by now almost reached. In 1978, the largest chips had 100K transistors; in 1987, the largest chips in production have a number of devices that is above 1M. Technology has allowed designers to integrate, on a single chip, more and more complex functions.

However, the productivity of an integrated circuit designer would have to increase by about the same speed as the number of devices. If this increase of productivity is not achieved, the design costs will not be within reasonable limits and the design of complex systems will be unfeasible.

The increase of productivity can be accomplished only with the use of sophisticated design aids. This problem is very well understood both in the industrial world and in academia. In fact, during the past eight years, the increase in university and industrial programs in Computer-Aided Design (CAD) for Very Large Scale Integration (VLSI) has been spectacular.

The two research cooperatives set up by American companies in the computer and semiconductor industry, the Semiconductor Research Corporation (SRC) and the Micro-electronics and Computer Technology Corporation (MCC), each have a large program in CAD. The Defense Advance Research Program Agency supports a number of important research efforts in CAD and the National Science Foundation has allocated more and more funds to fields in the area of design methodologies and computer engineering.

During the same time-frame, the CAD vendor community has grown from a fairly small number of producers to a large one, with the market size ballooning from a few million dollars to five hundred million in 1984 and an estimated 1.3 billion in 1986. The potential market is much larger since it can be demonstrated that the penetration of design tools in the electrical engineering community is far from being complete. The focus of product development has been shifting from less complex tasks such as layout editing and schematic data capture to more and more complex tasks such as simulation, design rule checking, automatic placement and routing.

During the years of tumultuous growth, little attention has been paid to *system integration*. The tools used by IC and PCB designers are often incompatible; they use different data bases and different user-interfaces. I believe that the key concept of the next few years both in university and in industry will be system integration, i.e., how to store and manage large data and how to integrate tools coming from different vendors and from internal developers.

The goal of this collection of papers is to review the *CAD system* concepts as they have been forming from 1981 to now.

The introductory paper by Newton and Sangiovanni-Vincentelli puts in perspective the field of CAD for VLSI describing the major components of an integrated system and linking design methods with software tools.

In the following group of papers, the design systems developed at universities are described. The first set of papers in this group appeared in a special issue on CAD for VLSI of the Transactions on CAD published in July 1981. These papers gave the foundations of the work carried out in subsequent years in this fast moving field as well as material for business plans of several companies in CAD.

The University of California at Berkeley and Carnegie Mellon University later became CAD centers of excellence for the Semiconductor Research Corporation and expanded their program considerably. The papers that appeared in 1985 in the Design and Test Magazine describe the evolution of the program of these two schools and point to the future of design systems.

The third group of papers includes the work carried out to build large design systems in industry. Emphasis has been given to papers that describe entire systems and their architectures rather than the single components. Thus, although this selection is by no means exhaustive of the CAD activities in industry, it represents an interesting sample.

The last group of papers describe the state of the art of design systems in Japan. There is a common belief that Japanese industries are definitely less advanced than American industries in the use and development of sophisticated design aids. I personally disagree with this assessment and the reason for including these papers is to make the engineering audience better aware of the excellence of the Japanese effort in CAD, especially in the development of advanced tools for placement and routing, testing and logic synthesis.

Alberto Sangiovanni-Vincentelli
Dept. of EECS
University of California, Berkeley

CAD for VLSI

1. Introduction

A. R. Newton and A. L. Sangiovanni-Vincentelli, CAD for VLSI, *Computers,* **Special Issue on Design Automation, pp. 38-60, April 1986.**

2. University Programs: an Historical Perspective

S. Trimberger et al. A Structured Design Methodology and Associated Software Tools, *IEEE Trans. on CAS,* **pp. 618-634, July 1981.**

J. Allen at al. VLSI Design Automation Activities at MIT, *ibid,* **pp. 645-654, July 1981.**

R. Dutton, Stanford Overview in VLSI Research *ibid* **pp. 654-666, July 1981.**

A. R. Newton et al., Design Aids for VLSI: The Berkeley Perspective, *ibid,* **pp. 666-681, July 1981.**

A.R. Newton et al. Design Aids for VLSI: A Perspective Revisited, *IEEE Design and Test,* **pp. 106-115, Apr. 1985.**

S. Director et al. A Design Methodology and Computer Aids for Digital VLSI Systems *ibid,* **pp. 634-645, July 1981.**

S. Director et al., Integrated CAD, CAM, and CAT of VLSI Circuits and Systems: The CMU Perspective, *IEEE Design and Test,* **pp. 87-100, June 1985.**

3. The Industry Programs: USA

A. Bose et al. A VLSI Design System, *Proc. of ISCAS 83,* **June 1983.**

T.J. Kowalski, D.J. Geiger, W.H. Wolf, and W. Fichtner, "The VLSI Design Automation Assistant: From Algorithms to Silicon" *IEEE Design and Test,* **pp. 33-43, August 1985.**

T. Rhyne, VLSI CAD Research at MCC, *IEEE Design and Test,* **pp.104-107, Nov. 1984.**

W. McCalla, et al., "ChipBuster VLSI Design System" *Proc. 1986 Int. Conf. on CAD,* **pp. 20-23, 1986.**

3. The Industry Program: Japan

S. Sudo et al. CAD Systems for VLSI in Japan *Proc. of IEEE,* **pp. 129-146, Jan. 1983**

O. Karatsu et al. An Integrated Design Automation System for VLSI Circuits, *IEEE Design and Test,* **pp. 17-26, Oct. 1985.**

Computer-Aided Design for VLSI Circuits

A. R. Newton and A. L. Sangiovanni-Vincentelli
University of California, Berkeley

Performance, cost, features, and time to market are the factors that separate winners from losers in the ASIC marketplace. Not coincidentally, these same pressures affect vendors of IC CAD systems, too.

Computer aids have been used for both the design and verification of electronic systems for many years prior to the introduction of commercial integrated circuits in the early 1960s. Such tools have found their way into virtually every aspect of the design of such systems, from IC process technology to the design of complex computer architectures. However, it is the IC and the complex electronic systems the IC has made possible that have made computer aids an indispensable part of the design of an electronic circuit or system. Not only are computer aids necessary for both the design and verification of integrated circuits today but, as the semiconductor processing technologies mature, computer aids will soon also provide key proprietary advantages as semiconductor and system design houses vie for the promising application-specific IC (ASIC) market of the next decade.

The pivotal technologies in future IC CAD systems include tools for *IC synthesis* tasks such as placement and routing, combinational and sequential logic synthesis, and architectural design; *design system management* tools, for the management of design versions and alternatives in a distributed computing environment, data dependency management, and efficient and flexible interfaces to new tools; and new *verification tools,* including physical and electrical rule checking, simulation, and formal verification techniques. In many cases, the new tools will take advantage of new multiple processors to improve their performance or will use evolving heuristic programming technologies, such as rule-based expert systems, to improve flexibility and encourage the evolution of the tool. With the recent rapid increase in compute power per dollar we have seen at the engineer's desk, the next few years promise spectacular progress in all of these areas.

In this article we review the CAD tools and techniques used to support the most common design styles. Nowadays, the field of CAD for IC design is very broad and it is not possible to cover all aspects of IC/CAD in a single article. For that reason, we focus on the techniques critical to both custom and ASIC design, the directions of present research and development for these areas, and future trends. A more detailed description of all areas, including an extensive reference list, may be found in Newton.[1] While the areas of testing and semiconductor process and device design involve extensive use of CAD and are also key technologies for IC design, these areas

Reprinted from *COMPUTER*, pp. 38–60, April 1986.

are dealt with in detail in other articles in this issue and for that reason are not reviewed here.

An historical perspective

Since the advent of the first IC, the evolution of computer aids for IC design has occurred in an ad hoc manner. In most cases, computer programs have been written to solve specific problems as they have arisen and very few truly integrated Computer-Aided Design (CAD) systems exist for the design of ICs. Most CAD systems currently in use for the design of complete ICs consist of a loose collection of programs, requiring a large collection of data formats and often requiring manual intervention to move from one program or computer system to another.

The first digital ICs were available commercially in the early 1960s, and in retrospect it is surprising how little the computer was used in the design of ICs prior to 1980. Early circuits were sufficiently small that mask patterns could be drawn by hand on rubylith, and then photographically reduced to generate the IC masks directly. However, for the verification of the function of the circuit, simulators proved quite useful. Hence initial work in the mid-1960s focussed on the development of semiconductor device analysis and circuit analysis techniques. These circuit simulators were originally developed for the analysis of nonlinear, temperature, and radiation effects in discrete circuits and it was not until the early 1970s that circuit simulators suitable for IC analysis became generally available.

As the complexity of the circuits increased, industry turned to the computer to store integrated circuit mask layout data (the arrangements of polygons that would be used to define transistors and interconnect on the final chip) and to produce the masks required for manufacture. Systems for layout digitization (where the layout is first drawn by hand on sheets of mylar and then entered into the computer using a tablet and a puck) and interactive correction of the layout data found extensive use by the early 1970s. However, it was not until the mid-1970s that programs for checking the physical layout rules for the circuit (LRC) began to find widespread use. These programs process the geometric descriptions of the layout and check to make sure that layout rules, such as minimum spacing between adjacent polygons or required enclosure of one polygon by another, are met.

By 1975 it became clear that computer aids were a necessity in the design of complex integrated circuits, both for physical and for functional design and verification. Until then, the layout of an IC and its transistor-level schematic diagram had been quite separate. In the late 1970s, computer programs became available for such tasks as extraction of transistor-level schematics from IC artwork data (recognizing transistors and interconnections from patterns in the artwork data), connectivity verification (comparing the transistors and their connections expressed in a schematic diagram with the connections extracted from the artwork data), and even extraction of gate-level netlists from the transistor list (recognizing the collections of transistors that form common logic gates). These programs were loosely coupled and were often incompatible with one another. All of the early tools were developed with a "batch" computing environment in mind; none of them addressed the problem of design data management (other than for the data they dealt with directly); and tools from different vendors typically used different input and output formats. The task of coordinating the tools and integrating them into a particular design flow fell largely to the IC house's central, or corporate, CAD group; and this task has traditionally been responsible for a lot of their headaches. In fact, even today the only fully integrated CAD systems that are in general use for the design of complex ICs are those for some highly specialized design approaches, such as the standard-cell and gate-array design styles.

Prior to, and in parallel with, the development of computer aids for IC design, a great deal of work has been ongoing to aid the digital system designer, particularly as applied to printed circuit board design using standard components. In particular, algorithms and programs for the optimal placement and routing of cells, logic simulation techniques, and test grading have resulted in sophisticated design packages. As the complexity of ICs and IC-based systems increased, these two worlds began to merge in the late 1970s and early 1980s. In addition, the IC industry saw the introduction of the first personal workstation-based interactive mask layout and schematic entry systems. The rapid drop in price/performance provided by these systems has had a dramatic effect on the IC design community, and has caused a great deal of confusion as well! The additional advantages of predictable response time, communication among designers, and improved user interface to the CAD tools, and the wide range of possible price/performance options has accelerated the acceptance of these systems. But the advent of the workstation had not been without its drawbacks. Early systems were often clumsy to use and did not live up to many of their claims.

Over the past few years, however, workstation-based CAD systems have improved dramatically and are now available on machines from a standard personal computer to advanced, color-display workstations with the power that a superminicomputer had just a few years ago. However, while the popular press promises the designer a "personal, integrated workstation," such an environment is still not cost-effective. The more useful workstations today are still relatively expensive—too expensive to provide a workstation per designer in most companies. Bringing the cost of these workstations down is an important challenge to the workstation vendor. In some cases the design systems run on proprietary hardware, where a full range of cost/performance and multi-user systems is not available, or use proprietary networking technology, making it difficult to integrate such systems smoothly into a complete office environment. Most engineers continue to require a computer terminal or PC with which to perform their work. However, there is no question that workstations for CAD are here to stay. With such rapid improvements in hardware platforms, it has been difficult to keep the CAD tools portable enough to keep up with the changes. Workstation-based CAD vendors have either committed their systems to a particular manufacturer or have chosen a portable operating system, such as Unix, in which to develop their tools. In that way, as long as Unix is available on the new hardware, the job of porting their software has been made considerably easier. Other techniques that have been used to ease portability while minimizing loss in efficiency include the use of portable programming environments such as Mainsail and Lisp.

With the advent of low-cost, high-resolution graphics, another portability issue has emerged—that of user interface. Early implementations of window-managed user interfaces have involved customization of the code down to the assembly language level. In some cases, hardware manufacturers have provided efficient interfaces that have a common program-level interface to a variety of display devices. However, the interface is often proprietary or, if defined in terms of a graphics standard such as CORE or GKS, the interface is often too slow or inadequate for a window-managed environment. Recent University developments, such as MIT Project Athena and the CMU Spice project, may provide the basis for future portability in this area.

The relationship between CAD and design

Since the first CAD tools were applied to the IC design process, designers have complained that CAD lags behind design. There are those who feel that such a situation is inevitable since, once a designer finds a problem for the CAD engineer to tackle, it takes some time for the appropriate CAD tool to be written, debugged, and documented. By that time, the designer has "moved on" to new designs and, with them, new problems for CAD. In the early days, designers were able to "work around" problems with the CAD tools at their disposal. Today, however, the job just can't be done without CAD.

For an ASIC design environment, a simplified designer's view of the role of CAD is illustrated in Figure 1. The designer, driven by the marketing need for a circuit that meets a particular cost, performance, and functional specification, works with system, logic, and circuit architectures to create a chip design. In that process, the CAD tools are used to evaluate tradeoffs and alternative designs, to construct specific circuit components, and to assemble and interconnect the components to form the final chip patterns. Once the IC mask patterns have been assembled, CAD tools are used to check the final layout and prepare it for the automated manufacture of masks. As the competition for designs increases, driven by the increasing number of companies in the ASIC business and by the high capital cost of a modern IC processing facility, there is increasing demand for designers to be able to differentiate their IC product from that of their competitors. Higher performance, lower cost, more features, or a faster time to market are all major factors which differentiate IC products in the ASIC marketplace. In the past, different companies have been able to provide such product differentiation through their IC fabrication technologies: the ability to pack more transistors on a given chip or to provide a higher switching speed per gate drove the designs and their advantage in the marketplace. However, the silicon planar process technology is maturing rapidly—significant gains in performance and density are becoming increasingly expensive and many companies are resorting to "joint ventures," often with former competitors in the United States, Europe, or Asia, to maintain their position in IC process technologies. Because of this decrease in the relative competitive advantage obtained from process technology, semiconductor companies and silicon foundries must emphasize other aspects of the design process if they are to compete effectively for the ASIC market. The two avenues available are in architecture—hiring better designers and system architects than their competitors, a task often difficult and certainly expensive—and in CAD.

Unfortunately, the perspective shown in Figure 1 is simplistic and incomplete. As illustrated in Figure 2, the design task involves three major components: CAD programs, support for specific design styles, and support for component libraries. On one hand, a lack of CAD support in any of these areas may result in a significant reduction in the competitiveness of the designers' final product. On the other hand, a significant proprietary advantage in any or all of these areas will maintain a company's position as a force in the marketplace.

The CAD tools area may be subdivided further into three areas: tools for circuit design or synthesis, tools for circuit verification, and tools for design data management and for managing the flow of the design process.[2] This last category of tools is of particular importance, for it provides the foundation on which the CAD system is build. If the design framework is inflexible and cannot adapt to new tools, new design styles, and changes in process technology, then the design system will soon become obsolete. It is also important in maintaining a competitive advantage in the design process since an open framework, which supports the addition of locally developed as well as commercial tools, can be used to provide a proprietary difference between one system and a competitor's system. For that reason, this important area is reviewed below.

As mentioned earlier, since design verification has received a great deal of attention in the past, most of the techniques and tools are relatively mature. The major research issues in the verification area concern improving the performance of the tools for large designs without sacrificing reliability of the results. The use of special-purpose hardware and new computer architectures is playing a major role here. In addition, new algorithms are being developed which exploit the properties of large circuits, such as the repetitive use of circuit structures. Many of the new techniques, while novel and requiring large engineering investments to achieve their potential, are relatively easy to duplicate and therefore cannot provide the foundation for a proprietary technology.

On the other hand, with a few notable exceptions, design synthesis systems for ASIC designs are far less mature and large gains in circuit efficiency and design time are still to be had. In addition, many of the state-of-the-art synthesis techniques involve far more "inspiration" than "perspiration" and, as a result, can form the basis of a proprietary and differentiating technology for ASIC design. Techniques for efficient synthesis (system design, register-level design, logic design, placement, routing, and array compilation) will provide a major focus for both university research and industrial competition over the coming years and, for that reason, they are emphasized in this article.

The second important area for differentiation is that of CAD support for different design styles, in particular for floorplan style (gate-array, standard-cell, macrocell, etc.) versus so-called "random" logic—that portion of a design that, unlike RAM, ROM, or datapath, cannot be cast into a straightforward and efficient regular layout style. Since designers are finding improved circuit design styles and layout styles continuously, it is essential that a CAD system be able to support a variety of design styles and adapt easily to new development in these areas.

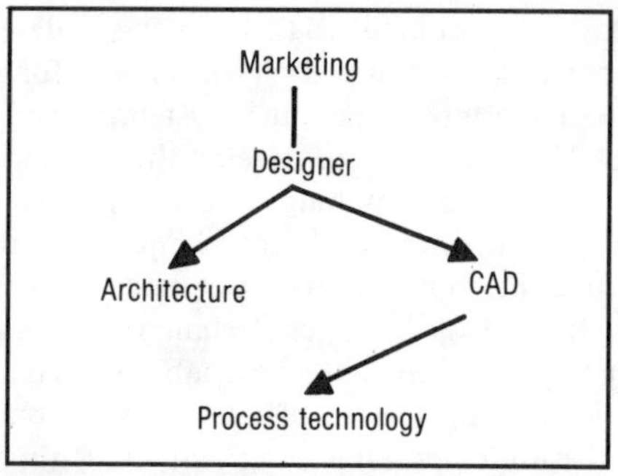

Figure 1. A simplified view of the ASIC designer's task.

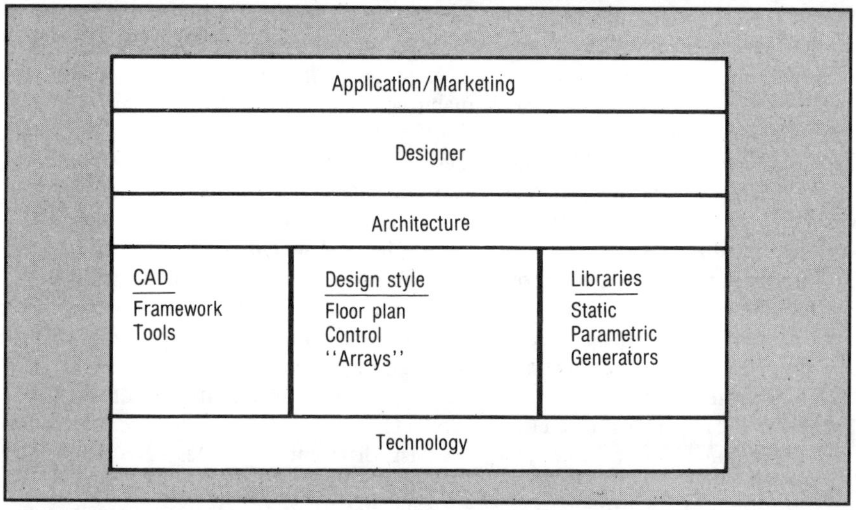

Figure 2. A more detailed classification of the ASIC designer's tools.

Managing CAD development

For many years, most companies have worked with a central, or corporate, CAD group supporting the entire company's, or business unit's, CAD needs. Occasionally, the CAD is distributed and a small team is responsible for all CAD aspects of a single design project. In either case, a major dilemma is always the "buy versus build" decision. Most companies would buy if they could and only build what they have to; but it is a common problem to find a tool that "isn't quite right" or to buy the tool only to find it does not perform as advertised.

Since the late 1960s, all large IC companies and most small companies have relied on the central CAD group to supply CAD support for a large number of design teams. It was often felt that since the computers themselves were expensive, often centralized corporate-level resources, the group which supports design aids on such a facility should also be centralized. The group would support common tools, such as circuit simulators and layout-rule checking programs, for the entire company. They would also develop new and innovative tools to support the particular design styles favored by their designers. Unfortunately, this often leads to a computer-center style of relationship between the IC design teams and the central CAD group. On the one hand, the designers felt that by having access to the CAD code themselves, with "just a few minor changes" a tool might do the job they wanted. On the other hand, the central CAD group was well aware that if they released the source to each design group, they would be asked—no, they would be expected—to support the resulting variant tools—tools they had no part in creating.

There have been many battles between a design team and the central CAD group. In recent years, design groups have often taken matters into their own hands by forming small CAD support teams for each design under way. These teams are responsible for bringing together the appropriate CAD technology for the design method in use and for building a CAD system to support the design. This generally involves obtaining tools from the central CAD group, buying some tools, and, generally as a last resort, building some tools of their own. While this approach often leads to a satisfactory solution for the design in progress, unless the design framework is managed carefully, it often leads to a CAD system of tools which is difficult to support and therefore cannot be used in other, similar design projects.

In many ways, the relationship described above is a no-win situation from both sides. However, a central CAD group has a very important role in a company that has a number of ongoing, state-of-the-art design activities. It is certainly the best place to support common, basic tools. It can also play an important role in the dissemination of information about new tools or a new use of an old tool. When one design team completes a successful design project, the central CAD group can often follow up on the use of CAD tools in that project and make the successes and pitfalls of the design style available to other, new IC design projects. Today, such information is often lost, rediscovered with each new design, or carried to the next project by the senior designers in the team.

But perhaps the most important role a central CAD group can play is in the maintenance of the design framework. In its simplest terms, this involves the

Finally, all ASIC systems require a library of primitive components, whether they be individual transistors, logic gates, or entire subsystems. These library cells may be invariant designs, such as the traditional standard-cell or gate-array building blocks; parameterized cells, such as those in the libraries offered by the silicon compiler companies; or sophisticated, module-generator-based libraries, where different cell topologies are generated on the fly as a function of the user's input description. If a designer is to compete in the competitive ASIC marketplace of tomorrow, he must be able to customize his CAD design environment in all three of these areas.

The CAD framework

Design data management. In the 1960s, database management was not an issue for IC design—the entire database often consisted of a box of punched cards and a hand-drawn roll of mylar that the designer carried with him. In the early and mid 1970s, as circuit complexity increased, proprietary and tool-dependent data formats were developed to represent particular classes of design data, such as mask layout data and transistor or gate-level netlist descriptions. Since most CAD programs were developed independent of one another and had their own input formats, coupling them together to form an integrated system for IC design involved writing translators to and from each program. In the worst case, for N programs, $N^2 - N$ translators would be needed. However, the CAD tools were evolving and their input formats were changing along with them. As a result, it was often necessary to keep a family of translators for each program, with each translator corresponding to a different version of the input data format. Maintaining such a family of translators soon became a CAD manager's nightmare! The number of translators can be reduced to a worst case of $2N$ by choosing a common, central format and translating to and from that format. A number of de facto standard formats evolved in the late 1970s to meet the need for a common format and different companies standardized internally on one format for each class of data. In the mid and late 1970s, a number of public-domain standard formats were adopted and the most successful examples are the CIF (Caltech Intermediate Form) for mask-level layout descriptions and SDL (Stanford Design Language) for gate-level netlist descriptions. While such formats provide a consistent way of storing the design data, there is no support for managing the data: Which copy is the latest version? Has the layout been changed since the schematic diagram

specification and support of standard interfaces between tools. Even today, these interfaces are often textual or binary interchange formats, such as the Calma GDS2 format for geometric data or the SPICE2 circuit simulation program input format for netlist data. Often there are many, occasionally synonymous, formats in use within a large organization—again resulting in a support nightmare for the central CAD group. New formats, such as EDIF and the VHDL, promise to reduce the complexity of this problem. However, a circuit design framework can be far more powerful than a standard text format. It can include both common data management tools that support interactive design styles, as well as a common user interface to all tools.

In the late 1970s and early 1980s, a number of new companies were formed, often by frustrated designers who saw the opportunity that low-cost computers provided to address these problems. These companies focussed on data management, integrated user interface, and a selection of supported tools. Often they focussed their efforts on a particular segment of the market. Logic schematic entry and associated logic-level tools proved to be the most successful as they addressed both custom and semicustom IC designers and board designers alike, therefore providing a much larger market to sustain growth. In contrast to earlier offerings, these systems provided interactive, graphical entry of the design data and they promised an interface to a wide variety of tools, as well as the ability to store and retrieve design data.

These companies have been joined by many others and the systems have broadened in terms of the design data they can deal with. At the 22nd Design Automation Conference in June, 1985, over fifty companies were offering CAD systems that ran on machines from personal computers to high-end mainframes. The systems showed a correspondingly broad range of capabilities. From the advertising literature, these new, commercial systems looked like fully supported solutions to the design managers' CAD dilemma. Unfortunately, a number of these systems did not live up to their promises and the designers soon discovered that the problems they had working with the CAD vendor were almost the same as those they had with their internal CAD group. Only now, they had less control over the solutions!

While a CAD group plays an important role in the design process, care must be taken to avoid the CAD tool under development becoming the focus of the group, rather than a successful chip on first silicon being the ultimate goal of the CAD group as well as the design group. Often, a central team is more concerned with squeezing the last microsecond out of a logic simulation run, rather than making the designer interface easier to use or offering the analysis options of most use during the design. Management plays the most important role here, by allocating some chip design responsibility to the CAD group. For example, by assigning a specific CAD individual to the task of supporting interactive routing aids for a specific, custom IC designer, a synergistic effect is achieved. The designer is pleased with the support he or she is receiving, while the CAD engineer sees the direct benefit of his or her work. The resulting tools designed in this way are generally far more useful than those designed from an initial "specification."

was updated? If I change this cell, which cells that use it will be affected? It is the ability to answer such queries that differentiates a true data management system from a simple data repository.

In parallel with this work, a number of companies developed conventional database systems for managing their IC design data. Often these companies were the large computer or system houses that had experience with the use of database management techniques for discrete digital system design. These record-oriented database management systems (DBMS) were developed to manage IC parts inventories, part location on standard printed circuit (PC) boards, and the connections among IC pins necessary to implement the logic schematic. These lists of connections, used to guide wire-wrap or stitch-weld machines, are generally referred to as *netlists*. While these companies found that the application of conventional relational, network, or hierarchical database management techniques was effective for structured, semicustom design styles like gate-array and standard-cell, these approaches were not successful for custom design styles or in situations where the underlying process technology and design style was evolving rapidly.

However, the same rapid increases in complexity that makes the use of conventional database management techniques difficult has made the need for a unified data management system critical, especially for full-custom or structured-custom design styles. No longer is the entire design process the responsibility of a small, tight-knit group; rather, teams of system designers, logic designers, circuit designers, and layout technicians must all work together and share the vast amount of data representing a modern IC-based system. For a state-of-the-art custom IC containing around 300,000 transistors, over 5G bytes of data is required to store the design. Within the next five years, the volume of data may grow by an order of magnitude!

The representations of IC design data, such as mask layout, schematic diagrams, documentation, and simulator input and output are quite diverse and new representations are being developed continuously. This evolution requires a flexible data management system which can adapt readily to new design methods. The use of conventional database management technology in this area has met with limited success.[3] The major limitations here involve problems specific to engineering applications, while today's database technology has often evolved from the business area. While specific differences between the needs of the business world and the needs of IC designers can be used to illustrate the problems here, simply solving these problems may not be sufficient. Rather, a completely different approach to the problem is necessary. These systems also exhibit low efficiency compared with the special-purpose solutions that have often been developed in-house. The requirements of a data management system for custom design include access methods for storing and retrieving geometrical data, multiple versions and design alternatives, back-out of nested transactions, support for workstation and network-based transactions, procedural attachment, and near-optimal performance with relatively cheap hardware.

An analogy can be used to explain where conventional database management fails for custom IC design. Many researchers have noted a strong similarity between the custom IC design process and writing computer programs—mask layout is akin to a binary image, symbolic layout is analogous to assembly code, and gates or modules are compared with lines of code in a high-level language. The sorts of operations the programmer performs on code are like those the IC designer would like to be able to perform on the IC design data. In fact, it is from such an analogy that the terms *silicon compiler* and *silicon assembler* evolved. Taking the analogy a step further, it is worth noting that programmers do not store their code in conventional DMBS systems. Rather, they have used conventional file systems (nowadays, often organized hierarchically) with tools to aid management of their code. In fact, these tools add to the programming environment many of the capabilities provided by modern database management systems. Over the past few years, a number of IC design data management systems have been developed based on this model and have been used effectively for custom IC design.

In recent years, the notions of procedural circuit design and rule-based expert system technology have emerged as key components in the design process. These techniques, coupled with the ever-moving boundary between entire systems, printed-circuit boards, and chips, have broadened the requirements for an integrated design system. What is needed to support this work is a flexible design and programming environment that allows a variety of approaches to design to coexist and permits system-level, logic-level, and circuit-level designers, as well as CAD algorithm developers to work together in a single, unified environment. The keys to such a system are common levels of abstraction and standard interfaces among them, as well as a powerful set of synthesis and verification tools which form the basis on which further research is carried out. Each object in such a system may be described by data (such as its mask layout), by a local procedure (such as a parameterized cell), by generic synthesis tools (such as a channel router or a placement program), or by a combination of all of these techniques.

As software systems continue to grow in size and complexity, programmers have turned to object-oriented approaches to code development and support. The next generation of workstations, with an order-of-magnitude increase in performance at the desktop for comparable price to workstations of today, will be a key factor in making such approaches practical and affordable outside the research laboratory. In an analogous way, IC designers are beginning to develop and use procedural descriptions of design components, akin to the objects in many of these languages. In addition, the database management community is directing its attention to the management of object-based descriptions of systems. From an IC design point of view, these three technologies will converge in the next generation of data management and programming systems for IC design. The interfaces to these systems will be indistinguishable from that of an object-oriented, message-based programming environment.

In summary, we believe that true open access to design data is an essential first step in the development of a data management framework.

Unfortunately, it is unlikely that a standard, object-based data management interface will be developed in the near future. There is still considerable research required to fully understand the issues involved before a suitable standard can be developed. In addition, competitive market pressures will continue to keep such interfaces proprietary. However, there is a need to move data from one design system to another. The design data represents the "life blood" of an IC design company. If a particular design tool does not function correctly under certain conditions, or a workstation or mainframe computer fails, the problem can generally be overcome and work can proceed. However, if the design data were to be lost in the middle of a large design project, the cost could be astronomical. Not only would the investment in design effort be lost to that point but such a situation would also cost valuable time and a market window might be missed. This is one reason why most IC design companies have resisted trusting all of their data management tasks to a single vendor, particularly if it is not possible to archive all of the data in a nonproprietary format. In addition, once a company has committed their data to a particular vendor's system, it is "locked in" to that vendor unless there is a way of migrating the data to another system.

Another need for data transfer arises in situations where more than one design system or design site is involved. For ASIC design, schematic capture and simulation may take place on a low-cost, customer operated workstation, while the actual silicon implementation of the design occurs at the ASIC vendor's plant.

To meet these and other needs, standard, textual interchange formats are being developed for IC design data. In some cases the definition has focussed on the data necessary to support a particular design style. For example, the EDIF (Electronic Design Interchange Format) has been developed by a broad base of semiconductor manufacturers, CAD workstation vendors, and system houses to address the needs of ASIC-based system development. In other cases, a broader charter is being attempted, such as the efforts to standardize hardware description languages and behavioral descriptions of systems. The VHDL (VHSIC Hardware Description Language) activity in the US and the HSL-FX effort in Japan are the most active in this area today.

In summary, we believe that true open access to design data is an essential first step in the development of a data management framework. Such access must occur at two levels—as an object-oriented, programming interface to all data in the database, and as a nonproprietary, standard textual format which reflects as much of the semantic content of such systems as is practical.

User interface. Over the past five years, the advent of low-cost, high-resolution bit-mapped graphics terminals and workstations has altered the economics of user interface dramatically. Low-cost artwork entry systems and schematic entry systems have evolved to compete with the more expensive systems and it has become cost-effective to have such capability at each engineer's desk. In the more advanced CAD environments, general-purpose computing functions such as computer mail and networked file systems are also integrated with the CAD interfaces. In most cases, the user interface is similar to that developed at Xerox PARC in the early 1970s and as embodied in the Apple Macintosh and Lisa families of personal computers—separate windows for each application, pop-up or pull-down menu-based command selection, and bit-mapped graphics interfaces to most tools. In the most recently announced systems, the command syntax is also common for all tools from schematic editing and mask layout to testing and report preparation. A typical screen from such a system is shown in Figure 3.

However, from a framework point of view, it important that the user interface be exposed to the CAD developer for the addition of new tools. Ideally, in a consistent object-oriented environment, the user interface (windows, menus, menu selections, etc.) would be treated as objects in the same environment as the data management system. With the exception of some experimental, in-house systems, there are no systems available today that provide such a uniform interface for the tool builder or CAD system developer.

Verification

To shorten the design cycle and to decrease design costs, it is crucial to eliminate as many errors as possible before manufacturing an integrated circuit. Verification tools, which help perform this task, were the first class of tool to be used in the IC design process. For this reason, the verification area is relatively mature and is only described briefly here. There are several different kinds of verification tools: structural verification tools, simulation tools, performance verification tools, and logic verification tools.

Structural verification. Structural verification is the task of verifying that the structure of a design—the arrangement of mask-layout shapes, the connections among those shapes or among the components of a design—satisfy a particular set of rules. For structural verification, the behavior of the components of the design is not considered, only their spatial relationships and connectivity.

Structural verification is subdivided into three areas: *layout-rule checking* (LRC), where mask geometries are verified to check if they satisfy a set of spacing, sizing, and enclosure rules; *electrical rule checking* (ERC), where the circuit schematic is verified to find electrical errors due to wrong connections of the devices; and *connectivity verification systems* (CVS), where a netlist description extracted from the layout is compared against the netlist description ex-

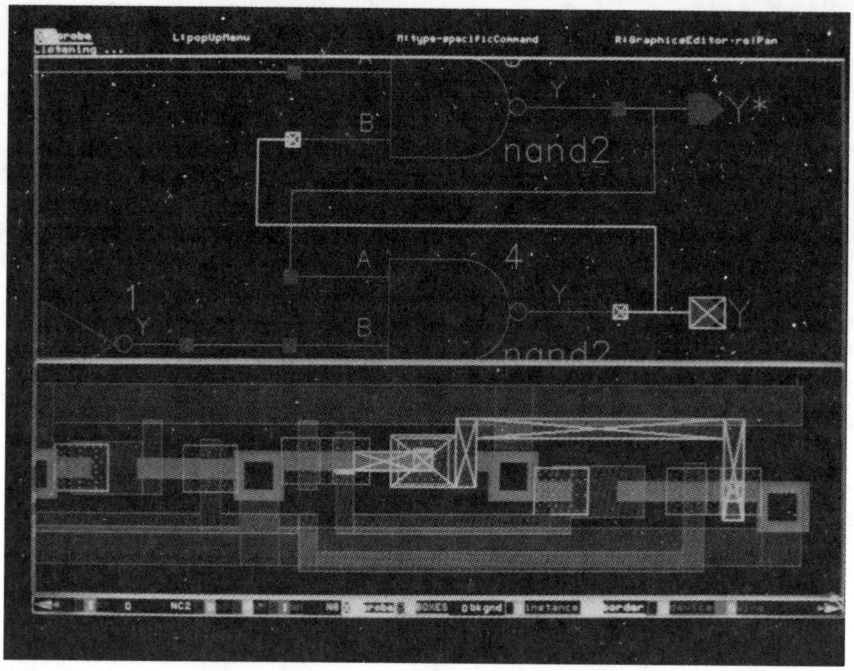

Figure 3. Typical CAD system user interface. (*Photo courtesy of SDA Systems.*)

tracted from an alternate description, such as the schematic diagram.

For custom and structured custom design styles, structural verification has made possible a substantial reduction in the design time needed to obtain functionally correct circuits. However, structural verification alone cannot provide a guarantee that the design meets necessary performance specifications.

For design styles where the circuit is constructed automatically from precharacterized and verified cells using computer programs, such as the standard-cell and gate-array design styles, or where large areas of the chip are constructed automatically using module generators, these techniques are used more as a final verification of the output of the CAD synthesis tools than as an active part of the iterative design process itself. For example, a design rule checker may be invoked after an automatic place-and-route system and the following symbolic spacing system have completed a layout, to verify that all the interconnections are properly spaced and that all the components of the design have been placed so that the design rules are satisfied. In our own experience with synthesis tools, we have found that a verification step after the synthesis step often detects obscure bugs that would have been very difficult to find otherwise. In addition, in systems where the user is permitted to modify the output of the synthesis system (for engineering changes or "green wires," or to improve the quality of the final output), a verification step is not only useful but essential.

Layout-rule checking. In some design methods, such as full-custom or structured custom, the mask geometries corresponding to the devices are entered manually by the designer. *Physical layout rules* or *design rules* specify the legal or illegal relationships among the polygons used in the IC mask-making and fabrication process to implement the final circuit. These rules account for necessary electrical separation of different components and signals as well as for imperfections in the mask preparation and manufacturing processes. If a designer enters the geometries of the masks such that they satisfy all layout rules, such as minimum spacing, minimum size, and minimum enclosure constraints, then the distortions occurring in the translation from the original drawings to the actual geometries in silicon should not reduce the yield of the design significantly.

Industrial rule-sets can be very complex, especially when the shapes of geometries are not restricted. In addition, electrical considerations may add to the complexity of the rules. For example, capacitive coupling between lines requires that long, parallel lines be spaced more conservatively than short parallel runs. For high-performance circuits, as the sizes of components continue to decrease, the importance of parasitic effects such as capacitance, resistance, and inductance is increasing and adding complexity to the electrical aspects of the rules.

Because of the size of IC designs today, a complete layout-rule check requires that millions of polygons must be inspected with several rules each. The running time of a batch LRC program is often of the order of days on a superminicomputer or mainframe. While early algorithms used for checking layout rules often showed $O(N^{1.5})$ performance, where N is the number of figures on the layer being checked, over the past decade researchers have been able to reduce the expected run time to almost linear in the number of mask shapes. Once the analysis has been performed, the errors have been reported, and the false errors have been discounted, the true errors must be corrected. Fixing such layout-rule violations is an expensive process in a batch checking environment because a layout-rule check must be performed every time the layout is modified to verify that no new errors have been introduced.

LRC programs have been developed in-house by large merchant and captive semiconductor companies including Intel, AT&T Bell Laboratories, Hitachi and IBM, and by vendors such as Calma, NCA, Phoenix Data Systems, Metheus, ECAD and SDA Systems. These programs can verify complex designs with complicated layout rules. Some offer the users a language with which new layout rules can be added rather easily. Over the past few years, in conjunction with VLSI design courses, several universities have developed LRC programs which are based on the Mead and Conway simplified lambda-based rules. Because the design styles used often require Manhattan geometries only (layouts comprised of rectangles

As ICs increase in size, the complexity of the design is often managed by increasing the regularity or repetition of cells.

only, with their edges aligned with the x and y axes, akin to the organization of streets in downtown Manhattan) and do not consider conditional rules, these programs are relatively simple and very fast.

A number of new approaches to LRC have emerged from this activity and have found application in industrial products as well. For example, the concepts of *hierarchical* and *incremental* LRC have been explored over the past few years and have been implemented in industrial systems. In incremental approaches, a background process checks the layout rules in the vicinity of each figure as it is added to the layout. Since manual layout is a slow process relative to the computing speed of a modern workstation, this is an effective way of using machine cycles that might otherwise have been wasted. It also permits errors to be corrected and rechecked in a tight, local loop so that the number of expensive, batch-mode checks can be reduced or even eliminated completely. However, incremental LRC must be applied so that temporary LRC violations, consciously introduced by the designer as an intermediate state, are not continuously reported, disturbing the editing session.

Hierarchical LRC takes advantage of the fact that cells are often used more than once in a large design. Once the insides of a cell have been checked, the cell is marked as "done" and then for each use, or instance, of the cell only the local context of the cell need be checked. In fact, it is not the hierarchical property of the design that really accounts for the savings but rather the repetition of identical cells or collections of figures. As ICs increase in size, the complexity of the design is often managed by increasing the regularity or repetition of cells in the circuit so that tools which exploit this fact often show large performance gains. In a modern, symbolic design system, where cells are often parameterized and many different variations can be created from the same master cell, or where a spacing system may adjust each cell differently to meet external constraints such as cell pitch, the advantage of hierarchical LRC is quickly lost.

The techniques used by LRC systems can be classified into three categories: region-operation-based, raster-based, and corner-based. An excellent review of these techniques can be found in Arnold.[4]

Once the mask patterns satisfy the physical layout rules, it is necessary to verify that they will actually implement a working circuit. The first step in this process is to recreate a netlist description of the circuit from the mask pattern data. This process is called *extraction*. Since the only information present in the mask layout data is the relative placement of shapes on different mask layers, mask-level operations must be performed to recognize individual components, such as transistors, capacitors, and nets. Since the types of operations necessary to recognize components from the layout are very similar to those used for checking the layout rules, it is not surprising that in most cases a LRC program forms the basis of an extraction program. The extraction program also determines parameter values for simulation, such as the sizes of the transistors extracted as well as related parasitic capacitance values. Depending on the design styles supported by the implementors of the extraction tool, the program may only extract gross components such as transistors, lumped parasitic resistance, and parasitic capacitance from interconnect to ground, or it may perform a very detailed extraction including interlayer and interfigure parasitic capacitance calculations and even solve two-dimensional field equations where necessary.

Like LRC, extraction can take many hours of computer time to perform if an accurate analysis of an entire chip is required. In addition, if coupling capacitances between parallel lines are extracted and RC-networks are used to represent interconnections, the amount of output data generated can be massive. While many of the industrially-developed extractors are used to obtain detailed parasitic information, those in general use in university design systems tend to favor speed rather than detailed analysis. In the most common university design styles for VLSI circuits, the layout rules and electrical requirements are constrained to reduce the likelihood of parasitic components altering the function of the circuit. Recently, the concepts of hierarchical and incremental extraction has been developed, making it possible to verify modifications to design interactively without sacrificing the level of verification that is needed in quality IC designs.

Connectivity verification and electrical-rules checking. The extracted netlist provides the basis for a number of additional checks. In particular, connectivity verification and electrical rules checking. Electrical rules checking programs evolved from simple implementations that searched the extracted netlist for ridiculously large or small transistors or direct short-circuits among the power supply lines, clock lines, and the ground line to programs which check for more complex relationships. Such rules include searching a local area of a circuit to be sure transistors related to one another in an electrical sense all have the correct sizes necessary for correct circuit operation. To increase the power of these programs and to follow the evolution of technology, new rules must be added to existing programs. Often, these rules were added by including new "hard-wired" procedures or data structures. Recently, a new breed of electrical rule checking programs has evolved based on the use of rule-based expert systems (see window). Using a rule-based approach, the addition of new rules becomes straightforward.

If a transistor-level netlist description of a circuit is available, either from manually-generated simulator input data or from a schematic entry system, the pattern of interconnections among those components and the pattern of interconnections obtained from the extracted netlist can be compared. This process is called *connectivity verification*. Connectivity verification systems (CVS) have been used over the past decade to improve dramatically the probability of functionally correct silicon on the first fabrication run. Connectivity verification was first used at the board level, for comparing a logic schematic input with a placed and routed board-level implementation of the circuit.

In general, the comparison involves a one-to-one correspondence between the circuit elements, such as transistors, and the nets in the two circuits, not a functional equivalence. Each circuit is represented by a graph, where the nodes in the graph represent either the circuit elements or the nets, and the nets or circuit elements, respectively, are represented by the arcs in the graph. In some cases, both the circuit elements and the nets are represented by nodes in the graph and the arcs simply represent the connections between them. The problem of determining that the two graphs are the same is equivalent to the graph isomorphism problem, a well known combinatorial optimization problem. The worst-case complexity of graph isomorphism is not known. No algorithm has been found with running time bounded by a polynomial in the size of the input, i.e., number of nodes and edges of the graph, but it has not been proven that this problem belongs to the class of NP-complete problems. However, good heuristics are available which can quickly detect if two graphs are isomorphic in most of the cases. If they are not isomorphic, the programs can then isolate the subgraphs that differ in the two netlists. This information is then provided to the user who tries to locate and correct the error.

Simulation. For circuits made from discrete components, breadboards (prototype boards with discrete components) were used extensively to check the functional correctness of the design as well as its performance. However, this approach does not work well for integrated designs since the parasitics on the breadboard are quite different from those on the IC and the thermal and electrical characteristics, as well as component matching properties of the discrete components, are also quite different from their IC counterparts. For this reason, electrical circuit simulation was one of the first CAD tools to be developed for IC design and has completely replaced breadboarding for analog and digital cell design. For large digital designs, breadboards are still used for software development.

Many different forms of simulation can be used for the verification of large digital integrated circuit designs at the various stages of the design process. They may be classified as behavioral (also called algorithmic or functional) simulators, register-transfer level (RTL) simulators, gate-level logic simulators, timing simulators, and circuit simulators.

Behavioral simulators are used at the initial design phase to verify the algorithms of the digital system to be implemented. Not even a general structure of the design implementation is necessary at this stage. Behavioral simulation might be used to verify the communication protocols in a multiprocessor design, for example.

Once the algorithms have been verified, a potential implementation structure is chosen. For example, a microprocessor, some memory, and a special-purpose input/output module may be chosen to implement the handshaking protocol mentioned above. An *RTL simulator* can be used to verify the design at this level. Only crude timing models may be available, since the exact circuit parasitics and other implementation details are not yet known. Useful information relating to congestion and hardware/firmware tradeoffs can be obtained from this level of analysis. A variety of RTL languages and associated simulators have been described in the literature.[5]

Depending on the design methodology and certain technology issues, a gate-level design may be undertaken, where each of the RTL modules is further partitioned into low-level logic building blocks, or gates. A *gate-level logic simulator* may then be used to verify the design. Sophisticated delay models may be introduced and testability analyses performed.

From the gate level, transistors and associated interconnections are generated to implement the design as an integrated circuit. Accurate electrical analysis can be performed for small parts of this design using a circuit analysis program or larger blocks may

> For circuits made from discrete components, breadboards were used extensively to check the functional correctness of the design and its performance.

be analyzed in less detail using a *timing simulator*. Once the integrated circuit layout is complete, accurate circuit parameters, such as parasitic capacitance values and transistor characteristics, may be extracted and used at the electrical level. These analyses may then be used to improve the delay models at both the gate and RTL levels to verify the circuit design using accurate timing data.

A number of simulators have been developed recently which span a range of these levels in thc simulation hierarchy. These simulators are called *mixed-level simulators* and allow different parts of a circuit to be described at different levels of abstraction. Not only does this approach permit a smooth transition between different levels of simulation (since the same simulator and associated input processor is used) but it allows the designer to take advantage of the time and memory savings available from higher level descriptions of parts of the circuit.

Excellent tools, either developed in-house by IC manufacturers or by tool vendors, are available at all levels of the simulation hierarchy. The parameters often used to judge the quality of a simulation tool are accuracy, speed, and flexibility. By flexibility, we mean the range of analyses supported by the program (time-domain transient, small-signal frequency-domain analysis, timing verification, fault simulation, etc.) and the range and quality of the component models it provides (both N- and P-channel MOSFETS, bipolar transistors, bidirectional switches, built-in registers, etc.). Because of the increasing size of IC designs, even the fastest simulators are not able to perform simulation as extensively as desired by today's designers. For this reason, several hardware accelerators have been developed for simulation and new algorithms are being explored to exploit multiprocessor architectures. (See sidebar, "Multiprocessors for CAD.")

Circuit and timing simulation. When accurate circuit models are available, circuit simulators provide precise electrical information (such as frequency response, time-domain waveforms, and sensitivity information) about the circuit under analysis. The majority of circuit simulators currently in use contain models for a wide range of active devices, including bipolar junction transistors, MOSFETs, JFETs, MESFETs, and diodes, and hence are largely independent of technology. For this reason, these programs must employ general algorithms for the solution of the set of coupled, nonlinear, ordinary differential equations which describe the integrated circuit and hence cannot exploit the special characteristics of a particular technology.

The most-used general-purpose circuit simulator is SPICE, developed at the University of California, Berkeley. This program has been adapted for use in many IC design companies and has formed the basis of a number of commercially available circuit simulators. The program ASTAP, developed at IBM and also widely used, is based on different algorithms than SPICE and has additional capabilities such as user-defined models and statistical analysis.

Without models whose accuracy is well matched to the expected accuracy of a simulation, the results of the simulation may not reflect the performance of the circuit under analysis. Recent work on modelling for MOS circuit simulation has focussed on the development of both analytic and semi-empirical models for MOS transistors which predict the characteristics of the devices accurately without requiring large amounts of computer time. With increasingly small geometries on ICs, signal delays and signal degradation caused by interconnection can dominate circuit operation. For this reason explicit models for interconnection are necessary for accurate simulation, and interconnect modelling is an active area of research.

Circuit simulation techniques can provide accurate waveform analysis for circuits of building-block complexity. However, as circuit size increases the time and memory requirements of a circuit simulation become prohibitive. On an IBM 3081/K computer, the average cost of a SPICE2 analysis is 0.3ms/MOSFET/clock/timepoint. For a 100,000-device circuit, with three clocks and for a single analysis of 10μs at 1ns steps, the computation time would be in excess of three CPU weeks, assuming the machine could even handle such a large problem. Nevertheless, the success of circuit simulation in design evaluation has been such that designers wish to continue to simulate large circuits at the level of accuracy provided by this type of program. A number of new techniques have been used to speed up conventional circuit simulation approaches.[6]

If simulation algorithms are tailored to specific technologies or applications, substantial speed improvements can be achieved. Many components of digital MOS or I^2L circuits can be considered unilateral in nature. This characteristic, as well as the facts that these families are saturating and hence accumulated voltage errors are lost at the extremes of signal swing, and that large digital circuits are relatively inactive at the gate level, are exploited in timing simulation. Timing simulators can improve simulation speed by up to two orders of magnitude while maintaining acceptable waveform accuracy. These savings are achieved by using node decoupling techniques in conjunction with simplified table look-up models for nonlinear devices.

Where a library of cells is used during the design, or when a group of transistors is used to implement a common function, such as a cell or logic gate, it is often possible to exploit the known structure of the circuit and use a simplified representation which maintains the essential characteristics of the cell at reduced computational expense. Such a reduced representation is called a *macromodel*; they are used in both circuit- and timing-level analysis.

While timing simulators are fast, they may be inaccurate for circuits containing tight feedback loops and large floating elements, particularly capacitors. Designers have often applied these simulators to problems which were not well suited for this type of analysis and have obtained incorrect answers. For this reason, converged relaxation-based circuit simulators[7] were developed. These programs evolved from the basic ideas of timing simulators but guarantee an analysis as accurate as the one pro-

Multiprocessors for CAD

As we approach the fundamental performance limits of uniprocessors, it is clear that only new, multiple-processor computer architectures can offer the large performance improvements needed to solve the complex problems of our time. Throughout the world, a great deal of research is in progress on the development of new, often unconventional, computer architectures for both symbolic and numerical processing. In addition, as large, loosely coupled networks of powerful workstations find their way into the design environment, CAD tools which can exploit unused machine cycles on these multiple-processor systems will also be developed.

However, multiprocessors will not be able to exploit the parallelism available in problems unless new algorithms are developed that are well suited to a multiprocessor environment. In the past, it has often been assumed that advanced compiler technology would be sufficient to translate a conventional computer algorithm for optimal use on a special-purpose machine. In the case of circuit simulation, even the most advanced compiler technology has shown poor speedup on pipelined machines (about 12-15 percent hardware utilization on the Cray 1).

There are several compute-intensive problems that face CAD of VLSI circuits today. In some cases, such as logic simulation and design-rule checking, new special-purpose machines have been designed to fully exploit the characteristics of the problem to be solved. In others, such as circuit simulation, existing general-purpose parallel processors have been used to speed up the simulation

vided by standard circuit simulators such as SPICE. The relaxation-based simulators exploit the unilateral nature of MOS devices, the inactivity of the circuit (latency), and the fact that two or more node voltages or branch currents of the circuit may change in time at different, sometimes very different, rates (multirate behavior). Savings in running time of up to two orders of magnitude have been obtained over standard circuit simulation programs, such as SPICE2, using this approach.

time. A careful analysis of the economic tradeoffs has to be made to decide whether to build a special-purpose machine or to use existing multiprocessors.

It is clear that the cost of the processors would be much lower for the case of general-purpose multiprocessors, while the performance of special-purpose machines could be orders-of-magnitude better. A "mixed" approach could probably offer the advantages of both strategies. In this case, the interconnection network and the basic software should be those offered by the general-purpose environment, while the particular compute-intensive tasks to be carried out by the single processors could be speeded up by the design of special-purpose accelerator boards to be used as coprocessors for the processors of the general-purpose system.

Several approaches have been followed in the past few years to develop CAD accelerators. An excellent review of the field can be found in Blank.[11]

We can classify the new research areas opened by the feasibility of designing special-purpose processors as well as by the availability of general-purpose multiprocessors into three areas: development of new algorithms for existing multiprocessor architectures, development of new architectures for existing algorithms, and development of new algorithms and new architectures. It is important to note that when using nonconventional architectures, existing algorithms which are considered less efficient when using conventional uniprocessors may become much faster.

However, for circuits that contain tightly coupled subcircuits or where many parasitic components are involved, the relaxation-based approaches alone may not perform as well as standard circuit simulators. Under these conditions, relaxation must be applied with great care to maintain the speed advantage over standard techniques. The key idea here is to solve the tightly coupled subcircuits with standard methods while the loosely coupled connections are dealt with relaxation. Most relaxation-based simulators in use today provide this capability. Automatic partitioning algorithms have been developed to partition large-scale circuits into collections of tightly coupled subcircuits.

These techniques alone cannot provide the speed that is needed for the detailed electrical analysis of VLSI circuits. Special-purpose hardware and multiprocessor systems are now being used to provide dramatic speed improvement for circuit simulation (see sidebar).

Logic and switch simulation. When the complexity of an integrated circuit design reaches the point where electrical analysis is no longer cost-effective, logic simulation may be used. Rather than dealing with voltages and currents at signal nodes, discrete logic states are used.[8] Only simple Boolean operations are required to obtain the output state value of a logic gate and these are generally the most efficient operations available on a digital computer. Rather than modelling the circuit at the individual transistor level, in a logic simulator transistors are grouped into logic gates wherever possible and a gate-level model is used. As in modern, relaxation-based simulators, asynchronous logic simulators exploit the inactivity in the circuit to improve speed by using event-driven and selective-trace techniques. The term *event-driven* refers to the fact that only logic gates whose input values change are processed and the term *selective-trace* refers to the technique used to find the logic gates whose inputs have changed. Rather than checking every gate at every timepoint to determine if its inputs have changed—an expensive process if most of the gates are not changing—when an output changes, a table containing pointers to the gates to which this output is connected is used to schedule the fanout gates for processing. As a result, the program selectively traces paths of activity through the circuit. With selective trace analysis and the above simplifications, asynchronous logic simulators are typically 10 to 100 times faster than the most efficient forms of electrical analysis.

Rather than using event-driven techniques, many of the early logic simulators were compiled simulators where the logic circuit was described in a programming language which was compiled directly to machine code and executed. Although this approach provides a very efficient mode of simulation, no compiled simulators accurately model asynchronous circuits. Compiled techniques are used, however, for higher level register-transfer level simulation and, of course, the bottom-level models in logic simulators are generally compiled. Recently, compiled simulators have been developed for the switch-level simulation of clocked, synchronous MOS circuits where the circuit is analyzed and switches that form distinct, combinational blocks are clustered as subcircuits and a corresponding program fragment is generated. These fragments are compiled and form the scheduled blocks of an event-driven simulator. An equivalent speedup can be achieved by using a gate-extraction program which recognizes specific patterns of switches and replaces then with an equivalent gate-level representation.

Register and behavior. While different description languages are described as RTL, PMS, or behavioral, it is often the case that the distinction is blurred in actual use. While the purpose of an RTL description is to describe a register-level implementation of a system, RTL descriptions are often used as input to synthesis systems where the structural information they contain may be ignored. However, the program structure in a behavioral or algorithmic description is often used as an initial hardware structure for implementation.

As the design representation becomes increasingly abstract, it also tends to become increasingly domain-specific. For example, while most designers can agree on what an AND gate is, at the system level a signal processing engineer will refer to a "sample time," the microcomputer designer talks of "four-phase clocks," and the data-driven system designer talks about "self-timed" modules. If a language selects any of these notions as its abstract representation of time, it will often not be used in the other application areas. The only way, therefore, to build a general behavioral language and simulator is to reduce the domain-specific notions to their lowest common denominator and to permit individual users to build libraries of domain-specific objects and operators. As a result, most behavioral-level simulators are implemented in existing concurrent programming languages, such as Simula and Modula-2.

Register-transfer level simulators may be classified by the manner in which they model time and the way in which they order the evaluation of blocks. In some simulators, all assignments correspond to clocked, register transfers. All logic between assignments is combinational and thus can be compiled on a sequential machine after a suitable static ordering has been determined. Such an ordering can be found using breadth-first search of the data-flow graph that represents the logic expressions. Since the logic is combinational, the graph must be cycle-free. This approach leads to a very fast implementation but cannot handle hazard detection or timing within a clock cycle and can only represent synchronous systems.

The other approach is to implement the simulator much more like a logic simulator, using a dynamic ordering based on the event-driven selective-trace technique described above, and scheduling code modules. The modules may be those specified by the user (code block or procedure boundaries) or may be extracted from the description between explicit delay assignment statements. If the system supports asynchronous design, specific delay constructs must be provided and these are used to determine locations for scheduling.

Register-transfer simulators, such as the ISP/S system developed at CMU, are over an order-of-magnitude faster than gate-level simulators for the same circuit, since they deal with fewer blocks and their model of time and signals is typically a somewhat coarser than in the logic case.

Performance verification. While simulation has been used successfully for the verification of large circuits, it cannot guarantee that certain timing specifications are met for all possible input combinations unless all such combinations are tried—an often impractical proposition. Performance verification techniques aim at the determination of the critical delays in a circuit independent of the input patterns.

Digital circuit design depends critically on the delay of the signals. For asynchronous designs, if the timing of the signals is not carefully considered, incorrect results may be obtained from an otherwise functionally correct design. For synchronous designs, the longest delay among the signals that flow from the primary inputs or latches to the primary outputs is called the *critical path*; it determines the period of the clock and hence the speed of the circuit.

To optimize the performance of a circuit, critical paths have to be identified and minimized. In addition, paths which are noncritical may be unnecessarily fast and consume extra power to no avail. This path optimization is becoming an increasingly difficult task as circuit size and complexity increase, but nevertheless more and more necessary because of the competition to build faster circuits. Simulation can be used, and has been used, to identify critical paths and to optimize circuit performance. However, detecting critical paths may involve the simulation of several thousand transistors for large ICs. Furthermore, pathological conditions may not be detected by simulation unless particular inputs are fed. *Timing verification* is a technique which aims at the determination of critical paths without performing simulation. This technique has played a very important role in the design of digital integrated circuits, in particular for bipolar gate-array technologies. The first timing verifiers were built around 1973 by IBM and other computer companies for the design of large high-performance computers.[9]

Other applications of timing verification have recently come to the attention of the design community. In particular, switch-level, or transistor-level, timing verifiers were developed at Berkeley and at Stanford. The task of accurately modelling delays at the switch level is often more difficult that at the gate level due to the complex coupling and distributed delay effects present in transistor networks. Most of the CAD vendors have developed timing verifiers with various degrees of complications and power. Both gate-level and switch-level timing verification is now available commercially.

Simulated annealing

Simulated annealing is a relatively new approach to the solution of combinatorial optimization problems. Since it was first proposed just a few years ago, this algorithm has found application to many CAD problems, from partitioning to gate-array placement, from floor planning to global routing, from topological optimization of regular arrays to logic optimization.

Heuristic algorithms are used to solve NP-complete problems approximately, i.e., to find "good" solutions which are "close" to the optimum. These algorithms explore a discrete space of admissible configurations, *S*, in a deterministic fashion. Often the search terminates at a local minimum. The local minimum reached can be quite far apart from the global minimum measuring their distance with the difference in cost. This is often due to the fact that heuristic algorithms are "greedy"; i.e., only moves which reduce "maximally" the cost are accepted.

Simulated annealing as proposed by Kirkpatrick et al.[12] allows "hill climbing" moves, i.e., moves that increase temporarily the cost of the solution to explore wider regions of the optimization space; but these moves are accepted according to a certain criterion which takes the cost into consideration—and not blindly. The controlling mechanism is based on the observation that combinatorial optimization problems with a large configuration space exhibit properties similar to physical processes with many degrees of freedom.

In particular, bringing a fluid into a low-energy, crystalline state has been considered similar to the process of finding an optimum solution of a combinatorial optimization problem.[12] Annealing is a well-known process to grow crystals. It consists of melting the fluid and then lowering the temperature slowly until the crystal is formed. The rate of decrease of temperature has to be very low—around

Accuracy is still a serious problem for timing verifiers, even though, for circuits designed with constrained design methods, timing verification can be within a few percent of the accuracy offered by circuit simulators. A method for detecting when the timing algorithms may have accuracy problems and for continuously changing the accuracy of the models used in the analysis is of key importance here.

> the freezing temperature. The Metropolis Monte Carlo method can be used to simulate the annealing process. It has been proposed as an effective method for finding global minima of combinatorial optimization problems. This method, when applied to combinatorial optimization, generates moves randomly and checks whether the cost of the new configuration satisfies an acceptance criterion based on temperature. If the cost decreases, the move is accepted. If the cost increases, then a random number between zero and one is generated and compared with
>
> $$f(\Delta c_{ij},T) = \exp(\frac{-\Delta c_{ij}}{T}) \text{ where } \Delta c_{ij}$$
>
> is the change in cost obtained by moving from configuration i to j and T is temperature, the controlling parameter. If the random number is larger than f, the move is accepted; otherwise the move is discarded. Note that the higher the temperature is, the more likely it is that a "hill climbing" move is accepted. Note also that "hill climbing" moves are less and less probable as the temperature decreases. A certain number of moves are generated and checked before a decrease in temperature is allowed. The initial temperature, the number of moves generated at each temperature, and the rate of decrease of temperature are all important parameters that affect the speed of the algorithm and the quality of the final configuration. Experimental results show that simulated annealing produces very good results when compared to other techniques for the solution of combinatorial optimization problems such as those arising from the layout of integrated circuits, at the expense of large computer time (a 1500-standard-cell-placement problem can take as much as 24 hours of a VAX 11/780).

Logic verification. Logic verification tools are also input-pattern-independent and verify that two design descriptions at gate- or structural-level and at the functional level are formally equivalent. In general, these techniques are much more expensive than simulation but their use increases the level of confidence in the design and therefore they become increasingly important as the complexity of IC design grows. At higher levels of the design process, verifying that a design implements the required function correctly is a serious bottleneck for complex IC designs and is an active research area.

In general, the functional verification step is carried out today by simulating the design with a set of input patterns which cannot guarantee that the design is correct, i.e., that the transformation from one level of the design hierarchy to the next has not changed its functionality. Formal verification techniques are input-independent and are designed to guarantee functional equivalence between two representations of the design at different levels.

Most of the formal verification techniques developed to date deal with behavioral-, structural-, and logic-level representations. These techniques have great potential for producing correct-the-first-time designs. In addition, verification is important for technology remapping, i.e., for the transfer of a logic design from a technology (for example, bipolar ECL gate arrays) to another (for example, CMOS gate arrays). In this case, the functionality of the new implementation has to be checked against the functionality of the former. Despite the importance of formal verification, only a few of the techniques developed in the past have been applied to industrial designs due to their complexity and computational requirements.

Most of the formal verification techniques proposed recently can be classified into three major categories: formal software verification techniques, semi-exhaustive logic simulation and logic comparison.

The most successful formal verification techniques are based on logic comparison and have been in use at IBM for many years. They are applied to designs that satisfy a set of restrictive assumptions. The main assumption is that a one-to-one correspondence between the memory elements, such as flip-flops, at the two levels of interest has been established. Then the problem of logic verification is reduced to the one of verifying the equivalence of combinational logic. This assumption holds when Level-Scan Sensitive Design (LSSD) or scan-path techniques are used in the design to make the task of testing the design easier.

This problem is very similar to the testing problem and, in fact, can be formulated as a redundancy identification problem for a compound circuit obtained from the two circuits under comparison by forcing their primary inputs to be the same and by tying their primary outputs to the input of a comparator (an XOR gate). If the designs implement the same logic function, then the output of the comparator is always zero. The verification problem is equivalent to proving that a stuck-at-zero fault at the output of the XOR gate is redundant. This problem can be easily shown to be NP-complete. Experience gained in coping with testing problems is instructive in this case. However, in the testing problem, the most common situation is that the fault under test is not redundant and a test for it should be found as soon as possible, while in logic verification the fault is expected to be redundant and we wish to come to that conclusion as rapidly as possible. This implies that efficient algorithms for testing may not be efficient for logic verification.

While some of the results can be used to formally verify a design, more research is still needed to improve the speed and the domain of applicability of the formal verification techniques. Recent results in multilevel logic optimization may be a key for developing faster techniques. Parallel processors can also be used to speed up this intrinsically very complex problem.

Synthesis

Synthesis is a crucial component of future CAD systems. The future com-

> The future competitive edge of IC design will most probably come from the use of effective synthesis, or design, tools.

petitive edge of IC design will most probably come from the use of effective synthesis, or design, tools.

A complete synthesis system should generate layout masks from a high-level algorithmic, behavioral or functional description of a VLSI system, a description of the target technology, a description of the design constraints, and cost functions. The design should be completed in reasonable time and with the quality a human designer could obtain.

Very few design aids are available to assist the VLSI designer at the algorithmic level. At this level, the designer describes the system by specifying its operations or functions without necessarily giving implementation details such as the hardware components needed to implement the system. Design at this level involves the translation of a required algorithmic-level specification into an architectural or register-transfer-level implementation. The architectural representation of the design includes components such as registers, memories, and processors, which specify the high-level implementation of the system.

Once the functional partitioning of the design is completed, estimates of the layout size, power-supply requirements, and speed of the high-level circuit blocks used to implement the various subfunctions are required. A chip plan must also be constructed to determine the relative placement of these building blocks. This chip plan is then further refined as the design proceeds. These tasks are often performed manually, perhaps with the help of the computer to perform bookkeeping tasks such as the storage of interconnection data.

Silicon compilers have been proposed to carry out the entire synthesis process. Since the task is so complex, early silicon compilers assumed that a target technology and a floor plan were chosen by human designers. In this case, the difficult steps of linking the high-level synthesis task to the layout problem was resolved by eliminating computer intervention. Among the most important contributions of the early research on silicon compilers is the development of *procedural design languages*. These languages are used to write programs which, when executed, generate in a flexible and possibly technology-independent way the layout of entire chips and/or of leaf cells, i.e., the basic low-level cells such as NAND gates, NOR gates, inverters, and register cells. The work on ICL at Caltech and the work on DPL at MIT are examples of such important contributions. Procedural design languages can be used effectively to generate the layout of regular structures such as ROMs, RAMs, PLAs and datapaths. In particular, the use of these languages eases the construction of parameterized and technology-independent module generators, i.e., of computer programs that generate the layout of a block such as a ROM, RAM or PLA, given a functional representation, such as a truth table.

Working designs have been produced with silicon compilers, but the quality of the design has always been a problem. While for a restricted class of designs, such as digital signal processors (DSP), the use of a fixed floor plan has been successful (e.g. the LAGER silicon compiler developed at Berkeley), and its use for less constrained applications results in inefficient utilization of area and poor performance. In addition, the structure of the control logic is often too rigid and not optimized, thus yielding a slow and large chip. J. Fox of GTE Laboratories has illustrated the pitfalls of an existing silicon compiler, MacPitts, originally developed at the MIT Lincoln Laboratories, by comparing a design for a telecommunications chip generated by a silicon compiler with a design obtained with the use of a standard-cell place-and-route system.

The present trend is to break the synthesis process into stages, and to use tools that optimize real estate and/or performance to go from one stage to the next. At first, attention was paid to the optimal generation of regular arrays such as ROMs, RAMs and PLAs. For example, early module generators have been built by VLSI Technologies Inc. and Silicon Compilers Inc. using a procedural design language. Simple routing techniques were also offered to connect the modules generated by these tools. These two companies were the first to introduce the concept of silicon compilation and procedural design languages in industry. While these concepts have now gained considerable attention in the industrial community and several other companies are offering procedural design languages (e.g., Silicon Design Laboratories and SDA) and module generators (e.g., Silicon Compilers Inc., SDA, and Seattle Silicon Compilers) a few years ago when these companies were founded, traditional designers expressed a great deal of resistance towards these new approaches to design.

The SILC silicon compiler under development at GTE Research Laboratories and the Yorktown Silicon Compiler (YSC) being developed at the IBM T.J. Watson Research Center are two examples of systems where layout optimization and efficient logic synthesis are introduced. While SILC addresses the problem of translating an algorithmic description of the system to be designed into an architectural description, YSC starts with an architectural description of the design, leaving the task of determining the architecture of the chip to the human designer. The Design Automation Assistant under development at ATT Bell Laboratories is the most recent entry in the automatic synthesis arena. This system is based on the work done at Carnegie-Mellon University for high-level synthesis and on the work done at Stanford for layout. One of the most interesting aspects of this system is the use of a knowledge-based expert system to carry out the translation of the behavioral-level description into a register-transfer-level description and to generate an optimized floor plan. Research work on silicon compilation is also carried out at the University of Illinois with the ARSENIC silicon compiler developed at the University of Illinois, Urbana.

The procedural design aspects of early silicon compilers have been neglected for some time. Recently Silicon Design Laboratories (SDL), founded by the developers of the PLEX system originally conceived at the AT&T Bell Laboratories for automatic synthesis of microprocessor-based designs, introduced a procedural design language for IC design which is available commercially. We believe that procedural design systems and knowledge-based expert systems are crucial for the synthesis systems of tomorrow.

In this section the three basic components of a synthesis system are reviewed: (1) physical synthesis or layout, including floor planning, partitioning, placement, routing and compaction; (2) logic synthesis, including combinational logic, sequential logic and algorithmic or behavioral synthesis; and (3) procedural design and module generation.

Physical synthesis. The layout of integrated circuits consists of the placement of the devices (cells) composing the design in a two-dimensional finite space and of the interconnection of the pins of these devices according to the schematic of the circuit to be implemented. The goal of this process is to complete the placement and interconnection of the design in the smallest possible area satisfying a set of design constraints (such as the ones posed on the position and size of the devices), a set of technological constraints (such as the ones posed by design rules and levels of interconnects), and a set of performance constraints (such as the ones posed by the timing of the logic to be implemented). This optimization problem is very complex (even simplified versions of it are NP-hard) and, given the number of modules to be laid out, has to be decomposed into smaller subproblems to be tractable.

The layout problem is traditionally subdivided into several stages. While these stages are often common to the various design styles such as gate arrays, standard cells and macrocells, their complexity may differ.

Automatic layout systems were introduced first for gate arrays and standard cells, since the layout problems associated with these design styles are in general simpler than the ones associated with the macrocell design style. Good gate-array systems developed both inhouse and by vendors companies are now widely available.

The bipolar discretionary wiring system developed at Texas Instruments was an early automatic routing system for gate arrays. The Engineering Design System of IBM was among the first complete place-and-route layout systems to be developed for gate arrays in the late 1960s, even though the first papers describing the algorithms used in the system were published in the early 1970s. In Japan, the first gate-array place-and-route system was developed by Oki and was reported in 1974. Silicon foundries such as LSI Logic offer their customers place-and-route systems optimized for their gate-array families. Mentor/CADI, Tektronix/VR Systems and Daisy have good place-and-route systems. Daisy has also a gate-array placement hardware accelerator which implements the simulated annealing algorithm (see sidebar).

For a while, gate arrays completely dominated the semicustom market, in part because of the wide availability of CAD tools for their design. Now standard cells are capturing the attention of a large user community. Interestingly, the concepts of standard cells and of gate arrays were developed at about the same time. However, automatic placement and routing tools for standard cells were developed earlier than the ones for gate arrays, for example, with the pioneering work at Philco Ford, Fairchild with the μMOSAIC system, and RCA with MP2D in the 1964-1970 time frame. Bell Laboratories developed a poly-cell (a synonym for standard-cell) place-and-route-system called LTX in 1973. The successor of this system, LTX2, was extended to gate arrays and to a limited version of macrocell place-and-route.

Macrocell placement and routing is particularly difficult. Japanese companies are the clear leaders in this area. NTT, NEC, Hitachi, Sharp, and Sony have all good working systems. In Europe, Siemens was among the first

> We believe that procedural design systems and knowledge-based expert systems are crucial for the synthesis systems of tomorrow.

companies to develop macrocell tools. However, their system is not in production use at this time. The GAELIC/COMPEDA system was developed by the University of Edinburgh and made available by a commercial company, but it has not found wide use. In the US, some companies such as Hughes aircraft have recently developed experimental systems for the VHSIC (Very-High-Speed Integrated Circuits) program of the Department of Defense. The SICLOPS system first developed by Preas and VanCleemput at Stanford has been improved at Sandia Labs. However, to the best of our knowledge, no complete system is currently in production use. Because of the importance that macrocell systems are bound to have in the future, we expect more companies to develop such systems.

In the academic world, Japanese and US universities have the lead in the development of algorithms and programs for placement and routing. Recently, universities have placed emphasis on the development of complete place-and-route systems for a variety of design styles. The PI system for macrocell place-and-route from MIT, the BBL (Berkeley Building-block Layout) system for macrocell place-and-route, and the ThunderBird standard-cell system, both developed at the University of California, Berkeley, are a few examples. The distinguishing feature of

> Floor planning can be used to verify the feasibility of integrating a design onto a chip without performing the detailed layout and design of all the blocks and functions.

these systems is the use of new and experimental algorithms.

Most of the work done on layout has been concentrated on optimizing area. Recently, performance (in particular, speed) has become a major concern. Coupling timing analysis with placement and routing has been proposed to influence the layout process of gate arrays with speed considerations. We expect this problem will receive more attention in the future.

In the following sections, we will review the various stages of the layout process: floor planning, partitioning, placement, routing, and compaction. The literature on layout methods is extensive and it is impossible to cover all technical aspects of this field. We direct the interested reader to Newton[1] and Soukup[10] for an extensive bibliography and more detailed description of the techniques.

Floor planning. Floor planning is the first stage in the layout of VLSI circuits. In this stage, the relative positions of the modules to be laid out are determined. Timing, power, and area estimations are the factors guiding the relative placement. Floor planning can be used to verify the feasibility of integrating a design onto a chip without performing the detailed layout and design of all the blocks and functions. This stage is typical of the less constrained design styles, e.g., macrocell.

At this stage, the aspect ratio of some of the modules may still be unconstrained. For example, if the control logic is implemented with standard cells, then the number of rows used for the modules is not necessarily fixed. Many rows will produce a block that is long and skinny; few rows will produce a block that is short and fat. Different aspect ratios correspond to better packing of the modules in the available area. As other examples, folding and partitioning of a PLA can be used to modify the aspect ratio of the module, or the number of bits used for row and column decoding in a RAM or ROM module can also modify their aspect ratio.

Another degree of freedom is the position of the pins of some of the modules. When standard cells are used for control logic, the signals may leave the block on any point of its periphery. According to their positions, the routing area may be minimized.

Because of the many degrees of freedom and of the uncertainties, the optimization is quite difficult. Very few tools have been developed for this task. Some are interactive and provide estimates for wiring length, channel congestion, area utilization, timing, and power dissipation.

Recently, the introduction of simulated annealing algorithms (see sidebar) has made it possible to develop algorithms where the optimization can be carried out with all the degrees of freedom mentioned above. A system developed at the IBM T.J. Watson Research Center and the TimberWolf package developed at Berkeley use the simulated annealing algorithm to produce a floor plan that not only gives the relative positions of the modules, but also aspect ratios and pin positions.

Automatic floor planning is more and more important as automatic module generators become available which can accept as constraints or parts of the cost functions, pin positions, and aspect ratios of the blocks.

Placement and partitioning. The placement problem involves the assignment of specific locations to building blocks of the layout. This includes the assignment of logic gates within a gate array, the placement of cells in a standard cell layout, or the placement of macrocells.

Placement algorithms may be subdivided into two basic categories: constructive placement and iterative improvement. *Constructive placement* algorithms build a placement from initial data such as the size and the type of the cells to be placed and the netlist. *Iterative improvement* algorithms start with a given initial placement (which can be given by the user, generated randomly, or obtained by constructive placement algorithms) and modify the layout to improve its quality. Interestingly, most of the placement algorithms can be used in the basic design styles (gate-array, standard-cell, and macrocell) with minor variations. Recently, placement programs have been developed based on simulated annealing. The results obtained are excellent and often better than the ones obtained manually at the expense of computer time.

The overall area occupied by the design is obviously dependent on the routing area. The estimation of the wiring area is one of the most difficult parts of a placement algorithm, especially for the macrocell case.

Partitioning is often used to decrease the complexity of placement. This task is certainly useful when thousands of objects have to be placed and the running time of the best algorithms often increases more than linearly (usually quadratically or cubically) with the number of objects. In this case, logic gates or functions are grouped together and assigned to blocks with fixed or variable dimensions. Then, the placement stage determines the actual positions of the components in the blocks and of the blocks on the chip. Note that partitioning can be and has been applied to all design styles. For example, the Engineering Design System of IBM uses partitioning as a pre-process to their gate-array placement program.

Partitioning is a most difficult task because of the many, possibly conflicting, criteria used to evaluate the quality of the result.

Routing. After the modules have been placed, the interconnections have to be completed in the available space. If routing over the cell is not allowed by the technology or by the design style, the area available for interconnections is the one which is not occupied by the modules. In the gate-array and standard-cell case, this area is fairly regular, consisting of a collection of rectangular regions alternating with rows or columns of cells. In the case of macrocell design style, the routing area is much less regular and can be organized in many different ways.

In general, it is convenient to decompose the routing region into rectangular regions, called channels. The way in which the routing region is subdivided may make the routing problem easier. This decomposition is called *channel-* or *routing-region definition* stage.

The routing stage, where the interconnections are laid out on the chip, then follows. This stage is in general broken down into two stages: global or loose routing and detailed routing. The *global routing* stage, sometimes called channel assignment, determines which channels the interconnections will go through. Finally the *detailed routing* stage determines the actual physical location of the interconnections inside the routing regions. If the region to be routed contains pins on two sides only, then effective detailed routing tools called *channel routers* can be used. Extensions to the basic tools can provide programs that can route channels with pins on three sides If the routing region has pins on four sides, then a switch box router can be used. In general, channel routers have the best results in terms of the area used to complete the interconnections. Thus routing regions with fixed pins on more than two sides should be avoided. Channel definition and the ordering with which the channels are routed has a great impact on this issue.

Very effective channel routers have been developed over the years, while detailed routers for more general regions are still not totally satisfactory. Global routing for all design styles is a difficult problem that still needs work to solve the most complex layout problems. Combining global routing and placement into a unique optimization step is an active area of research and we expect new important results to appear soon. Technological advances in processing are making multiple layers of interconnections a reality. New routing tools are needed to take advantage of this degree of freedom that makes the macrocell design problem very similar to the printed-circuit board problem.

Compaction. In any design style, basic circuit cells have to be designed and verified, either for each design as in the case of full-custom, or once for gate arrays (where the designers responsible for the structure of the gate array has to design the basic cell) and standard cells (where those responsible for the libraries have to design the cells). These cells may be simple logic gates, such as NAND gates, NOR gates, and flip-flops. Alternatively, they may be cells used in regular arrays, such as the one-transistor cells used in a programmable logic array (PLA), which do not perform a complete logic function alone. The most commonly used aid for the physical design of a cell is a mask-level digitization and interactive correction program.

Some programs allow direct symbolic layout entry, using either fixed-grid or relative-grid schemes. With the fixed-grid symbolic approach, the grid is designed to ensure all basic layout rules are satisfied upon data entry. For relative-grid schemes, it is necessary to modify the layout such that all layout rules are satisfied. Programs which carry out this operation are often referred to as *compaction programs* since they also attempt to reduce the area occupied by the circuit.

Once a symbolic layout has been entered into the computer, it may be compacted by adjusting the size of non-critical components (such as interconnections) under the constraints imposed by the physical and electrical layout rules of a given technology. Hence, this approach allows symbolic layouts to be updated more easily than physical ones as design rules or technology change. The FLOSS program, developed at RCA for the compaction of circuit cells, paved the way for the development of transistor-level compaction programs for ICs.

Compaction was proposed many years ago, but it has not been as widely used in IC design as expected. We believe this situation is due to the limitations of early compaction systems, which did not allow upper-and lower-bound constraints, did not take into account important electrical considerations, and made it difficult to express and edit constraints. Recent research has been focussed on developing more general algorithms and systems and we expect symbolic layout and spacing programs to become a common tool in the designers' kit.

Combining global routing and placement into a unique optimization step is an active area of research and we expect new important results to appear soon.

Logic synthesis. The synthesis of a circuit—deciding how to partition the logic, in what form to implement specific pieces of the logic, and what layout style to use for implementation—is still a largely manual process. For digital circuits separated into datapath and control circuits, the control logic portion of the chip is often the most time-consuming piece to design. It is generally on the critical path for timing and, because of limits in design time, is often implemented in a very inefficient way. Automated synthesis of the control logic blocks of a chip, optimized for speed and area, provides one of the major challenges facing CAD today.

PLA-based synthesis of control logic. Programmable logic arrays are perhaps the most popular structures for the implementation of two-level logic functions. Most modern VLSI microprocessors include large PLAs to implement the datapath control, as well as a variety of smaller PLAs for controlling other activities on the chip. Other chips such as memory management circuits often consist almost solely of PLAs.

Many PLA layout generators have been written based on simple translations of the Boolean equations into layout. However, a straightforward implementation of the logic entered by the designer may result in PLAs which are large and, as a result, have poor performance in terms of speed and power.

It is clear that PLA optimization is necessary to obtain an effective implementation. The optimization steps involved in the transformation of combinational logic into the layout of a PLA include (1) *logic-level* optimization, which aims at the reduction of the number of product terms needed to implement the function, (2) *topological* optimization, which aims at the elimination of unused space inside the core of the PLA (e.g., folding and simple partitioning), and (3) *layout and circuit* optimization, which attempts to perform optimal sizing and placement of drivers, loads, core cells, and additional ground lines.

Over the past few years, a great deal of attention has been paid to logic minimization of two-level logic. When the logic function is implemented using a PLA, logic minimization both reduces the area occupied by the PLA and improves its electrical performance. The algorithmic complexity of complete logic minimization is very high and so approximate logic minimizers are used when medium and large logic functions have to be minimized. MINI, developed at the IBM T.J. Watson Research Center, was the first efficient heuristic logic minimizer to provide quality minimization. Recent research on approximate logic minimization algorithms has produced an efficient new logic minimizer, called EspressoII-C. EspressoII-C has been found to be very effective in minimizing complex logic functions while consuming a reasonable amount of computer resources.

Once the logic minimization is complete, topological optimization can be performed to minimize the area of the core occupied only by interconnect which does not contribute directly to the implementation of the logic function. The objective of folding is to determine a permutation of rows and/or columns of the array which permits a maximal set of column pairs to be implemented in the same column or row of the logic array.

Much research has been done on folding and other aspects of PLA optimization at various institutions including Berkeley, IBM, Stanford, Katholieke Universitat at Leuven, and the University of Illinois.

Heuristic programming

Over the past two decades, many algorithmic-based tools have been developed for the analysis of ICs. However, many of these tools must process a large number of "special cases," which are exceptions to the basic algorithm. It is often the processing of the special cases that dominates the run time. In other situations, a simple algorithm may not be known for a particular problem but a collection of simple heuristics may be used, such as for checking for more complex electrical-rule violations in the design. In both cases, these heuristics are often difficult to express in existing algorithmic programming languages. These problems are amenable to the relatively new field of heuristic programming.

Over the past few years, a number of both synthesis and verification tools have been developed which rely on the use of such an approach in the form of expert systems.[13] In general, it is our conjecture that if a satisfactory algorithmic solution to a problem is known it should be used. When such a solution is not known, or where a problem has many "special cases" that dominate the run time of the algorithm, the problem is a good candidate for an expert-system-based approach. In the coming years, we expect to see new heuristic programming environments developed which will also allow the tool developer to express an algorithm efficiently.

A number of key areas where heuristic programming approaches have been applied successfully to CAD problems to date are outlined below.

Connectivity verification. Almost all connectivity verifiers can handle the straightforward problem very efficiently with fast, heuristic algorithms. However, most of the time in these programs is spent handling the special cases.

There are many special cases that can degrade the performance of the basic algorithms. Two such cases are terminal permutability and parallel paths. For some circuit elements, the terminals are logically and/or electrically equivalent and are allowed to permute. The inputs to the basic logic gates (NAND, NOR, etc.) and the source and drain of MOSFETS are examples of such situations.

Identical and almost identical parallel paths (as in bit-slice circuits and RAMs) also present a problem to many connectivity verifiers. If the paths are identical, the algorithms currently used cannot distinguish between the paths and may not process them. Also, if two paths have only small differences (as in ROMs), since only local effects are taken into account, connectivity verifiers also may not be able to distinguish between them.

Whereas standard connectivity verifiers process circuits locally (looking at individual elements and nodes, and their immediate neighbors), in the rule-based approach the circuit is processed in a global manner (collections of elements and nodes). As a result, both of the problems mentioned above are handled naturally in the rule-based connectivity verifiers. The designer's hierarchy from the schematic may be used to build patterns that match equivalent groups of elements in the layout. These patterns are rules for a rule-based system, with the circuit description being the data.

Advanced electrical-rules checking. A peer-group review is one of the major checkpoints of the VLSI design process. During this phase, a designer will have his work reviewed by one or more experienced designers who study the schematics and layouts of a design, their mission being to "flush out bugs" that may have been overlooked by the original engineer and to provide feedback to him on his work. To date, this is an area where no effective algorithmic approach has been found short of complete simulation at the circuit level. Even then, the design critic may find errors that circuit simulators cannot detect.

Recently, sequential circuits have been implemented using regular arrays. In particular, PLA-based finite-state machines (FSM) have been used in the design of several microprocessors and telecommunication circuits. Such circuits use a PLA to implement the combinational part of the logic and the secondary outputs are fed back to the secondary inputs of the PLA via clocked latches. For a given set of primary (external) inputs and a required set of primary outputs, the objective is to (1) choose the number of secondary outputs to be fed back, via the latches, as inputs and (2) assign values to these outputs (logic-"1" or logic-"0") for each state specified in the FSM description, such that the total area occupied by the combinational logic and/or the critical-path delay through the PLA are minimized. The "textbook" approach to this problem, originated in the days of discrete SSI circuits, is to choose the number of secondary variables so as to minimize the number of latches used. By doing so, one minimized the number of expensive IC packages needed for latches. To reduce the amount of combinational logic, various heuristic schemes were used. The most common approach was to used a "distance-one" state encoding for adjacent states. Unfortunately, such a state-encoding strategy does not work well for PLA-based FSM designs.

To solve this problem optimally for PLA-based designs, an efficient approach to state assignment is needed. Many algorithms have been proposed in the past to perform optimal state-assignment. However, the results obtained were not satisfactory because of the complexity of the algorithms suggested or of the poor electrical performance of the PLA used to implement the resulting combinational part of the FSM.

Further work is needed to partition a large FSM into smaller machines where the intermediate output values are encoded. Though this process may result in more stages of logic, in many cases it is expected that the increase in clock speed that can be achieved using the small PLAs will more than compensate for the additional clock cycles caused by more stages. Once again, algorithms have

The items turned up during this review cycle take many forms. In some cases they are very simple problems. Perhaps a new designer may not understand the implications of a design rule or needs some guidance as to the more practical ways of implementing a logic function. In other cases, extremely subtle problems are found which can elude even the most careful worst-case simulations. Charge coupling, MOS capacitor inversion time constants, charge sharing on dynamic buses, voltage swings on bootstrapped nodes, and problems due to clock undershoot, overshoot, overlap, and skew are some examples of the latter.

Over the past three years, a number of circuit design critic programs have been implemented using rule-based systems to perform these checks.

Routing. Detailed routing is quite complicated if the region to route has pins on four sides (the two-dimensional routing problem). In addition, most of the available detailed routers are constrained to route vertical connections and horizontal connections on two different layers to simplify the structure of the algorithms. Many objective functions are used to evaluate the quality of the results by designers. For example, the number of vias used to complete the routing can be important to improve the reliability of the chip and net length is important to optimize the speed of the chip. Thus the detailed routing problem in two dimensions seems to be a natural application of expert systems and some early work in this area from CMU looks very promising.

The algorithms used to route gate arrays rarely provide 100 percent interconnection completion, and several hours—if not days—of designer time are needed to complete the interconnections left out by the algorithms. An expert system developed at NEC uses a set of rules, obtained by capturing the designer expertise in performing the rip- and-reroute operations, to complete the routing. In addition, this expert system can call on algorithms to evaluate the application of a particular rule.

Multilevel logic synthesis. One of the first systems for automatic optimization of multilevel logic, LSS of IBM, could be considered a prototypical rule-based system. Recently, a full-fledged rule-based expert system has been developed at GE Microelectronics Center to optimize combinational logic for a specific target technology. The system starts with a description of the logic to optimize and performs a series of algorithmic steps to produce a starting point to which a rule-based expert-system applies a series of local transformations. The knowledge base can be easily enriched by the user through a rule-generation module that automatically encodes new rules and inserts them in the knowledge base. A control module directs the application of the rules to the data. In particular, the module determines in which sequence the transformations are to be applied, based on evaluations of the effectiveness of the transformation.

Behavioral synthesis. A rule-based expert system was built at CMU to translate effectively a behavioral representation of a design into a structural representation. The rule base was built by successive approximation, with input from designers who criticized designs that were obtained with the tool. The first rule set used contained 70 rules; the latest version of the tool contains more than 500 rules.

The conclusion of the CMU researchers has been that while the algorithmic approach can be more effective if the cost function is well specified, the expert system approach offers more flexibility and a better environment to capture the often difficult-to-express intent of the designer.

> In recent years, the notion of procedural circuit design has emerged as a key component in the design process.

been proposed in the past, but the technological constraints and objectives which drive the decomposition have changed drastically so as to make the existing algorithms inappropriate.

Multilevel synthesis. As seen in the previous section, a great deal of work has been done to implement combinational logic in optimal, two-level form using the PLA. However, some control logic has a two-level representation which can have as many as 2^n product terms, where n is the number of primary inputs of the logic, even after minimization. In addition, even if a two-level representation contains a reasonable number of terms, there are cases in which a multilevel representation can be implemented in much less area and generally as a much faster circuit. In fact, a two-level logic representation can be viewed as a special case of general multilevel representations. Hence, a general framework for control logic design should offer multilevel synthesis tools which are able to select a two-level implementation wherever the two-level form is more effective in terms of area and/or speed. To be able to explore the design tradeoffs, such a system should offer a variety of both electrical design styles (e.g., Domino logic, static CMOS) and layout design styles (e.g., Weinberger arrays, gate matrices, standard cells, and gate arrays) alternatives.

Several systems are being built for the design of control logic using multiple levels of logic. The precursors in multilevel logic synthesis are the IBM Logic Synthesis System (LSS), which has as target technology a variety of gate arrays and has been extended to standard cells and to CMOS dynamic logic; the IBM Yorktown Silicon Compiler, which has Cascode voltage switches as its target technology; and the UC Berkeley MAMBO system, which uses Domino Logic. AT&T Bell Laboratories with the FDS system, NTT with the Angel system, and NEC and Hitachi with the POLARIS system were all developed with standard cells and gate arrays as target technologies. Only recently, electrical consideration have been taken into account during the synthesis process; MAMBO is an example.

For multilevel design, there are two basic approaches to the logic optimization step: (1) *global optimization,* where the logic function is refactored into an optimal multilevel form without considering the form of the original description (e.g., the Yorktown Silicon Compiler, part of Angel, and FDS); and (2) *"peephole" optimization,* where local transformations are applied to the user-specified (or globally optimized) logic function (e.g., a part of Angel, LSS, and MAMBO).

Both these techniques have good properties, but much research still needs to be done to improve the basic algorithms used by both. In particular, the mathematical aspects of multilevel logic synthesis are not as well understood as those of two-level synthesis. We expect a great deal of research in this area and some system to become generally available in the next few years.

Synthesis from high-level behavioral description. The translation from a behavioral description to a register-transfer level description involves architectural decisions that have great impact on the final quality of the design. The difficulty stems from the large number of constraints, design objectives, and design configurations to consider. In addition, it is very difficult to evaluate at this stage the effect of an architectural decision on the speed, power consumption, and area of the chip.

A key issue in behavioral-level synthesis is the selection of the language used to describe the design. The language must be concise and have high-level constructs to express compactly the intent of the designer. High-level programming languages such as Modula, PL/1, and concurrent Pascal have been used to describe designs at behavioral level. Special-purpose languages such as the ISPS language developed at CMU and the MacPitts language have also been developed. These languages are powerful enough to allow the description of existing computers in only a few pages.

Many approaches have been tried at CMU, some involving algorithmic approaches such as EMUCS and Facet, some involving knowledge-based expert systems such as the Design Automation Assistant (DAA), and the Sugar system under development.

The MacPitts input description is compiled into a datapath and control by replacing the language constructs with hardware. The control part is expressed as a finite-state machine and implemented with a Weinberger array. The datapath is synthesized using basic one-bit units called organelles. Operations which are specified in the input as mutually exclusive are implemented in parallel, i.e., hardware is generated for all the operations which can be done in parallel, while operations which are not mutually exclusive are implemented using as much as possible hardware already implemented for other operations. It is clear that if the input description specifies more operations as mutually exclusive, the execution of the operations is faster but more silicon is used. Thus, the user has some control over the hardware generated by the silicon compiler. As pointed out before, the fixed floor plan and design style for the control unit has resulted in silicon implementations that are not as compact and efficient as human designs.

Procedural design and module generation. In recent years, the notion of procedural circuit design has emerged as a key component in the design process. The use of procedure rather than just graphics (or data) for describing IC designs was pioneered at Caltech and applied later at MIT in the DPL project. These efforts, and others, have inspired a large number of projects at university and industrial sites for in-house use. The term "silicon compiler" is often associated with procedural design and a number of new companies are advertising silicon compilers. Unfortunately, the majority of systems offered so far do not offer the general user a procedural design capability. Rather, they can be characterized as cell-based systems where the circuit building-blocks are parameterized cells. These cells can be assembled in a variety of ways, depending on the design style the "compiler" is using.

True procedural design, where the IC designer can write programs which, when executed, produce layouts, is still of key importance to the productivity of the custom IC designer. Silicon Design Labs offers a procedural design system based on their "L" language, which is an evolution of the "I" language developed at AT&T Bell Laboratories.

Many of the early procedural design systems had limited success for a number of reasons:

- The relationship between graphics and procedure was not exploited sufficiently. Graphics and procedure are generally treated as disjoint descriptions of the design. The graphical mask layout was the result of the procedure, rather than an active part of it.
- Verifying the correctness of a procedurally generated design is generally performed at the mask layout level. At that point, no correspondence between the active procedure that created an object and its geometrical layout is maintained. As a result, debugging the design is very difficult.
- Often the designs produced by the early systems did not achieve high density, high performance, or meet power requirements.

Another way of reducing the potential for errors, as well as ensuring the technology-parameterization necessary at the lowest level of design, is to have the procedural design system generate symbolic layout rather than detailed geometry. A spacing program, or *constraint solver,* can then be used to guarantee a layout-rule-correct design and to convert the symbolic layout to mask artwork.

As described in this paper, computer aids have been used for both the design and verification of electronic systems for many years prior to the introduction of commercial ICs in the early 1960s. These tools have found their way into virtually every aspect of the design of such systems, from IC process technology to the design of complex computer architectures. Today, it would not be possible to design a complex IC without CAD tools and we believe soon these tools, for data management, verification, and synthesis, will be as significant as the underlying semiconductor technology in differentiating products in the marketplace.

The use of CAD in IC design is now a very broad and very deep subject. While it was not possible to go into detail in this paper, we have indicated the history of CAD for IC design, the state-of-the-art, and the present directions for future work. The reader is directed to Newton [1] for a more detailed presentation of this area, with an extensive reference list. □

Acknowledgments

This paper could not have been written without the many years of work in the area of CAD for ICs in collaboration with our excellent graduate students and with our colleagues in the microelectronics and computer architecture area at the Department of Electrical Engineering and Computer Sciences and Electronics Research Laboratory of the University of California at Berkeley, and in particular with Don Pederson, who continues to provide support and encouragement. We wish to acknowledge the support of the Semiconductor Research Corporation, the Defense Advanced Research Projects Agency, the member companies of the Berkeley CAD/CAM Consortium and the Berkeley Micro-Affiliates programs, and the MICRO project of the State of California, who have all contributed to the development of our research laboratory.

References

1. A.R. Newton and A.L. Sangiovanni-Vincentelli, "Computer-Aided Design for VLSI Circuits," Electronics Research Laboratory Memorandum No. UCB/ERL 86/16, Univ. California, Berkeley, Feb. 1986.
2. A.R. Newton, "Computer-Aided Design of VLSI Circuits," *Proc. IEEE,* Oct. 1981, pp. 1189-1199.
3. R.H. Katz, *Information Management for Engineering Design,* Springer-Verlag, Berlin, 1985.
4. M.H. Arnold, "Corner-Based Geometric Layout Rule Checking for VLSI Circuits," Rep. No. UCB/CSD 86/264, Computer Science Division, Dept. EECS, Univ. California, Berkeley; also, Kluwer Academic Publishers, to appear.
5. A number of references may be found in *Computer,* Vol. 7, No. 12, Dec. 1974.
6. G.D. Hachtel and A.L. Sangiovanni-Vincentelli, "A Survey of Third Generation Simulation Techniques," *Proc. IEEE,* Oct. 1981, pp. 124-1281.
7. A.R. Newton and A.L. Sangiovanni-Vincentelli, "Relaxation-Based Circuit Simulation," *IEEE Trans. Elec. Dev.,* Vol. ED-30, No. 9, Sept. 1983, pp. 1184-1207. See also *SIAM J. Scientific and Statistical Computing,* Vol. 4, No. 3, Sept. 1983; and *IEEE Trans. CAD of ICs and Systems,* Vol. CAD-3, No. 4, Oct. 1984, pp. 308-331.
8. *Computer Design Aids for VLSI Circuits,* ed. D.O. Pederson, H. DeMan, and P. Antognetti, Sijthoff and Noordhoff, 1981.
9. R.B. Hitchcock, Sr., "Timing Verification and the Timing Analysis Program," *Proc. 19th IEEE/ACM Design Automation Conf.,* June 1982, pp. 594-604.
10. J. Soukup, "Circuit Layout," *Proc. IEEE,* Oct. 1981, pp. 1281-1305.
11. T. Blank, "A Survey of Hardware Accelerators Used in Computer-Aided Design," *IEEE Design & Test,* Vol. 1, No. 3, Aug. 1984, pp. 21-43.
12. S. Kirkpatrick, C.D. Gelatt, M.P. Vecchi, "Optimization by Simulated Annealing," *Science,* Vol. 220, No. 4598, May 13, 1983, pp. 671-680.
13. E.A. Shortcliffe and R. Buchanan, *Rule-Based Expert Systems,* Addison-Wesley, 1984.

Part II: University Programs: An Historical Perspective

A Structured Design Methodology and Associated Software Tools

STEPHEN TRIMBERGER, JAMES A. ROWSON, CHARLES R. LANG, MEMBER, IEEE, AND JOHN P. GRAY, MEMBER, IEEE

Invited Paper

Abstract—**The problems encountered designing very large scale integrated circuits (VLSI) are fundamentally different from the problems encountered in the design of small scale integrated circuits. The differences require a new methodology of design for the new large scale circuits, and the new design methodology requires a new set of tools. The computer-aided design work at Caltech has progressed from a recognition of the inherent differences and has produced a new design methodology and a set of tools which attack the new problems in integrated circuit design.**

I. Introduction

THIS PAPER is intended to give the reader an introduction to the ideas and tools that underlie VLSI design research at Caltech. These ideas and tools have evolved not only from the faculty and students, who have been designing integrated circuits since 1970, but also from industrial representatives from Caltech's Silicon Structures Project. This project, in existence since 1977, staffed in part by sponsor's representatives on one year assignments at Caltech, has been a valuable source of insight into the problems being faced and about to be faced by VLSI designers.

The paper is divided into three parts: a description of the design philosophy, a summary of important tools built around this design philosophy, and a discussion of results and conclusions.

II. Design Philosophy

VLSI technologies have the capability to produce chips containing a hundred thousand transistors [15]. With current design techniques, it takes around 60 man years to design, and another 60 to debug, such a chip. At the theoretical limits, VLSI chips will contain nearly 10 million transistors. Without some method for reducing the complexity of design, a 10-million transistor chip would take somewhere near 6000 man years to design.

These numbers clearly indicate that there is a widening gap between what VLSI technologies can produce and what system designers can design. At present, the only chips that approach the available density are memory chips, and those only because of their extremely regular patterns. With less regular systems like microprocessors, designers are having difficulty merely *completing* complex designs.

The *structured design methodology* of Mead and Conway [17] in use at Caltech is an approach to VLSI system design that attacks the problems of complex designs. Elements of this structured style are apparent in many successful designs. The structured design methodology has two major parts: hierarchy and regularity. Hierarchical techniques have long been used to design complex systems [13], [24]. Hierarchies are used to partition designs among the design team members. Also, common parts of a design can be factored out and specified only once. By introducing regularity into a system, the design problem is reduced in complexity as subunits are replicated many times and connections between units are simplified. Traditionally irregular control structures have their regular counterparts in ROM and PLA. Wiring strategy and regularity are addressed from the start, eliminating inefficient and costly routing.

The structured design methodology is similar in concept to structured programming [7]: the design proceeds in a top-down manner in which the problem is decomposed and refined. The designer is limited in the kinds of structures he may use to implement a function. The advantage of restricting the structures is that the design can be implemented more quickly and reliably. Proofs of correctness are easier to produce and more of the design can be automated.

In the words of Dijkstra [8]: "Testing can be used to show the presence of bugs, but never to show their absence." A major goal of computer-aided design at Caltech is to produce correct designs without the need for checking. This goal of *correctness by construction* permeates the design methodology and tools at Caltech. Emphasis is placed on tools which generate designs to be correct in some sense, rather than on tools which check designs after they have been generated. These tools can be very specific to the design methodology in use at Caltech. The mating of the tools to the design style allows more powerful tools to be constructed and leads to much greater productivity from the designers for whom the tools work.

A. The Separated Hierarchy

Hierarchical design at Caltech is predicated on a restricted hierarchy called the *separated hierarchy* [21]. The

Manuscript received November 5, 1980; revised January 22, 1981.
The authors are with the Computer Science Department, California Institute of Technology, Pasadena, CA 91125.

Reprinted from *IEEE Trans. Circuits Syst.*, vol. CAS-28, no. 7, pp. 618–634, July 1981.

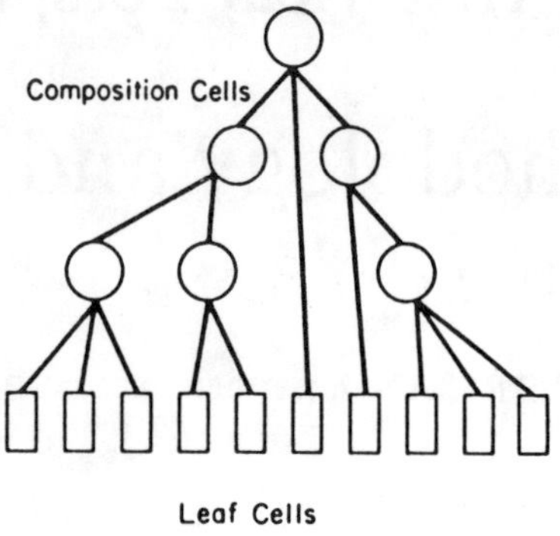

Fig. 1. The separated hierarchy.

separated hierarchy, shown in Fig. 1, has two different kinds of cells, *leaf* cells and *composition* cells. A leaf cell is defined *only* in terms of primitives. No instances of other cells are allowed. A composition cell contains *only* logical interconnections of instances of other cells, no primitives allowed. The tools needed for manipulating these two kinds of cells are vastly different.

Composition cells in the separated hierarchy form a *representation* independent language for specifying a design. A representation is one particular view of a design. There are many possible representations including mask geometry, stick diagram, schematic diagram, English language description, behavioral description and so on. Leaf cells, containing only primitives, must be specified for each representation. The same composition cells can be used for all representations because they contain no primitives. In order to deal with a composition cell within a particular representation, a *composition rule* for each representation is specified. A composition rule is just an algorithm that will produce a definition of any legal composition in the corresponding representation. A legal composition could be as simple as "No two outputs may drive the same node."

A set of design tools built around the separated hierarchy provides mechanisms for designing leaf cells in many different representations, a composition rule for each representation, and a way to design composition cells. One problem inherent to a multiple representation system is that of consistency checking. If a designer specifies a particular cell twice, i.e., as geometry and as circuit schematic, some method for guaranteeing consistency between the two representations is needed. Composition cells, being representation independent, need not be checked for consistency. Composition rules must be consistent between representations, but this consistency need only be checked once per representation. Leaf cells clearly need to be checked. The hierarchical decomposition of the design leads to relatively simple leaf cells, making consistency checking (even by eye) relatively simple.

B. Algorithmic Design

Large chips closely resemble large programs in their variety and complexity. The need for powerful design systems to make these chips pushes us into the most powerful forms we have: algorithms represented by programming notations. The algorithm may be very simple, such as a series of rectangle placements. The use of a programming language allows placement of features to be done in a relative way, so that if one item or cell moves or grows, others follow. The procedures may be very high level, such as routing of bus connections, deforming cells to interface without routing, designing a PLA, or computing the width of power wires according to the current required.

Cells can be parameterized so that they can adapt to the environment in which they are instantiated. For example, a transistor may change size to react to a change in load capacitance. The list of parameterizations is as long as the number of designers involved. Adaptable cells designed in this parameterized fashion will "live" longer and be more useful as building blocks in later designs which may have different requirements. They also allow decisions about detailed characteristics of the cell to be delayed until later in the design cycle.

As an example, consider a register cell in a dual-bus system. This register may be required to read data from or write data to either of the two buses in the system [11]. Depending on intended use, the cell register may require refresh and preset logic. These registers may be used many times on each chip, and each usage, or instantiation, of the cell may require a different subset of the six available functions.

Since the cells are programs, we need not design all 64 possible register implementations and select the appropriate one as required. Instead, we can generate *one* register cell program that can remove unused circuitry and thereby generate any particular register instance as it is required. Fig. 2 illustrates this technique. Fig. 2(a) shows the six function register cell described above. It is possible for the same parameterized cell to be generated without the preset capability and without access to the lower bus, in which case the smaller register is produced, as shown in Fig. 2(b). The user of the register can specify its configuration with a list of the desired functions for the register: a behavioral description.

The recognition that the difficulties of complexity management overwhelm the problems of geometry generation leads to a new set of constraints on the design tools. Design tools must address the problems induced by the complexity of the design at all levels. Cells defined algorithmically can adapt to changes in their neighbors, thereby restricting small changes in the design to small amounts of effort by the designers as they incorporate those changes. Programming languages provide the necessary versatility, although the precise characteristics of an ideal design language have not been determined.

C. Descriptions of the Design

It is possible to identify three domains of design description [4] which must be addressed in a finished design. The three descriptions are: the behavioral, the description of the function of the design; the structural, the description of the form of the implementation; and the physical, the description of the physical implementation of the design. There are many possible representations of a circuit in each description, and judicious choice of representations is important in tool design.

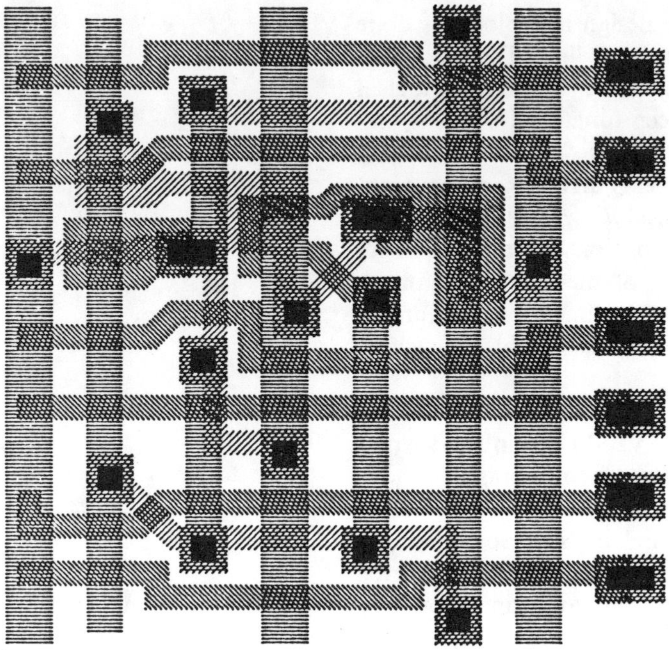

(a)

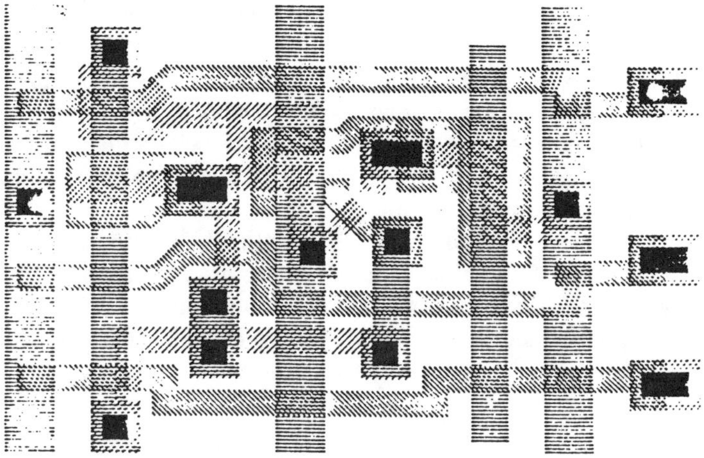

(b)

Fig. 2. (a) Register cell with all functions. (b) Minimal register cell.

1) Behavioral Description:

Integrated circuit designers are in the business of building behaviors. A part which does not have the desired behavior is not acceptable no matter how clever the design. In this sense, designing integrated circuit hardware is similar to designing software and the problems associated with the two are often similar.

The global solutions to the problems in these two kinds of designs are similar. Designs are structured, hierarchical, and divided functionally into meaningful pieces. Tools are needed which help convert the high-level description of the behavior into a low-level implementation description.

Since we are concerned with the behavior of the circuit, work on design tools is influenced by rigorous disciplines and notations for describing circuits that can be guaranteed to be correct in some sense. For example, the self-timed discipline [23] can guarantee correct sequencing based on interconnection rules. Another good example is signal restoration in CMOS, ordinarily assured by checking, which can be guaranteed by rules of construction [20].

2) Structural Description:

A design is not merely a behavior. Designing integrated circuits is a mapping of a behavior onto a physical structure. There are fundamental differences between software design and integrated circuit design. One cannot blindly implement a function as a chip without addressing physical implementation issues. Integrated circuit design deals with inherent massive parallelism; software traditionally does not. Traditional software solutions to problems may not be proper in silicon. The designs, as well as the design methodology, must address the physical aspects of the medium.

Integrated circuit design has a severely limited communication space which shares the computation space on the two-dimensional surface of the silicon. Thus communication costs are high in silicon, but very low in software. In order to effectively use the silicon area, a design system must take into account these properties of integrated circuits.

The structural description is a description of the logical connection of blocks in the system. Hierarchical decomposition of the behavior of the design into blocks is done along both geometrical and functional lines. The functional and geometrical hierarchies must match because the logical connection, the interface between functional units, is precisely along the geometrical interfaces.

The hierarchical decomposition is driven by a very high-level part-behavioral, part-physical *floorplan*. The floorplan is a general functional decomposition strategy which includes a wiring strategy as part of the decomposition. A good floorplan recognizes the two-dimensional nature of the silicon chip and addresses physical problems such as global wiring.

3) Physical Description:

The physical characteristics of VLSI chips introduce new difficulties in complexity management not present in programs: the production of a geometrical structure under many constraints of topology and physics. Some systems have been built which implement function on an integrated circuit without regard to the inherent communication limitations of the silicon [5], [19]. These systems work well for small scale designs, but are overburdened with communication overhead with larger designs [26]. For large designs, the physical constraints of the silicon medium are so restrictive that they must be addressed early in the design. Systems which deny the physical nature of the silicon implementation cannot hope to effectively use the silicon medium in large designs.

Correctness by construction has been applied to the geometrical design task. Stick diagrams are used to generate geometrical layouts without the need for design rule checking. Designs can be produced much more quickly, and more iterations on a design can be quickly done to obtain an acceptably optimized layout.

D. Design Process Flow

This section describes the design process as practiced at Caltech. This design process embodies the structured methodology described above. This section also provides a framework for the design tools described in Section III.

The design process has two distinct parts: design and implementation. Design proceeds top-down, with global decisions made first. The implementation then proceeds bottom-up, where constraints from low-level implementations are propagated to higher levels in the design hierarchy.

The design process is divided into five pieces: architectural design, where the design is partitioned into functional pieces and the general floorplan of the design is decided; cell estimation, where cell interfaces—size and interconnection—are decided; cell detailing, where detailed cells are laid out; chip integration, where cells are assembled into chips; and preparation for fabrication, where the finished design is converted into a form suitable for fabrication equipment. Of course, the design process may iterate in any of the loops seen in Fig. 3.

The high-level top-down architectural design is still predominantly a human task. In order to achieve a good design, the designer must not only concern himself with functional decomposition, but with wiring strategy as well. The product of the architectural design is a floorplan, a

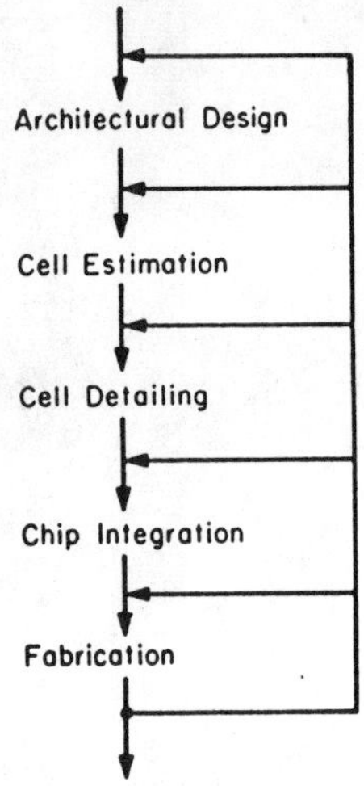

Fig. 3. Design process flow.

tiling of the plane with functional units. The floorplan includes a rough specification for each of its elements, including constraints, interfaces and desired features. At this point, expected critical paths, both in area and speed can be estimated. This estimate is important in order to properly direct later optimization work.

Cell estimation is an important precursor to cell detailing. In the cell estimation phase, approximate geometries are tried for the blocks defined in the floorplan. If these designs cannot be made, or if optimization of the design is too difficult, the floorplan may have to be altered. Stick diagrams are a useful notation in the cell estimation phase, they allow both structural and geometrical information to be expressed in a highly readable form. The cell estimation phase proceeds until all interfaces between cells are reconciled.

In the cell detailing phase, final designs are produced for each low-level cell using graphic or program oriented design aids. Detailed cells may be generated in the form of hard mask geometry, malleable sticks or algorithmically defined cells. Cell detailing is the start of the bottom-up implementation of the design.

Cell detailing has typically been given the lion's share of the CAD vendor's effort. In large systems, most cells are not in the critical path for speed of area use and therefore do not require optimization. Systems which automate the layout task are useful and desirable even if the result is not as good as hand layout. The fact that it is faster than hand layout is enough to make it worthwhile. A cell design must be consistent with the floorplan and it should have the ability to interface to high-level assembly tools.

Correctness by construction leads to the exploration of means by which classes of errors are avoided. Therefore, symbolic layout or stick diagrams are an important means for the generation of detailed cells as they guarantee designs free from geometrical design rule violations. In addition, since the stick diagram geometry is malleable by the very nature of the representation, sticks provides a good interface to the chip integration phase that follows.

The chip integration phase is the one in which cells are assembled, according to the floorplan, into a finished design. The assembly task is extremely difficult if the floorplan is badly done or if proper attention was not paid to the cell interfaces. Much work has been done at Caltech on chip assemblers and silicon compilers, which, working with a given floorplan and properly defined cells, will assemble low-level cells into complete integrated systems. These design systems have been very successful, able to quickly produce many different chips in one class of floorplan.

Programming language based systems are preferred tools for the chip integration task because of their versatility. Powerful compositions can be easily defined in this algorithmic manner. Properly defined compositions allow changes to individual cells to be made without requiring a change in the way those cells are composed into a system. This leads to a very versatile design method. New ideas and optimizations, as well as bug fixes, can be made without requiring large changes to the composition of the system.

The final step in the design process is preparation for fabrication. This is typically a batch process, requiring a large processor for vast amounts of time. The hierarchical nature of Caltech designs enable this phase of the design to proceed faster than traditional methods. Plotters, mask generation programs, and design rule checkers exist which take advantage of the hierarchy in the Caltech designs to speed up the processing enormously.

III. Design Aids

Caltech has developed aids ranging from pattern generation software to silicon compilers. Many of the tools have been tested by extensive use in designing large chips.

A. Overview

The design aids at Caltech are being developed almost entirely on a DECsystem-20 computer. The DEC-20 has a number of design stations consisting of an LSI-11 driving a text terminal, a four-color pen plotter, a color frame buffer for graphics output and a Xerox "mouse" pointing device for graphics input. The graphic workstation is shown schematicly in Fig. 4.

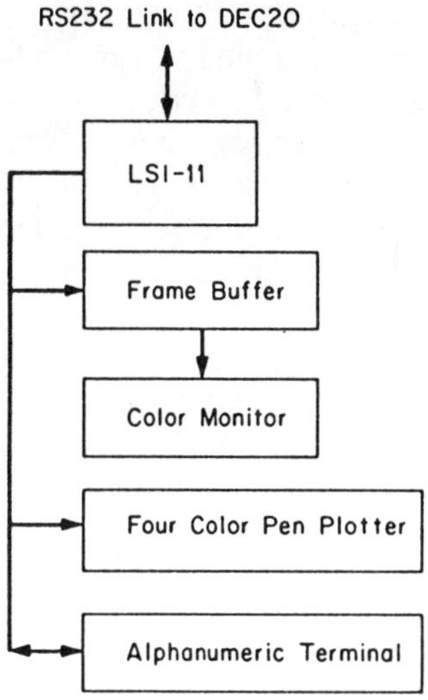

Fig. 4. Graphic workstation.

The computer science department has access to an IBM 3032 computer belonging to the Caltech Computer Center. This machine is used for large scale batch-mode processing and use is made of the additional plotting facilities on the IBM machine. The department has also recently procured a Digital Equipment Corporation VAX 11/780. Some of the software has been transported to that machine, and much more will be developed for the VAX environment.

1) Data Structure Overview:

Caltech has forsaken the standard approach to a unified set of tools: that of the all-encompassing restricted-access database. There are inherent problems with the database approach. Much of the effort in a database is in access control, file storage and optimization of memory usage. These are precisely the tasks performed by an operating system and executive. Rather than reimplement these functions, Caltech has decided to use the externally supported, highly debugged facilities provided with the DECsystem-20 computer.

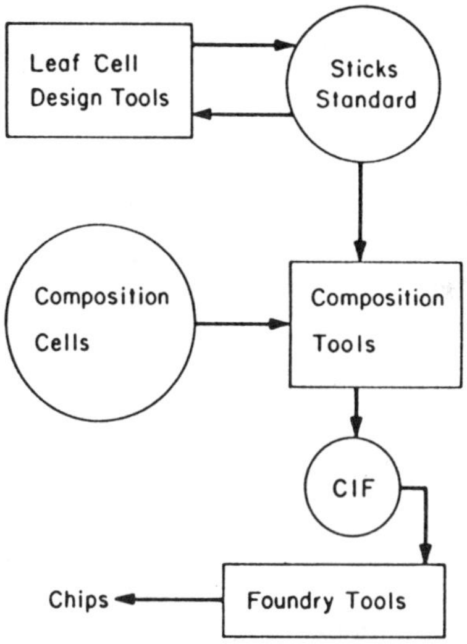

Fig. 5. Tool communication overview.

A database allows tools to communicate, but only after conformance with the predefined database model. This model is usually extremely restrictive, since it was designed before the tools which use it. Extensions to the database are disruptive and costly. Interface to the database is typically a considerable fraction of the effort required for the development of a new tool. In a university, where innovation and experimentation are necessary, being locked into a stagnant database is surely a death knell.

Rather than limit future design systems with an outdated data description language, simple interchange standards allow the tools to communicate to achieve compatibility (see Fig. 5). Each set of tools: leaf tools, composition tools, and foundry tools, is a collection of loosely coupled programs communicating through a standard file format. Thus a tool may communicate with other design systems while still maintaining any internal representation necessary or desirable for its application. Such a data structure allows easy incorporation of new forms of data and new storage techniques. In a sense, we are following our own chip design advice by organizing our software to communicate through simple well-defined interfaces.

The Caltech Intermediate Form (CIF) [25] is the standard interface to all the "foundry" (fabrication) tools. CIF is a hierarchical geometrical description, and contains commands to construct rectangles, wires, polygons, and circles as well as facilities to define and call symbols. The Sticks Standard [27] is the standard interchange form for systems which deal with symbolic layout and is being used as input to composition tools. The Sticks Standard is a symbolic layout description, and contains cells described with components, interconnections, and constraints on positions of the components in the stick diagram. These standards provide the means for communication in a precisely specified language, they clarify the essential concepts of the form of data they describe, and they stimulate tool development around the standard.

2) Tool Overview:

Tools developed at Caltech can be grouped into three categories: those that help design leaf cells, those that help design composition cells, and those that help interface to fabrication. The leaf design tools aim at producing the primitive functional units used in a chip design. Leaf cells contain only primitive elements: polygons, rectangles, and wires. Composition cells contain only instances of other cells, either leaf cells or other composition cells. The interface to fabrication is needed because of the large number of designs that Caltech produces, both for research projects and class projects.

An overall view of the tools is illustrated in Fig. 6. A set of "foundry tools" underlies a system of "compilers" which produce full silicon systems based on leaf cells designed using "cell design" tools. These sets of programs are not tightly bound into a single system, they communicate through standard data interchange formats.

The split between leaf cell tools and composition tools is deliberate. By separating the two, we not only match the structure of the separated hierarchy, but gain a great deal of independence and flexibility. The independence allows tools to be built with little detailed knowledge of other tools. Flexibility is manifested in the ease with which new tools and new design approaches can be added to the overall system.

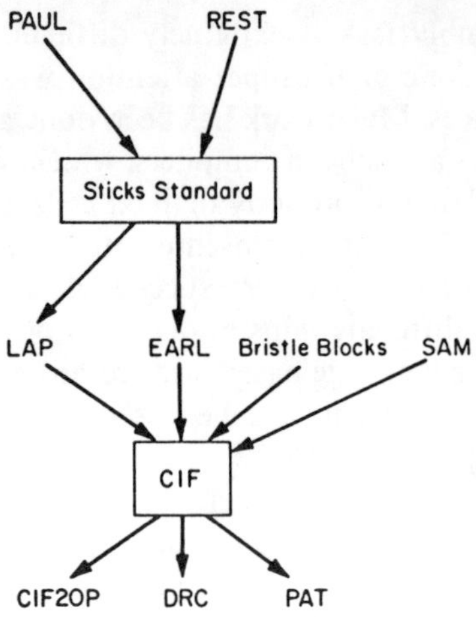

Fig. 6. Tool overview.

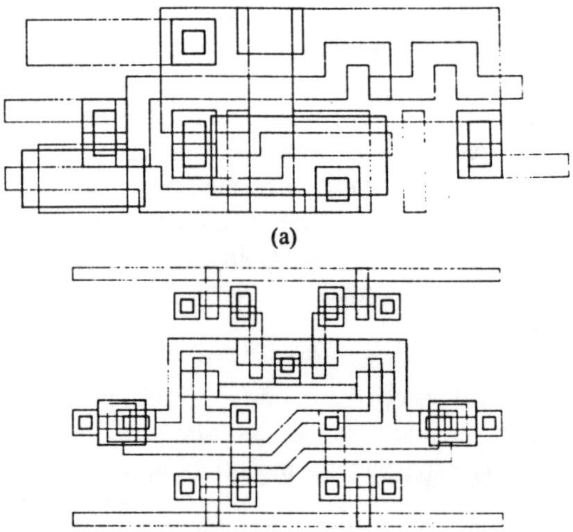

(a)

(b)

Fig. 7. Leaf cell examples. (a) Super buffer cell. (b) Static register cell.

B. Leaf Cell Tools

A leaf cell is generally a simple functional unit that is implemented using only primitive elements. The set of leaf cells represent the semantic units the designer can use to describe a chip. Deciding which pieces of a design are to be made into leaf cells is a tradeoff, matching a desire to have it be a self-contained functional unit with some limit on design difficulty. Too small, and the leaf cell won't be complete enough to specify an understandable function. Too big, and the designer won't be able to understand the relationships among all the parts, making it difficult to understand and modify. (See Fig. 7.)

In chip design, the design of a leaf cell is basically graphical in nature. The cell designer must specify a mask geometry which implements the desired function.

There are three methods used at Caltech to design leaf cells: interactive graphics, graphical languages, and a combination of both. All three have their advantages. The interactive graphic systems provide a very good human interface, allowing rapid on-line design. Graphical languages provide a less interactive environment because designs have to be "executed" before a plot is produced. But they give the designer more power, in the sense of variables, expressions, conditions and loops, than the interactive system. Hybrid systems have some of the advantages of both of these systems.

The three leaf design methods are embodied in the following design tools: REST, LAP/PAUL, and SAM.

1) REST (graphics):

Richard's Editor for STicks (REST) [18] is a leaf cell design system based on symbolic layout techniques of [29] and closely related to the work of [10]. REST runs on the high-resolution color display with editing functions handled by code in the display processor. The symbolic layout manipulation is done in the DECsystem-20. Since REST is a leaf design tool, it does not support a design hierarchy. REST interfaces to the Sticks Standard which can be read by silicon compilers for chip assembly.

The REST system deals with a form of symbolic layout known as *stick diagrams*. A stick diagram is a notation

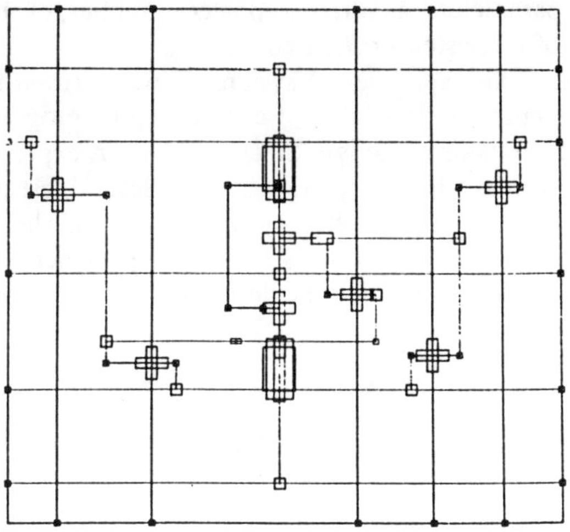

(a)

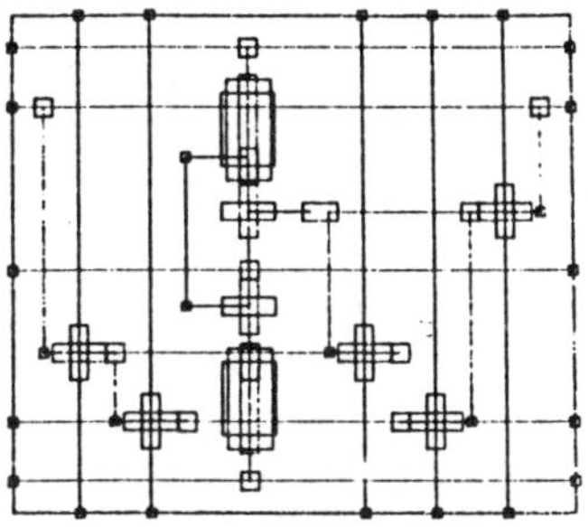

(b)

Fig. 8. (a) Uncompacted stick diagram. (b) Compacted stick diagram.

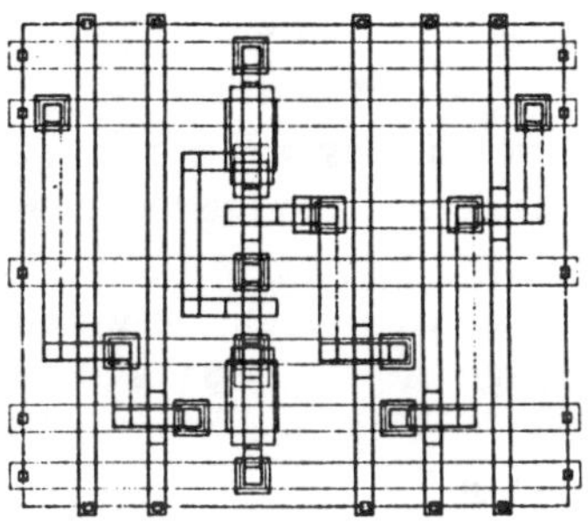

Fig. 9. Layout generated from stick diagram.

midway between transistor diagrams and full mask layouts. Stick diagrams specify more geometrical information than a transistor diagram in that the relative positions of transistors and wires are meaningful, but less than a full layout in that the absolute positions of transistors and wires are not meaningful. This intermediate position has many advantages. The designer can specify his layout in a very sketchy manner, with no regard to exact positioning, yet still have control over the relative topology of the layout. Component recognition and circuit compaction free the designer from worries related to details about mask making. Contrast the sticks drawing, of a static register cell in Fig. 8(b) with its derived layout in Fig. 9. The stick diagram includes device sizes and connectivity information, which are essential for performance estimation and simulation. This abstract specification allows the designer to concentrate on the system design. Stick diagrams were designed to be deformable, making them a preferred input form for silicon compilers and chip assemblers (see Section III-C.)

The compaction algorithm in REST compacts along the coordinate axes one axis at a time. The algorithm is simple and very fast, providing a unified, highly interactive man–machine interface. The user can control the direction of compaction, which axis is compacted first, and add constraints to the graph. REST does not make topological changes. The results of compaction are illustrated in the contrast between an uncompacted stick diagram input by the designer, and the stick diagram resulting from compaction in Fig. 8.

REST has been used successfully in conjunction with other design tools at Caltech to produce working designs. Its usefulness will expand as additional composition tools become available.

2) *LAP/PAUL (procedures)*:

There are many advantages to using an expressive language to design integrated circuits (see Section II-B). Almost any programming language can be used to design mask geometry simply by adding some procedures to output primitives of the Caltech Intermediate Form (CIF). A simple example of such a procedure is "Rectangle" which simply outputs the CIF code for a rectangle. The coordinates are the parameters of the procedure. Most programming languages allow those parameters to be expressions, enhancing the adaptability of the procedure. A more complex procedure might be "PLA" to generate a programmed logic array. The parameters of this procedure are the coordinates of the origin, number of minterms, inputs, and outputs, and the names of the files holding programming for the AND and OR planes. There is no limit on the complexity of the design generated by the embedded procedures: the PLA procedure may optimize the PLA without any changes needed by the PLA user. These procedures can be called by the designer in the midst of a program that "computes" the chip. Such procedures create a kind of "embedded language" that can be used as a design language.

LAP [14] and PAUL are two such design languages, both embedded in the programming language SIMULA. LAP produces CIF files, PAUL produces Stick Standard files. The full power of the SIMULA is available to the user in either of these languages. LAP has been in use since 1977 at many installations in research, education, and industry and has been a major chip design system for both leaf and composition cells. Over the years, it has been enhanced with a PLA generator, several simple routers, mask lettering generator and Sticks Standard to CIF translator. PAUL is a recent development designed to allow the generation of stick diagrams without the use of expensive graphics equipment, and to give the user the power of the SIMULA programming language with which to generate the stick drawing.

3) *SAM (graphics and procedures)*:

Layout languages provide users with the capability to algorithmically define cells. But the specification language is so nonintuitive that it is impossible to debug a design in that language. One must plot the design. Interactive graph-

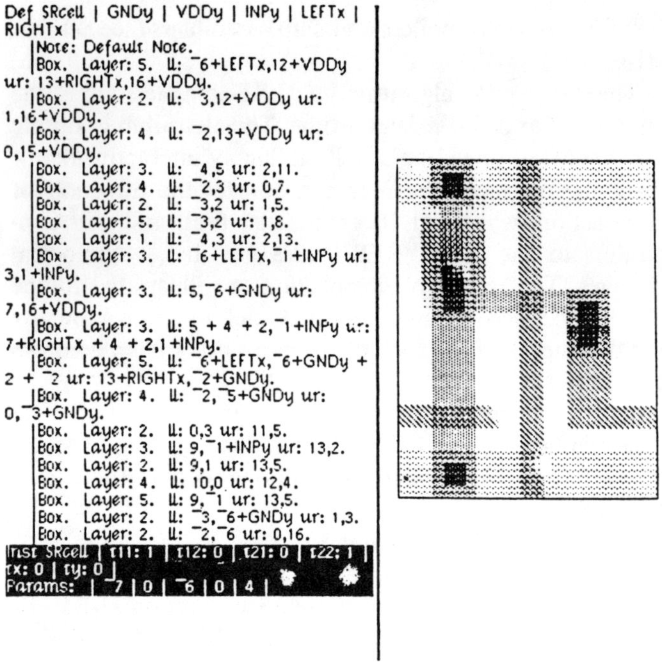

Fig. 10. SAM display.

ics systems, on the other hand, allow the user to debug in the form in which he sees the design, but severely restrict the language he may use to express the graphics. For example, he cannot express loops or conditionals. What is really needed is a single interactive system that combines layout language and graphic modifications to the data. SAM [28] is just such a system.

SAM provides the user with a two-part viewing window on the display as seen in Fig. 10. The left side shows the program view of the design under edit, the right side shows the graphics view of that cell. The user may move the viewing location in either window and may make edits to the data in either window. When the design is changed in either window, the change is reflected immediately in both windows.

The data displayed in the windows are "pictures" of the data structure. The data structure is the base form, the program view and the graphic view are merely different ways of looking at the base form of the data. When either the graphic bitmap form or the program character-string form is needed for display it is generated from the data structure. When the user makes what appears to be a modification of the data in either window, the commands are translated into calls on procedures in the data structure to carry out the action. The data structure makes the modification and causes both displays to be updated. The two views are kept consistent because they are both refreshed from the same data in memory.

SAM allows a designer to create algorithmically defined cells easily, cells defined as a program rather than a list of graphics. These cells are parameterized as well. When an instance of the cell is made, parameters are passed to the cell which specify desired attributes of that instance—for example, cell size or the locations of the connectors. The parameterization can be used with the program features of looping and conditional execution to process properly such input parameters as driving power, number of bits of length of a register, or PLA coding.

SAM is an experimental system. It was written to test the concepts and feasibility of combining the program and graphic methods of design. Since SAM is an experimental system, it is not in use as a design tool. Several test designs were made with SAM with encouraging results. The knowledge gained from those tests is directing further work in this area and has affected the development of several other new tools.

C. Composition Cell Tools

The problem of VLSI design is largely a problem of composition. It is a problem of organizing separately designed and debugged modules so that they work together correctly. The process of composition is at least as error prone as the process of leaf cell design, and the means for discovering errors in composition are poor at best and extremely costly. Composition errors are more likely to persist until late in the design process.

In order to make correct composition possible, the designer must be freed from responsibility for all the myriads of details that must be specified. This emancipation is often accomplished by imposing structure through the design tools. By imposing some structure, the design tool is able to perform the mechanical aspects of the design, leaving the more intuitive portions of the design to the user.

Once the composition tool is debugged, the designer is also freed from worry about whole classes of mistakes. By accepting a small set of restrictions brought on by the imposed structure, the designer is relieved of most of the particulars of the design, and the tool can be trusted to do them correctly. Once the user has gained sufficient confidence in the tool, much of the post-design checking like geometrical design rule checking can be bypassed. "Correctness by checking" has been replaced by "correctness by construction."

Composition tools come in many forms, and under many names, but they all trade designer restrictions for degrees of computer automation. An example of a composition tool is the common PLA generator. This tool imposes the structure that the design will be in the form of a PLA. The input is generally in the form of some form of logic equations or AND and OR plane code and all of the layout details are handled by the program. It is of course unreasonable to expect every design to be implemented efficiently as a single giant PLA. However, the PLA generator can be used to automatically implement appropriate pieces of a large design. Drawing on software analogies, more general composition tools are often termed "chip assemblers" or "silicon compilers."

Caltech has three powerful composition tools currently under development: Bristle Blocks, SLAP/Earl, and SPAM. These are described in the sections below.

1) Bristle Blocks:

Bristle Blocks [11] is a silicon compiler that specializes in the construction of *datapath* chips. A datapath chip consists of data processing elements connected by and com-

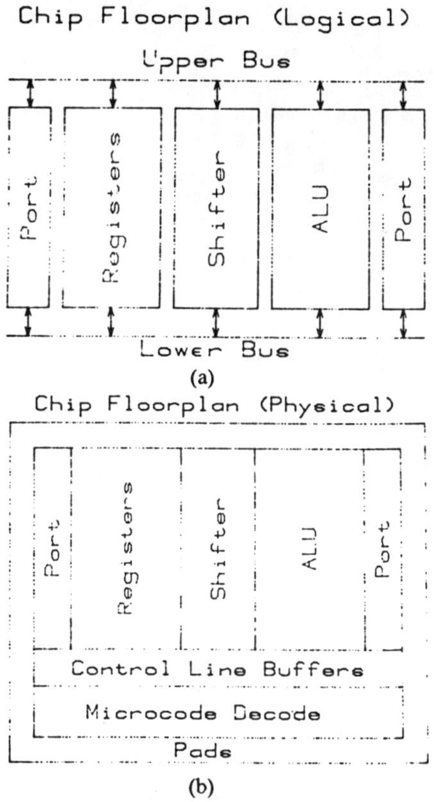

Fig. 11. (a) Chip block diagram. (b) Chip floorplan.

municating across data busses. Typical data processing elements include register files, arithmetic/logic units, and shifters. The chip is controlled by a microcode word which is decoded on-chip to drive each of the individual control lines of the processing elements. The functional block diagram for a Bristle Blocks datapath chip is shown in Fig. 11. Notice that the physical floorplan of the chip is identical to the functional floorplan. Like many other automatic layout systems [5], [19], Bristle Blocks imposes a generic floorplan in return for ease in automating the layout. While not all designs can fit into the high-level "datapath" floorplan, those that can are implemented as efficiently as hand layout.

The primary characteristic of Bristle Blocks cells is that they are programs rather than data (see Section II-B). Rather than designing cells as wires and boxes, the cells are designed as programs which, when executed, generate the required wires and boxes. This allows the cells to perform computations and participate in the design of chip.

Because of the need to minimize area, the cells composing the datapath processing elements will not all be the same size, which means that the cells cannot be interconnected by simple abutment. Analysis has shown that global chip area can be reduced and global chip performance increased by stretching the cells to connect by abutment instead of placing the cells and routing between them. See Fig. 12. Bristle Blocks makes use of stretching to assure that all cells are of uniform height. Bristle Blocks leaves the mechanics of stretching to the individual cell: each cell is designed with the means for stretching built into it. Fig. 13 shows a stretched register cell.

The input to Bristle Blocks consists of the cell definitions as programs and a high level description of the chip. The high level description is in terms of calls to the cell programs, passing the appropriate parameters for the chip to be designed. These parameters typically consist of behavioral information, such as the conditions required for the register cell to load information from a bus. The description for a fairly large chip is approximately three pages long. To build the chip, Bristle Blocks executes the cell programs as specified in the description to create the datapath portion of the chip. Bristle Blocks stretches cells to be of uniform height and abuts them. The datapath timing and control information is used to add control line buffers, parallel load shift registers and a semioptimized instruction decoder to drive the datapath. Finally, Bristle Blocks adds bonding pads and wiring to complete the chip and provides documentation about the locations of pads and signals.

Bristle Blocks has been operational since 1978 and has undergone several improvements and enhancements. The user may now direct the compiler to insert LSSD registers for testability [9]. Recent enhancements allow a much more general floorplan to be compiled, enabling the system to compile entire multiple processor computer systems. Bristle Blocks can compile systems with circuit densities comparable to hand design which are well beyond the capabilities of current integrated circuit fabrication technologies to implement. The process of compiling a sixteen bit datapath chip from the high-level description took fifteen minutes of elapsed time on the DECsystem-20/60 computer. The resulting design was 3600 by 4250 lambda (lambda equal to 2.5 μm) and contained approximately 13 500 transistors.

2) *SLAP/Earl*:

SLAP/Earl are two implementations of a system tied closely to the separated hierarchy.

Composition cells specify interconnections between instances of other cells. Any of a wide variety of notations can be used to describe composition cells. The notation familiar to most designers is a *netlist* form. A netlist is simply a list of instances along with a list of logical connections between instances.

Logical connections are actually made between *connectors* on instances of cells. Each cell has a list of connectors that define the *interface* to the cell, the means by which the outside world may interact with the cell's instances. The connectors are the only available attachment points to an instance of the cell. By explicitly declaring the interface, interactions with the cell can be controlled, localizing the behavior of the cell and simplifying the task of verifying correct usage.

One particularly important representation for VLSI design is the geometrical representation. The composition rule presented here is only one possible algorithm and certainly not the best, merely one of the first tries. This rule is based on the structured layout ideas developed in Mead and Conway [17]. In particular, the layout is designed to be a regular array of instances of cells designed to *abut* perfectly. Perfect abutment implies that all the connectors along the shared edge (see Fig. 14) are designed to be at the

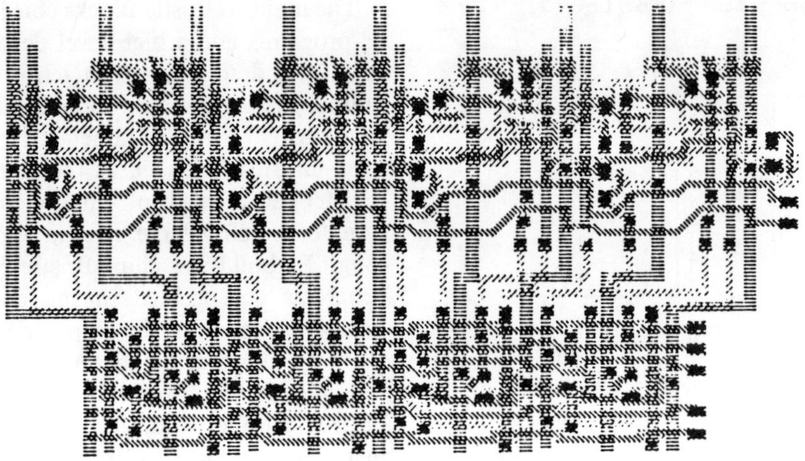

(a)

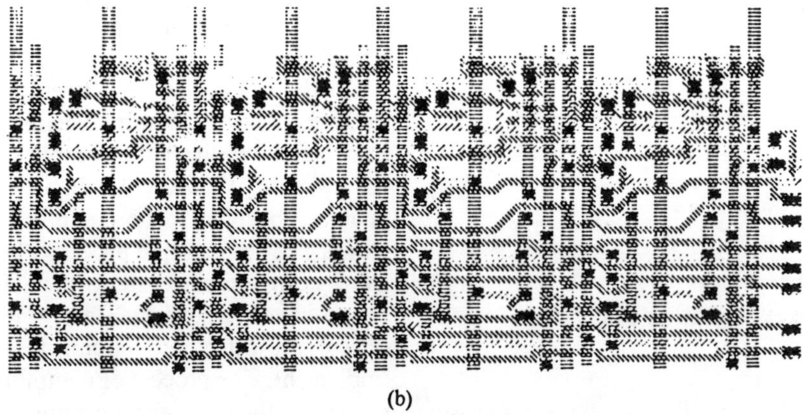

(b)

Fig. 12. (a) Connections made by routing. (b) Connections made by stretching.

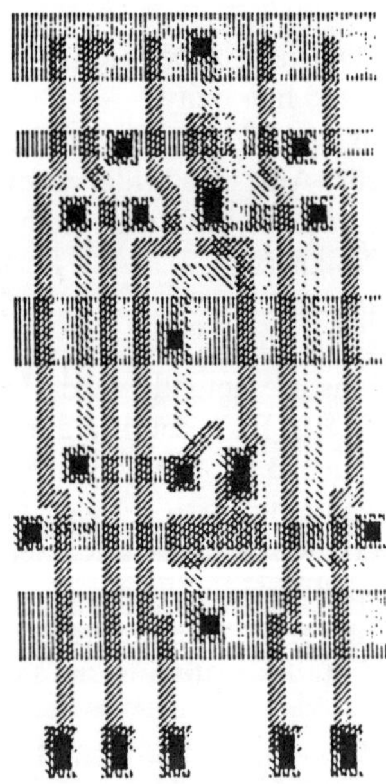

Fig. 13. Stretched register cell.

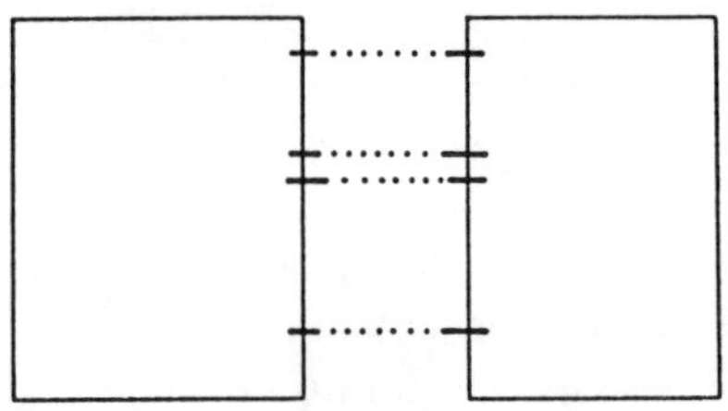

Fig. 14. Connection by abutment.

same position so that abutment will perform the logical interconnection desired, i.e., all the connectors on the right edge of instance *A* will connect to corresponding connectors on instance *B*.

This method for structuring the layout design fits very nicely into the separated hierarchy. Composition cells, containing only logical interconnections of instances of other cells, are logically very similar to the abutment style shown in Fig. 14. An automatic mapping can be made between the netlist description of a composition cell and an abutting cell layout description.

To implement geometry composition, interconnection is defined to be *superposition* and is accomplished through *stretchable cells*. Every logical interconnection between connectors on instances results in those two connectors being placed at the same physical location. In order to guarantee that this placement can be accomplished, all cells must be able to increase the distance between any two adjacent connectors. This use of stretchable cells is a slight generalization of the identical notion in Bristle Blocks.

The composition rule must perform some algorithm that satisfies all the constraints introduced by the interconnec-

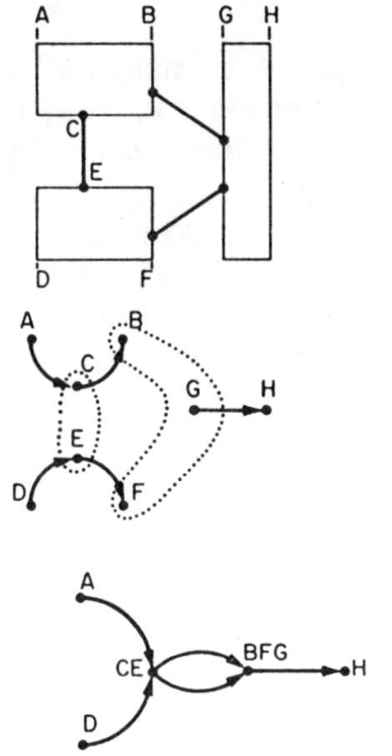

Fig. 15. Solution graph for horizontal direction.

```
------------------------------------------------------------------------
|                     TYPEA|                   TYPEA|         MINEY|
|          EENY            |        MEANY           |              |
|                          |                        |              |
|                          |                        |              |
------------------------------------------------------  MINEY      |
|        TYPEB|             LARRY|          TYPEB|                  |
|   CURLY     |      LARRY       |         MO    |                  |
|             |                  |               |                  |
------------------------------------------------------------------------
```

Fig. 16. Example SPAM floorplan output.

tions. By restricting cells to be rectangular with all connectors lying on the boundary, the basically two-dimensional layout problem is split into two one-dimensional problems. Each dimension is translated into a directed graph and solved using graph techniques.

The nodes of the geometry solution graph represent *equivalent* sets of connectors. A logically connected group of connectors is represented by one node in the graph. The weighted, directed arcs represent minimum separation constraints between connectors resulting from constraints inherent in the leaf cells. Fig. 15 illustrates the resulting graph. Notice that the graph must be acyclic, but there may be multiple arcs between the same two nodes, possibly with different weights. The graph is solved with an algorithm that is very similar to the algorithms used in stick diagram compaction [10].

The SLAP system was embedded in SIMULA. Currently, work is in progress to produce an interpretive implementation, called Earl, which will provide a concise composition cell description language. The Earl language is based on list manipulation primitives that allow simple construction and interconnection of lists of instances and/or connectors. A major design project to build a tree connected parallel processor [3] is currently underway using the Earl system as a primary design tool.

3) SPAM:

Structure, Placement and Modeling (SPAM) [22] is a system that can be used to describe a hierarchical design which can then be simulated at any level of detail. The system provides a concise method for describing composition cells. SPAM deals with a structural description from which a physical positioning might be generated using Earl. In addition, the behavior of a composition cell can be described. An entire design can be simulated to any desired level of detail by choosing which cells are the "leaves" of the simulation, i.e., at which points in the hierarchy the behavioral descriptions are used instead of the behaviors of its parts.

Since individual physical devices are not important in SPAM, the task of describing the structure of a design is primarily the task of *interface specification*, the specification of the organization of communication between modules. Cell names, pin declarations, and pin typing are all the parameters necessary to completely describe the structure of a leaf cell. Typed pins allows a more complete structural verification. SPAM provides six types: *Input, Output, IO, Power, Ground, and Clock*. When the hierarchy is instantiated, every connection of two pins is checked for legality.

Composition cells must specify an interconnection of instances of other cells. Cell instances can be declared individually or as arrays. The interconnection of instances is done only through abutment, which can be described either graphically or textually. No individual pins are mentioned, whole edges of rectangular instances are interconnected. Thus "the left of *X* abuts the right of *Y*" implies that all the pins on the left of *X* connect to all the pins on the right of *Y*. Since it is occasionally desired that some pins be left unconnected, a clause can be added to abutment declarations to omit some connections. From the interconnection description a floorplan can be derived, as in Fig. 16.

Until it is possible to "compile" function into designs it will be necessary to use simulation to assist in the verification of behavioral integrity. The SPAM environment favors designs produced and verified in a top-down manner. That is, an initial chip description will be produced and simulated. When the results are deemed correct, the design is broken up into components and each module is given a behavior. The resulting network is simulated and the results are compared to the original test. If the same results are produced, then this refinement has been successful. The process continues until the leaf cells have been described.

Simulation in SPAM is highly interactive. SPAM contains a four-value, combination event-driven [12] and clock-driven [2] functional simulator. The two simulation modes may be used in the same simulation run in different cells. The user may examine any node at any time. The user interface to the simulator is similar to the interface to a symbolic debugger for a programming language.

Once a circuit description has been compiled, the user may request a *documentation workbook* consisting of a hierarchical map of the entire circuit, an interface specification diagram for each cell definition, and a floorplan diagram for each composition cell in the description.

D. Silicon Foundry Tools

The most important aspect of a silicon foundry is its interface to the designers it serves. Designers communicate their designs to the foundry using the Caltech Intermediate Form (CIF) and receive from the foundry bonded dice and appropriate documentation. Designs sent to a foundry are specified as the actual geometry of the desired devices on the chip. Acceptance or rejection of the chips is determined by the performance of standardized test patterns. The designers are insulated from any detailed knowledge of the foundry's process, but require assurance that their designs have been fabricated correctly.

In the past, Caltech has acted as its own foundry, accepting chip designs described in CIF and performing the necessary steps to produce pattern generator tapes. The mask generation and chip fabrication services were then purchased from commercial houses. The advent of the Multi-Project Chip (MPC) program, administered first by Xerox Palo Alto Research Center [6] and currently by USC's Information Sciences Institute [31], has eliminated the need for Caltech to perform these functions in house. In the future, it is hoped that commercial organizations will develop to provide this service.

A number of tools have been developed at Caltech to aid in the transformation of individual chip designs to fabricated chips. Not all of these tools are restricted to foundry use, many are useful in the design process as well. In particular, interactive checkplotting programs are of great use to designers. At the top level, there are check plotting facilities and a design rule checker. These act as a filter to prevent the unwitting fabrication of designs which cannot work due to the presence of design rule violations or other errors which preclude fabrication. Once individual designs are deemed acceptable, they must be packed into one or more dice. Test circuits and patterns, alignment marks and scribe lines must be added to suit the process line on which the designs are to be fabricated. Finally, the image of the die must be converted to a format from which masks can be generated.

1) Check Plotting Facilities:

At several points, the image of the designs, or of the entire die, must be viewed to detect any gross errors. If one had complete confidence in the accuracy of the tools that manipulate the artwork, check plotting might not be necessary. However, the cost of mask making and fabrication is high enough to make such checks worthwhile.

Two plotting tools have been developed at Caltech for the purpose of plotting CIF files. The first of these, an interactive program named CIF20P, was originally developed for use by designers in verifying their layout designs. It has become useful in assembling multiproject chips as well. It accepts any legal CIF file as input and, in response to commands by the user, plots various portions of the file's geometry on a variety of output devices. The primary virtue of this program is its fast response. It is essential that the designer be able to make small, localized modifications to a design and view the result quickly. Interactive views of individual cells and areas of the design must be available to the designer within minutes. This fast feedback is an important characteristic that allows small changes in the design to be viewed immediately regardless of the overall size or complexity of the layout.

A second plotting program, ALEX, is a Fortran program intended for batch use. Like CIF20P, it can utilize a variety of graphic output devices. ALEX is used primarily for making large plots on a Calcomp four-color plotter. Users can specify a window within their designs and can control which symbols and which layers are to be plotted.

A third check plotting program provides verification that the pattern generator tapes have been made correctly. This program is called CHKPLT and it takes as input both Electromask and Mann 3000 pattern generator (PG) formats. It plots the PG instructions as they would be encountered by the PG machine, verifying that no obvious errors have occurred in the generation of the PG instructions and also showing that the instructions have been sorted in the order appropriate for the particular PG machine. This is the last check made of the data before it is sent to the mask making house.

2) Design Rule Checking:

While many of the design tools developed at Caltech endeavor to produce correct artwork by construction, until these tools gain the confidence of designers and are shown to be error free, design rule checking will remain a necessary part of the design process. A significant improvement in design verification has been made recently at Caltech by the development of a hierarchical design rule checker (DRC) [30]. The ideas and motivations for this tool are described by McGrath and Whitney [16].

Rather than taking the traditional approach of doing geometrical operations on a fully instantiated description of a design, this DRC makes use of the symbol hierarchy found in the CIF description to eliminate as many redundant comparisons as possible. The correctness of geometry inside a CIF symbol is checked only once, regardless of how many instances of that symbol are made. The environment in which each instance of a symbol is found is remembered so that a particular set of symbol interactions is only checked once. This technique gains a great speed advantage on regular chip designs without restricting the complexity of the geometrical shapes taken as input. Design rule checks can be done on arbitrarily complex rules.

This approach contrasts other attempts to design efficient DRC programs which generate a rasterized picture of each mask using a one lambda resolution and perform design rule checking using this representation [1]. Such an approach gains performance at the cost of restrictions on the form of the input and on the complexity of design rules which can be checked.

As much of the design intent as is available in the CIF description is used to reduce the redundant checking of duplicate geometric situations. Of course, some of the design intent is lost in describing the design in CIF. Some extensions have been made which allow more of the structural data associated with a design than is normally found in CIF to be made available to the DRC. Also, some conventions have been imposed on the input CIF file. Primitive elements have been defined for such objects as transistors and contacts. Since many design rules control the legal construction of these primitive elements and their interactions with other objects, considerable simplification and performance are gained if these objects are identifiable in the design description. Some of the most obscure and difficult design rules are related to the construction of transistors. Since these are irreducible structural objects, they may be defined as primitive elements without reducing the flexibility available to the designer. Primitive elements can be checked for correctness prior to their use by the designer and only the interactions of primitive elements with other objects need be checked, avoiding the need to check their internal parts.

These techniques minimize the number of geometrical spacing and width checks that must be performed between primitive geometrical shapes. The use of the design hierarchy can eliminate the high complexity inherent in the design rule checking of fully instantiated designs. The number of false or spurious errors is also reduced. Since few redundant checks are made, few redundant errors are generated. The dramatic reduction in the number of shapes compared is shown in Figs. 17 and 18. The first shows a small design, the second shows those shapes that were actually compared with other shapes during the check. For clarity, only the polysilicon and metal layers are shown in the figures. When design rule checking, of course, all layers are processed.

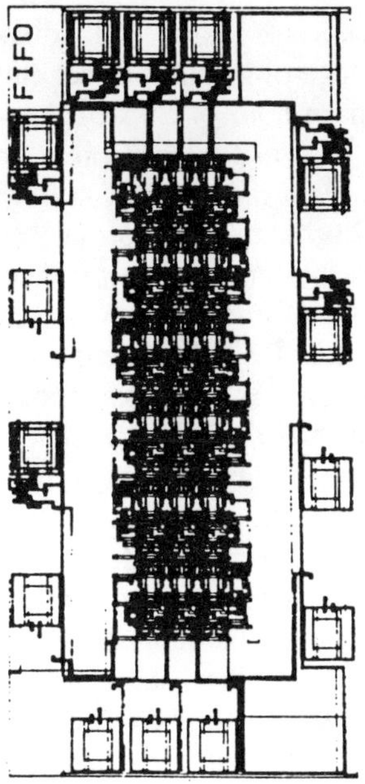

Fig. 17. Design rule checker example.

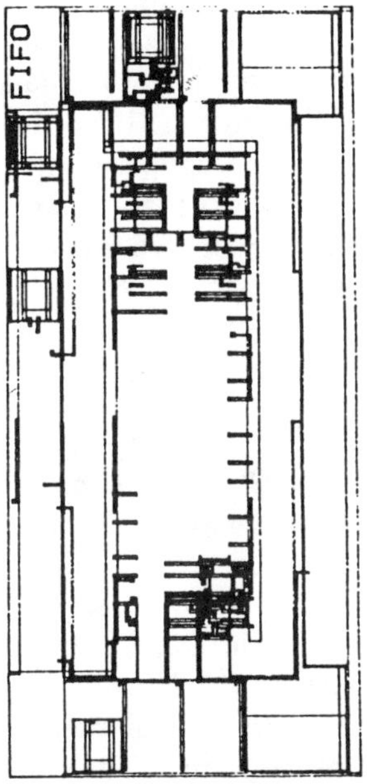

Fig. 18. Shapes actually examined by DRC.

The extraction of the design topology, a side effect of DRC, can provide an important verification tool for designers in determining whether or not they have implemented the circuits they intended in the artwork. When two shapes are later checked against design rules, their membership in the same or in different nets can play an important role in detecting whether or not an error truly exists, reducing false errors. The ability to compare the designer's topological data with that found by the DRC is not part of the DRC function but it will hopefully be available at Caltech in the near future.

3) Assembling Multiproject Chips:

A set of three programs have been developed at Caltech to automate the process of assembling a collection of independent project designs into a single die image. The projects are input as individual CIF files with no restrictions on the use of legal CIF commands. The result is a single CIF file containing all the projects placed inside a rectangular die, including scribe line geometry, mask labeling, and fiducial marks.

The first step in the process verifies that each project file contains legal CIF commands and extracts the bounding box of the geometry. The second step is the packing of the projects into a single rectangle representing the size of the die. When a trial packing has been accepted, the location and possible rotation of each project on the die are saved for use in the third and final step.

Lastly, the packing information and the original CIF files are used to produce a single, large CIF file representing the entire die of the multiproject chip. Fiducial marks are added and each layer is labeled with the name of the multiproject chip and the mask identifier. Process line and mask house dependent information is read from a file describing what type of fiducial marks, scribe lines, layer labeling are necessary for the organizations who are to make the masks and process the wafers. Test patterns for checking the lithography, circuits for checking electrical characteristics and alignment marks are added to the die merely by including the appropriate CIF files as additional projects.

At the completion of this last step, a CIF file is available for conversion to pattern generator format. To document the placement of projects on the die, a plot is made of the die with outlines of each project and their names included.

The entire process of taking individual CIF files and producing the image of a multiproject chip in this fashion requires less than an hour of elapsed time.

4) Translating CIF to Pattern Generator Formats:

A Fortran program named PAT was written at Caltech to convert CIF 2.0 data to the formats required by Mann and Electromask pattern generator machines. This program removes the CIF symbol hierarchy by recursively replacing instances of symbols with their corresponding geometrical primitives as it moves through the input file. PAT fractures all the primitive CIF shapes, including arbitrary nonself-intersecting polygons, into the set of rectangles available on the PG machines.

To insulate the user from the details of chip fabrication, CIF files specify the actual geometry of devices on the chip as they are returned to the designer. To translate from the fabricated geometry to the actual mask artwork requires knowledge available only to the silicon foundry and which the foundry would prefer to keep confidential. These characteristics of the process affect the relationship between the artwork and the fabricated geometry. PAT performs the translation of fabricated artwork to drawn geometry by bloating and shrinking the features on the individual mask layers before the format conversion is done.

PAT optimizes the ordering of the data on the PG tapes to minimize the time required by the PG machine in making the reticles. The optimization algorithm is easily modified to accommodate variations in different machines. The PG output data produced by PAT can be visually checked by plotting it with CHKPLT. This provides a picture of the final data as it is sent to the mask house.

IV. Conclusions

The structured design methodology and algorithmic approach to design tools appear to be a good fit and are working well. The emphasis on structures which allow implementation of the desired function in a manner suitable to integrated circuit implementation has been and continues to be a great success. A great advantage has been gained by building tools closely related to the design methodology. These tools are built with assumptions about the general form of their input. These assumptions lead to simpler implementations of tools and more powerful design systems. Powerful design systems have allowed us to quickly produce complex chips.

A. Successes

The major successes have been in four areas: methodology, chip assemblers, standards, and silicon foundry software.

There is no doubt that adopting the structured design methodology and rigorously structuring a design produces great benefits. These benefits include shorter design times, better design management, easier checking, and increased probability of correctness. Over a period of time, work at Caltech has produced some crude measurements of improvements. For example, it is possible to achieve an order of magnitude speedup in delivery of artwork over conventional designs.

The invention of the chip assembler has contributed significantly to reduction in design time. Paradoxically, there is even evidence to show that compiled layouts consume less area than hard-packed designs. This could be a result of the compiler inexorably applying area optimizations where a designer may be less careful. Additional savings come from global optimizations made possible by the quicker turnaround of the assemblers. These area reductions may be as large as twenty percent.

The somewhat radical view at Caltech which promotes helpful, yet unpretentious standards in lieu of an all-encompassing database has allowed nearly unconstrained work to progress on a number of tools independently, yet has allowed the tools to be interfaced in a well-defined manner. This has contributed significantly to the productivity of the group as a whole.

Although superficially less interesting, the development of the basic CIF software (foundry software) has been a notable success. It is in continued use at Caltech and other institutions.

B. Difficulties

No major difficulties have been found with the structured design methodology, although refinements continue to be made. The major criticism of correctness by construction is that it interferes with optimization of the circuit. It is true that work at Caltech has addressed the issue of correctness at the expense of area and time optimization. It seems clear that some area or performance penalty should be incurred in structuring a design. After all, the largest determinant in systems performance is design at the highest level of abstraction. However, it is difficult to make convincing arguments without quantifiable evidence. At this point, we are generating measures and collecting data.

The Caltech methodology moves the responsibility for correctness to the construction programs. The feedback paths traditionally provided by verification programs have been replaced by restrictions on the design structures. This trading of verification feedback for the simplicity gained by automating details of the design task, is precisely the tradeoff made in the software community for compiled languages over assembly languages. Elimination of feedback paths may diminish confidence in the resulting designs. It is not yet known if verification tools can be eliminated totally or if they can be replaced by simpler verification tools.

The major difficulties in tools have been in those areas not compatible with the directed thrust of the department, specifically, checking tools. The difficulties with checking tools appear to stem from the current state of tools at Caltech, which only recently embraced the correctness by construction philosophy for all phases of design. Previously, tools existed only for selected parts of the design process, and much checking was necessary for the remaining design tasks.

C. Future Work

To date, emphasis has been on physical and structural design. In the future, we will explore temporal and functional approaches to design. Inevitably, this will center around models and notations appropriate to each design domain.

Compilation without regard to the physical aspects of the implementation medium appears to lead to poor use of the medium. This may be due to the low-level target representation for the circuit, transistors. Preliminary work has shown that it is much easier to convert a behavioral description of a circuit into a physical device by compiling to a known set of low-level structures, somewhat analogous to redefining the low-level primitives for a programming language compiler. Compiling to transistors appears to be extremely difficult, compiling to datapaths, memory arrays, PLA's and decoders may be more easily implemented. Future work will explore reasonable goal structures for compilers along these lines.

Many of the techniques in use at Caltech produce a great savings in design time. However, there are, as yet, no good measures of that savings. Future work will be done to test the cost-performance of the design methodology in general, as well as specific instances of it. Subject to test in the near future are sticks diagram systems and self-timed logic.

D. Open Questions

There are many areas of computer-aided design which have not been addressed at Caltech. Some of these may come under investigation in the near future, some may not.

The Caltech standards work has proved to be very successful. However, it is not immediately apparent what further standards are needed. It is not reasonable to define standards in the hope that tools will arise that need those standards. More experience will reveal the need for further standards as well as needs for changes in those currently in use.

Many books and papers have been written on the organization of design teams for software. These are usually part of a software design methodology. Caltech has done no work on the organization of design teams using the structured design methodology for integrated circuitry, although it is becoming clear that that organization must be different than that currently in use in the integrated circuit industry. The design team organization will be influenced by the role of the computerized design aids to be used, which, of course, are ultimately dependent on the design methodology.

Caltech Intermediate Form

The Caltech Intermediate Form (CIF Version 2.0) [25] is a means of describing graphic items (mask features). Its purpose is to serve as a standard machine readable representation from which other forms can be constructed for specific output devices such as plotters, video displays, and pattern-generation machines.

The form is a fairly readable text file, to simplify combining files and to maintain machine independence. The text form allows unlimited precision of numbers without requiring large amounts of storage for small numbers. The intermediate form is not intended as a symbolic layout

```
( CIF created by STKCIF from STK file: invert.stk  1981-01-15  12:12:40 );

DS 1 250 4;   9  TRANSISTOR ;
L ND;    B  16,16 0,0;
L NP;    B  32,8 0,0;
DF;

DS 2 250 4;   9  RESISTOR ;
L ND;    B  8,40 0,16;
L NP;    B  24,32 0,16;
L NI;    B  20,44 0,16;
C 3 R 0 -1;
DF;

DS 3 250 4;   9  BUTTING ;
L NC;    B  16,8 2,0;
L NP;    B  12,16 -4,0;
L ND;    B  16,16 6,0;
L NM;    B  24,16 2,0;
DF;

DS 4 250 4;   9  GB ;
L NC;    B  8,8 0,0;
L NM;    B  16,16 0,0;
L ND;    B  16,16 0,0;
DF;

DS 5 250 4;   9  INVERT ;
C 1 T -12,-33;  (TRANSISTOR);
C 2 T -12,5;   (RESISTOR);
C 3 R 0 1 T 20,-17;  (BUTTING);
C 4 T -12,-49;  (GB);
C 4 T -12,49;  (GB);
L NM;
W 12  -34,49  -12,49  34,49 ;
W 12  -34,-49  -12,-49  22,-49  34,-49 ;
L ND;
W 8  -12,-11  20,-11  20,-19 ;
W 8  -12,-33  -12,-11  -12,5 ;
W 8  -12,-49  -12,-33 ;
W 8  -12,5  -12,49 ;
L NP;
W 8  -34,-33  -12,-33 ;
W 8  20,-33  34,-33 ;
W 8  20,-33  20,-19 ;
DF;

C 5;  (INVERT);
E
```

(a)

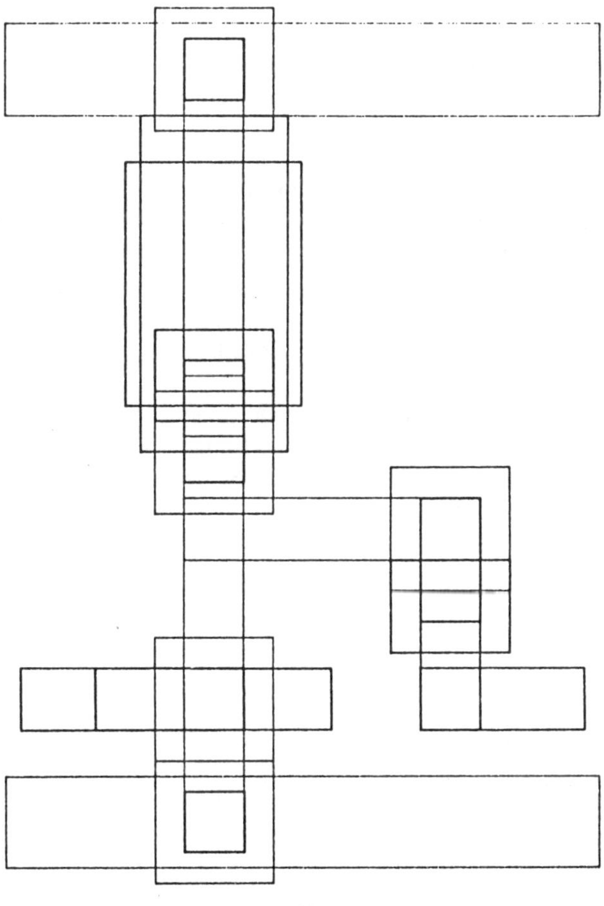

(b)

Fig. 19. Simple NMOS inverter. (a) CIF code. (b) Plot.

language: CIF files will usually be created by computer programs from other representations, such as an embedded layout language or an interactive design program.

CIF thus serves as the common denominator in the descriptions of various integrated circuit projects. No matter what the original input methods, the designs are translated to CIF as an intermediate, before being translated again to a variety of formats for output devices or other design aids.

CIF is a hierarchical geometrical description (see Fig. 19), and contains commands to set the layer, construct boxes, wires, polygons, and circles, and to define, call and delete symbols. A symbol is a set of geometry and calls on other symbols, with an identification number and a scaling from CIF units to the symbol's internal units. Calls on symbols can be nested to any depth and can be mirrored, rotated and translated to the desired position.

CIF has found widespread acceptance in the university community and in several industrial locations. A large number of tools have CIF interfaces, giving LSI designers a common language in which to communicate their designs. The Defense Advanced Research Projects Agency is sponsoring a *silicon foundry* to read CIF and generate chips [31], [32].

Acknowledgment

We are extremely grateful to the sponsors of the Silicon Structures Project: Burroughs Corporation, Digital Equipment Corporation, Fairchild Corporation, Hewlett-Packard Company, Honeywell Incorporated, International Business Machines Corporation, Intel Corporation, Motorola Corporation, Sperry-Univac Corporation, Xerox Corporation, and the National Science Foundation for their support of our computer-aided design effort. We would also like to express additional thanks to the individual visitors from those sponsors who have contributed immensely to the success of the project.

The work on design aids has benefited from complementary research in the "Submicron Systems Architectures Project" under Defense Advanced Research Projects Agency sponsorship.

We are also grateful to the good people at Caltech who helped put this paper together: D. Johannsen, C. Kingsley, G. Lewicki, R. Mosteller, R. Segal, C. Seitz, and T. Whitney.

References

[1] C. M. Baker and C. Terman, "Tools for verifying integrated circuit designs," *Lambda Mag.*, vol. 1, no. 3, 1980.

[2] G. Bell and A. Newell, "The PMS and ISP descriptive system for computer systems," in *Proc. AFIPS SJCC*, 1970.

[3] S. A. Browining, "The tree machine: A highly concurrent computing environment," Ph.D. dissertation, California Inst. of Technology, 1980.

[4] I. Buchanan, "Modelling and Verification in Structured Integrated Circuit Design," Ph.D. Thesis, University of Edinburgh, 1980.

[5] K. A. Chen, M. Feuer, K. H. Khokhani, N. Nan and S. Schmidt, "The Chip Layout Problem: An Automatic Wiring Procedure," Proceedings of the 14th Design Automation Conference, 1977.

[6] L. Conway, A. Bell and M. E. Newell, "MPC79: The Large-Scale Demonstration of a New Way to Create Systems in Silicon," *Lambda Magazine*, vol. 1, no. 2, 1980.

[7] O. J. Dahl, E. W. Dijkstra and C. A. R. Hoare, *Structured Programming*, Academic Press, New York, 1972.

[8] E. W. Dijkstra, "Notes on Structured Programming" from [7].

[9] E. B. Eichelberger and T. W. Williams, "A Logic Design Structure for LSI Testability," Proceedings of the 14th Design Automation Conference, 1977.

[10] M-Y Hsueh, "Symbolic Layout and Compaction of Integrated Circuits" Ph.D. Thesis, University of California at Berkeley, UCB/ERL M 79/80 Memo, 1979.

[11] D. Johannsen, "Bristle Blocks—A Silicon Compiler," Proceedings of the 16th Design Automation Conference, 1979.

[12] D. C. Ko, "BUILD User's Manual," Burroughs Corporation Mission Viejo, 1979.

[13] A. Koestler, *The Ghost in the Machine*, Chicago, Henry Regency Co., 1967.

[14] C. R. Lang, "LAP User's Manual," California Institute of Technology Computer Science Department Technical Report #3356, 1979.

[15] W. Lattin, "A 32-Bit VLSI Micromainframe Computer System," ISSCC, New York, Feb 1981.

[16] E. J. McGrath and T. E. Whitney, "Design Integrity and Immunity Checking: A New Look at Layout Verification and Design Rule Checking," Proceedings of the 17th Design Automation Conference, 1980.

[17] C. A. Mead and L. A. Conway, *Introduction to VLSI Systems*, Addison Wesley, 1980.

[18] R. Mosteller, "REST—Stick Diagram Editing System," MS Thesis, California Institute of Technology, 1981, in preparation.

[19] G. Persky, D. N. Deutsch and D. G. Schweikert, "LTX-A System for the Directed Automatic Design of LSI circuits," Proceedings of the 13th Design Automation Conference, 1976.

[20] M. Rem and C. A. Mead, "A Notation for Designing Restoring Logic Circuitry in CMOS," Proceedings of the Second Caltech Conference on VLSI, 1981.

[21] J. A. Rowson, "Understanding Hierarchical Design," Ph.D. Thesis, California Institute of Technology, 1980.

[22] R. Segal, "Structure, Placement and Modelling," MS Thesis, California Institute of Technology, 1980.

[23] C. L. Seitz, "System Timing," from [17].

[24] H. J. Simon, "The Architecture of Complexity," Proceedings of the American Philosophical Society, vol. 106, no. 6, 1962.

[25] R. F. Sproull and R. F. Lyon, "The Caltech Intermediate Form for LSI Layout Description," from [17].

[26] I. E. Sutherland and C. A. Mead, "Microelectronics and Computer Science," *Scientific American*, September 1977.

[27] S. Trimberger, "The Proposed Sticks Standard," California Institute of Technology Computer Science Department Technical Report #3880, 1980.

[28] S. Trimberger, "Combining Graphics and Layout Language in a Single Interactive System," in *Proc. 18th Ann. Design Automation Conf.*, 1981.

[29] J. D. Williams, "Sticks—A New Approach to LSI Design," MSEE Thesis, Massachusetts Institute of Technology, 1977.

[30] T. Whitney, "A Hierarchical Design Rule Checking Algorithm," *Lambda Magazine*, vol. 2, no. 1, 1981.

[31] D. Cohen, "MOSIS-The ARPA Silicon Broker," Proceedings of the Second Caltech Conference on VLSI, 1981.

VLSI Design Automation Activities at M.I.T.

JONATHAN ALLEN, FELLOW, IEEE, AND PAUL PENFIELD, JR., FELLOW, IEEE

Invited Paper

***Abstract*— A large part of the M.I.T. research program in very large-scale integrated circuits (VLSI) deals with design automation. Industry today is facing a "design crisis" that requires substantial advances in the state of the art in design aids. University research such as that at M.I.T. is aimed at making such advances. The design of an integrated circuit is viewed as a series of transformations among various domains of representation for the design. Many computer tools which assist in those transformations have been developed at M.I.T. There appear to be two new design methodologies emerging which are radically different from design styles currently in use in industry.**

Manuscript received October 10, 1980; revised January 17, 1981. This work was supported in part by the Air Force under Contract AFOSR 4-9620-80-C-0073 and in part by the Advanced Research Projects Agency of the Department of Defense and monitored by the Office of Naval Research under Contract N00014-C-80-0622.

The authors are with the Department of Electrical Engineering and Computer Science and with the Research Laboratory of Electronics, Massachusetts Institute of Technology, Cambridge, MA 02139.

I. Introduction

RECENTLY, industry's ability to fabricate large-scale integrated circuits (IC's) has outstripped its ability to design such complicated circuits, except for the special case of memories which have a great deal of repetition. In this sense, industry today faces what might be termed a "design crisis." This crisis will become more severe as the era of very large-scale integration (VLSI) is entered, and unless substantial advances are made in the state of the art of design, the potential of VLSI will not be realized.

In many industrial firms, the response to this design crisis has been to develop better design automation tools, intended to support design styles currently in use. In other cases, attempts have been made to develop radically new design styles felt to be better able to cope with very

Reprinted from *IEEE Trans. Circuits Syst.*, vol. CAS-28, no. 7, pp. 645–654, July 1981.

large-scale designs. However, the introduction of radical changes in the midst of ongoing designs is very difficult.

If radical new ideas are needed, university effort may be more appropriate. Along with a few other universities, M.I.T. is responding to this need with a VLSI research program. M.I.T.'s program includes projects in semiconductor materials, devices, processing, submicrometer patterning, large-scale circuit theory, design automation, complexity theory, and systems architecture. This paper describes some aspects of one of these areas, namely design automation.

Anyone who develops design-aid tools, such as those described below, must have in mind a model of the design process into which his tools are to fit. The model which is understood at M.I.T., and either explicitly or implicitly lies behind most of the results achieved, is discussed in Section II.

Recent results are discussed, in the context of this model, in Section III. It is possible, even relatively early in this work, to identify two trends which, if successful, will lead to design styles that are so different from those prevailing in industry today that they should be considered "revolutionary." These are discussed in Section IV.

II. Model of the Design Process

The process of designing an integrated circuit is thought of as a series of transformations of two types. One is a transformation of the design through several domains of representation. The other is the process of adding detail to a design, within any given domain, by specifying how a part of the design is to be built up from smaller parts connected together. Among the domains believed to be important are the functional, block-diagram, logic, electric-circuit, and layout domains. A more complete discussion of the domains appears below. Adding detail within any one domain constitutes refining the system's architecture. Both types of transformations involve making decisions about aspects of the design that were previously discretionary.

At the beginning of a design it is important to specify the requirements without unduly restricting design. That is, those aspects that are mandatory should be given, and all others, which are discretionary, should be avoided. During the design process, these discretionary aspects are fixed. This same comment applies to the design of any part of the circuit, and to descriptions in any domain. This distinction between what is mandatory and what is discretionary is similar to the dichotomy between competence and performance in software, the first referring to the results of a calculation, and the second to how these results are achieved.

A. Domains of Representation

A given integrated circuit can be described in many ways, corresponding to different domains of representation.

Functional: This domain is an informal one, in which formal languages do not appear to exist. The object is to describe the purpose of the design including all aspects that are mandatory, such as the function to be calculated, timing constraints, and power-dissipation requirements. Often a verbal description is adequate. Sometimes a computer program which produces the same result is useful, perhaps accompanied by a statement that the internal structure of the program is not mandatory.

Hierarchy can be introduced in this domain by specifying in some way how the function of the design can be achieved by connecting together two or more simpler functions. In some cases this can be done by reference to the internal structure of a program, or a verbal description. In case the purpose of the design, and perhaps its architecture, are already known, then this domain might be omitted.

Block Diagram: Descriptions in this domain may show either data flow, control flow, or both. The individual blocks generally correspond to hardware modules, and normally recursion is not allowed, although the same effect may be achieved with iterated structures.

In these two domains, and in all the domains described below, the description of a given module can be in terms of the input–output, or "black-box" or "boundary" behavior, or it can be in terms of internal details. For example, a discrete Fourier transform may be specified in terms of input and output signals as a function of time. Another description may be in terms of connections of functional units such as adders, multipliers, and delays. Both descriptions are important. The boundary behavior is what is, for this module, mandatory. The internal description reflects discretionary design choices that have been made. The internal description specifies the boundary behavior of the submodules (what is mandatory for them) and their interconnection. The equivalence between external and internal descriptions of the same module forms the basis upon which notions of hierarchy are built.

In general, a description of the IC in either the functional or the block diagram domain consists both of the input–output description, and the way that this behavior is to be realized in terms of subordinate modules. In turn each of these modules is described both in terms of *its* input–output behavior, and as an interconnection of other modules. These hierarchical ideas apply to all the domains. The internal description of a module may be given in another domain. If a module has no internal description, then the design is incomplete. Ultimately this hierarchy stops when the internal description is in terms of mask geometry, which is primitive.

Three points are important about this hierarchical picture. First, hierarchy appears within a *single* domain of description. Second, each module is required to have *both* an external description, for use in analyzing the module in which it appears, and an internal description. Third, a given design may have different hierarchies in different domains, although there are obvious advantages from using the same hierarchy.

Many engineers use hierarchy and modularity in block diagrams or computer programs. In these domains hierarchy suppresses unnecessary detail, simplifies system design through a "divide-and-conquer" strategy, and leads to more easily understood designs that are more readily

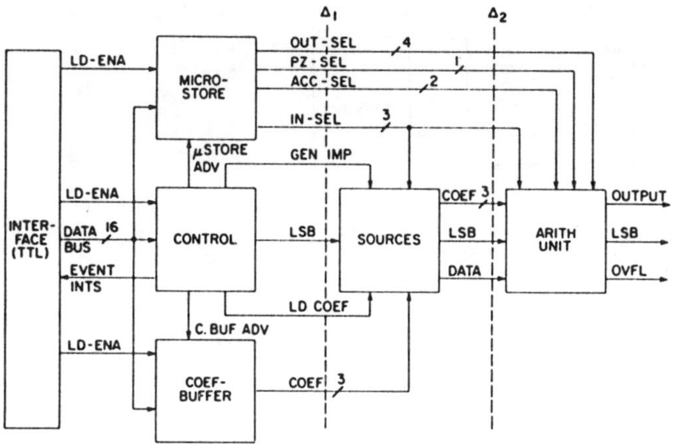

Fig. 1. Digital speech waveform synthesizer block diagram. The synthesizer is a hardware VTM. An arithmetic unit consisting of a time multiplexed digital resonator (second-order section) is configured by the microstore and excited by noise and impulse sources. Buffering permits the updating of the filter coefficients at a 5-ms interval.

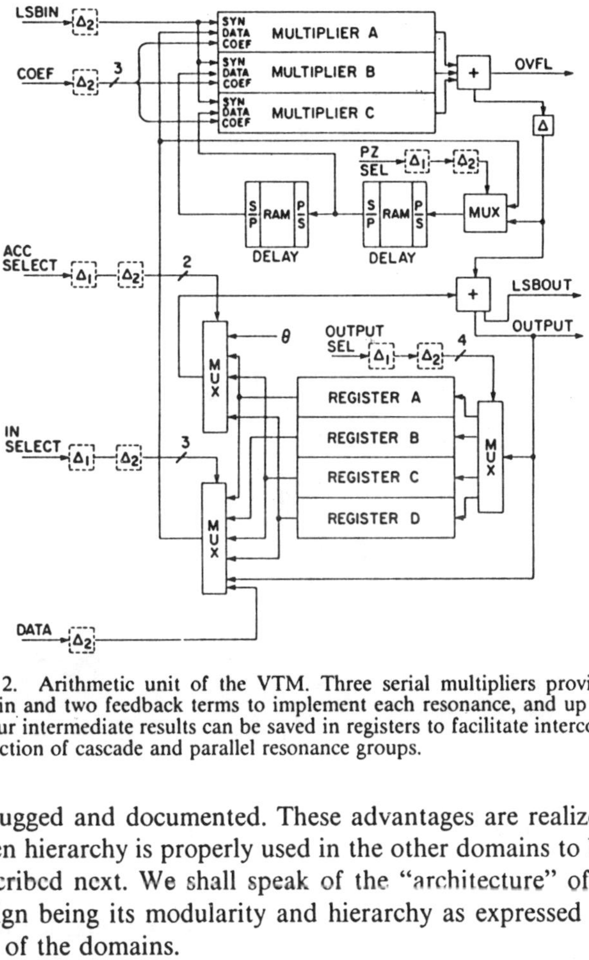

Fig. 2. Arithmetic unit of the VTM. Three serial multipliers provide gain and two feedback terms to implement each resonance, and up to four intermediate results can be saved in registers to facilitate interconnection of cascade and parallel resonance groups.

debugged and documented. These advantages are realized when hierarchy is properly used in the other domains to be described next. We shall speak of the "architecture" of a design being its modularity and hierarchy as expressed in *any* of the domains.

An example of hierarchy in the block-diagram domain is shown in Fig. 1. A speech synthesis vocal tract model (VTM) currently under design at M.I.T. is shown composed of six modules. One of these elements, the arithmetic unit, is shown in Fig. 2, where one set of three multipliers, together with two delays, is time-multiplexed thirty-two ways to achieve the effect of thirty-two second-order difference equations which simulate the effect of thirty-two resonances in the vocal tract model. Proceeding down the hierarchy, Fig. 3 shows the modular structure of one of the delay RAM's, and Fig. 4 illustrates how this memory may be further analyzed all the way down to details of the layout. In particular, Fig. 5 shows a typical storage column. It is interesting to note that although all thirty two such columns are essentially the same, each one has a different column selector five-bit address, and hence the address comparator in each column must be specialized for a particular address. Using parameterized cell specifications, these variations are easily produced by simple program "calls" to the basic cell type.

Logic: Another domain of interest is that of unilateral logic containing, say, NAND, NOR, or other combinational-logic gates, and state-preserving elements such as JK flip-flops. Subnetworks containing such components and perhaps other subnetworks may be usefully regarded as modules that can appear in larger networks.

Logic networks are constructed according to rules that are intended to prevent undesirable effects such as hazards or bus contention. These rules are syntactic in nature, in that they are independent of the particular function being implemented. A logic network or block diagram that does not obey the appropriate syntactic rules is not well formed and cannot express any particular function. A separate, semantic, question is whether the module implements the desired function. The distinction between syntactic and semantic correctness is useful in other domains as well.

If two or more logic networks implement the *same* functionality, they may be said to be "equivalent." An important research topic (for any of the domains) is to sharpen the idea of equivalence and develop criteria for selection from among several equivalent networks. In the logic domain techniques such as logic minimization serve this role.

Switching Networks: This domain is particularly important for MOS technologies, since pass transistors are not well modeled by unilateral logic gates. In this domain individual transistors are modeled by voltage-controlled switches, and pullups may be abstracted away very easily. This domain can handle large scale designs, since so little information is retained about each transistor; on the other hand the representations are very close to the topology part of electrical-circuit representations. Fig. 6 shows a switching-network representation of one of the modules of Fig. 5, together with some of the interconnections. This is the so-called three-transistor dynamic RAM cell. If electrical parameters such as capacitance and transistor sizes are put on Fig. 6, it becomes an electric-circuit module.

Electric Circuit: This domain deals with topology and also device parameters such as capacitances, resistances, and transconductances. Alternatively, geometric information about the active devices (length and width of gates in MOS circuits), together with parasitic capacitances and resistances, may be used.

Stick Diagrams: In this domain, the orientations of wires and transistors, and their relative positions on the IC, are given, but neither their absolute locations nor sizes. This domain is, therefore, somewhat simpler than a full layout, and automatic means based on particular design rules can be used to "flesh out" the wires and devices by giving them

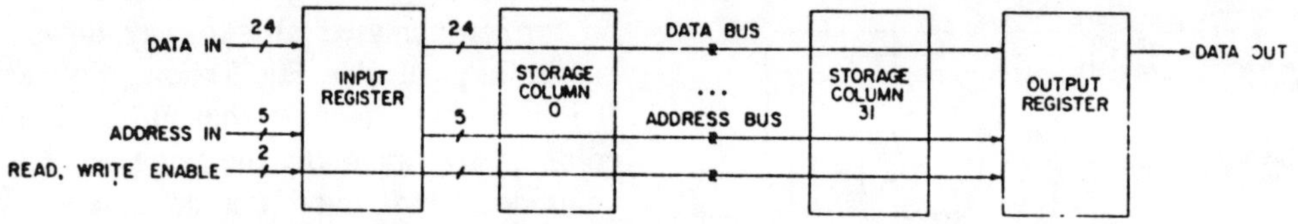

Fig. 3. Delay RAM for the VTM arithmetic unit. Input and output registers couple to 32 storage columns (one for each resonance) each containing 24 bits.

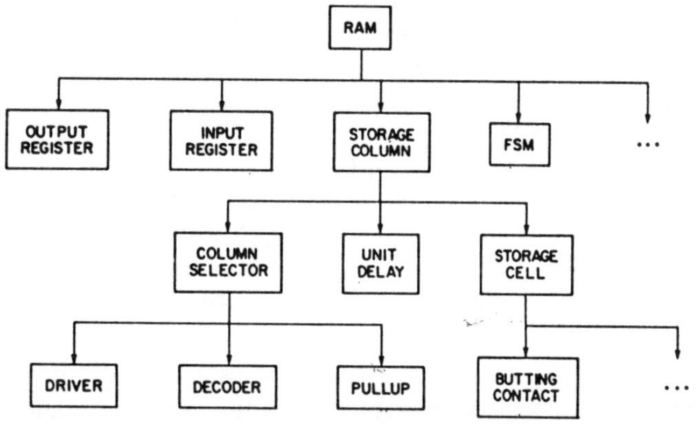

Fig. 4. Hierarchical typed dependency graph of the delay RAM. Design modularity at each level of the hierarchy is clearly shown, with emphasis on the RAM storage column.

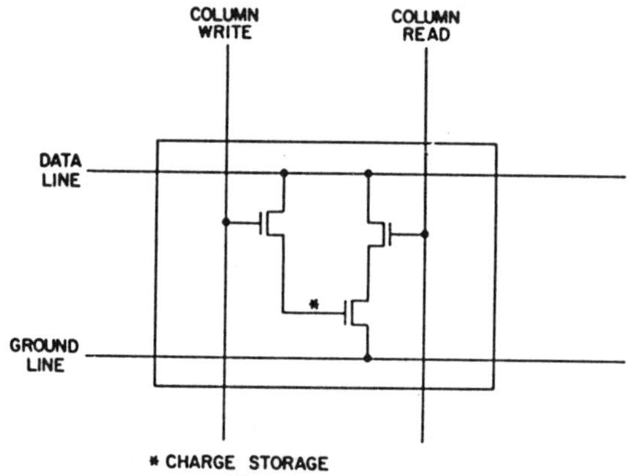

Fig. 6. Storage cell for use in the storage columns of the delay RAM's. This three-transistor cell is automatically refreshed by the serial nature of the computation.

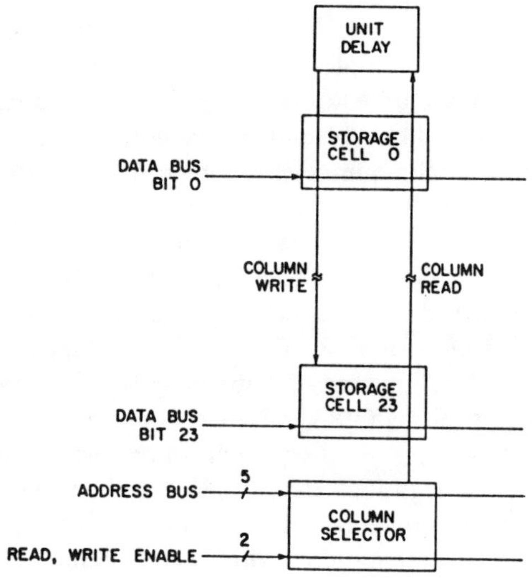

Fig. 5. Delay RAM storage column. Twenty-four storage cells are controlled by column address selection and read/write signals. The unit delay is used as part of a state machine to control read-before-write sequencing.

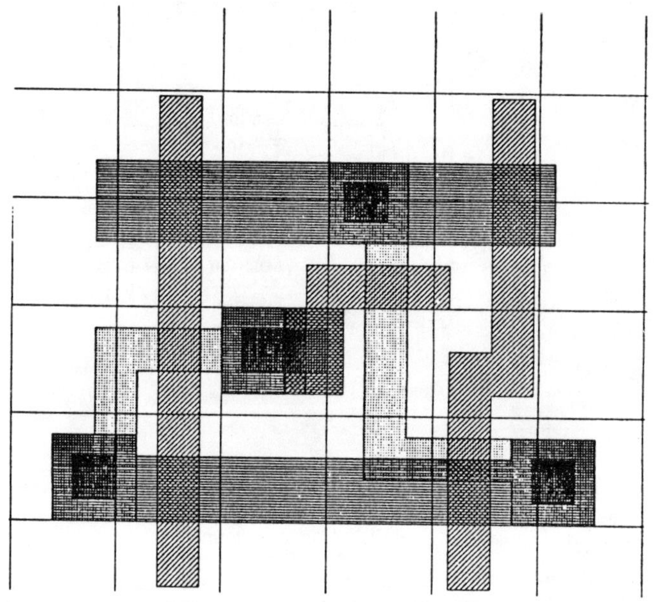

Fig. 7. Mask layout of the RAM storage cell for a silicon-gate NMOS process.

specified or default widths. Stick diagrams may be created by either a human or a computer program, and may be displayed for human verification and editing, and "compiled" into a layout by a computer. This illustrates the idea that, in each domain, a description of part of a design may be generated or read either by a human or by a program.

Layout: This domain is, for most design-automation work, the last one to consider. It is the domain from which physical masks are made, and usually represents a designer's last convenient chance to verify his design. There are many syntactic-type checks to be made—for adherence to geometric design rules, for power-bus sizing, for noise margins, impedance ratios, multiple threshold drops, and the like. Then there are the semantic checks, to verify that the functionality is the same as that of the same design as viewed in another domain. Perhaps this is done by simulation of an electrical circuit or logic network produced by analysis of the artwork. Fig. 7 shows a layout of the RAM cell from Fig. 6.

Other domains in which the design of an IC can exist include the mask set, and of course, the final fabricated chip.

In summary, the following domains of description have

been discussed:

Functional
Block diagram
Gate logic
Switching network
Electric circuit
Symbolic layout
Layout.

These have many features in common. All are able to express the architecture of a design—that is, the modularity and hierarchy. All are able to be used directly by humans or by computer programs. All of them admit the possibility of a given functional requirement being satisfied in more than one way. A brief discussion of the design-automation tools required at these levels is given next, followed by a description of how some existing design styles fit into this framework.

B. Design-Automation Tools

Computer tools to aid IC designers can be developed to work within any one domain, or among two or more domains. Needed for each domain are:

1) An internal data format that can be used to represent a particular design.

2) Human-oriented notation for a designer to use in specifying a design or a portion of a design, or for editing a design.

3) Human-oriented display of the design.

4) Test for equivalence between two designs.

5) Criteria and programs for optimization, i.e., selection from among two or more equivalent designs.

6) Simulation capabilities.

7) Syntactic tests for well-formedness independent of the functionality.

In addition, synthesis tools are needed which start with a description in one domain and return a description in another that is "closer" to the final layout. Finally, analysis tools which perform the reverse transformation are necessary. Semantic checking is then done by analysis followed by equivalence checking, or by simulation.

Such tools should support a hierarchical design style, since otherwise the complexity inherent in VLSI will preclude their use on massive designs. Thus for example, a hierarchical logic-node extractor will, when developed, offer substantial advantages over one which ignores the modularity and hierarchy present in a design. These advantages are both in faster execution because identical calculations can be avoided, and in preserving for the designer the architecture that has been specified, presumably because it reflects the way he thinks about the circuit.

In addition, such tools should preferably deal with appropriate performance criteria and their relation to technology constraints. This is often a difficult requirement to meet. An example is a placement and routing algorithm that, through calculation involving process parameters, checks to verify that the long interconnections it makes do not introduce intolerable signal delays. Another is an automatic PLA layout program that performs process-dependent power-bus sizing and speed optimization.

C. Design Styles

The structure of the "design space" represented so far requires three dimensions of discussion—one dimension for the domain used; another the architecture as expressed in the various hierarchies; and the third the design tools required. This structure is static in the sense that no indication appears as to how, by what path, a given design is done. Design styles differ in the exact paths used through this design space, and different design styles place different demands on the software needed.

A standard-cell approach is possible, in which predesigned modules are described thoroughly in several domains—layout, logic, electrical, functional, and so forth. The challenge then is to implement the design using such modules exclusively. Important tools are placement and routing algorithms, and, for cell generation, circuit extraction and simulation. Less important are design-rule checking and interconnect verification at the entire-chip stage, since the final layout is prepared by computer programs and therefore may be regarded as "correct by construction."

An interactive-graphics style is possible, in which small circuits are laid out and interconnected graphically. Important tools here are design-rule checking and logic node extraction for the entire chip, along with logic simulation.

A promising automatic style involves expressing the algorithm in the form of a finite-state machine. While such a representation is very general, it is not guaranteed to lead to efficient realizations of given logical requirements. However, implementation of the FSM in terms of one or more PLA's is straightforward, and entirely automatic methods, including minimization heuristics, are possible. In this style there is little if any need for logic extraction or artwork analysis, but many higher-level tools dealing with interconnections of FSM's are required.

Other design styles, too numerous to discuss here, can be described in terms of the design space.

III. Examples of M.I.T. Activities

In this section, we describe a variety of IC design activities at M.I.T., in the context of the previous discussion of the design process. Initially, the focus has been on individual tasks in the overall design process, such as architectural specification, layout, or topology extraction. This approach provides for detailed experience with domain representations, inter- and intra-domain transformations, and design styles, leading to a perspective on overall design system requirements. Top-down system design is avoided, at least for the present, and multiple approaches can be explored without rigid system specifications. Overall system designs are expected to evolve naturally under these circumstances, and we anticipate that several exploratory interactive systems will appear shortly. In the summer of 1980, most of our design tools were installed on one computer and loosely integrated to gain experience with their collective interaction as experienced by users.

Strong interest exists at M.I.T. in the specification of algorithms and their corresponding implementation architectures. This interest derives from difficulties with the dual specification nature of algorithms. Ideally, we would like to specify *what* an algorithm does (its competence, or semantics) independently from *how* it does it (performance). Unfortunately, algorithms as usually encountered confound these two attributes, since the detailed statement of an algorithm always assumes an architecture to interpret the algorithmic statements. At M.I.T., two efforts have focused on means to manipulate competence and performance separately.

Miranker [1] explored formal means to manipulate performance while retaining the algorithmic competence or intent. Starting from a high-level hardware description language, statements are manipulated in the language from serial to parallel form (and vice versa) whenever permitted. These local performance changes can be seen as intra-domain transformations that allow the designer to explore performance issues while preserving the algorithmic competence. Once a performance change is made at the hardware-design-language level, its consequences can be observed by transforming to a low-level description language where statements are in one-to-one relation to the final physical design, and hence allow space and time estimates of the design to be made. The formal theoretical apparatus of this thesis provides a basis for architectural design exploration in an interactive design system of the future.

Another study by Sussman and Steele, (in part leading to Steele [2]) seeks to pull apart algorithmic competence and performance, so that different performance strategies can be erected on the same competence representation. The desire to construct such an algorithmic decomposition is a fundamental long-range goal of computer science, and the consequences of this work are very broad. The need for this understanding is particularly strong in the area of integrated-circuit design where there is great architectural freedom. The construct used to represent the competence of an algorithm is the constraint network, which defines those relations that must be preserved regardless of the performance of any algorithm utilizing this competence. Most fundamental laws of science are normally expressed as constraints, e.g., Maxwell's Equations, Kirchhoff's Laws, or Ohm's Law. Other examples are sets of simultaneous equations, which arise in many areas. These equalities are neutral with respect to how they are solved. Sussman and Steele have shown that in general, many performance strategies can be constructed on the same competence constraint network, and hence performance can be readily varied. To date, several application domains have been explored for constraint relations [3], [4], and Steele [2] has dealt with the language issues inherent in the formal representation of these networks. The study of the "solution" of a system, for specified unknowns, from its underlying constraint network, demonstrates that constraints permit powerful inferencing to be made. Such techniques of constraint propagation can be used, in several design domains, to alter a design subject to specified constraint maintenance. Hence a geometrical layout can be modified to any of many equivalent layouts with respect to the design-rule constraints imposed by the fabrication process. This equivalence, of course, while maintained for design rules, does not necessarily pin down equivalence in other representation domains, such as the electrical circuit entailed by a layout, since different layouts, all satisfying the design rules, may implement different electrical circuits. The constraint ideas are seen as a powerful, fundamental tool for algorithmic representation and correct design.

Both of these techniques aimed at the exploration of performance strategies can provide useful interactive design assistance. At this stage, there is little inclination at M.I.T. to provide automatic means for architectural generation from a high-level functional specification. Instead, the human designer provides the global control for these programs, and is free to interpret the results at any stage and to impose direction on the evolving design.

In the logic domain, work at M.I.T. has centered on the simulation of switching network representations of an IC design. This network can be specified either by use of a language for the topological interconnection of transistors (viewed as ideal switches) or by transformation from a given mask layout specification by a connectivity or topology network extraction program. The switching network has been chosen as the appropriate representation for MOS logic simulation rather than a network of (unilateral) gates because of the prevalence of pass transistors, busses, and other structures whose logical behavior is best captured by a switching network. Two programs have been written for logic simulation. One, developed by Bryant [5]–[7], can process input derived either from a special input language (appropriate to a designer's "stick figure" representation) or from the mask layout. Another program, by Terman [8], [9], is for mask-derived networks only. Both simulators are event-driven for efficiency, but differ in the detailed means by which unknown node states are propagated and resolved. These simulators work on both clocked and unclocked networks, and include macro capability appropriate to hierarchical design techniques. The selection of the switching network as the appropriate level for logic simulation can be seen as an important exercise in abstraction and domain representation, since the key elements needed for proper logic simulation are retained without the encumbrance of other irrelevant detail.

The desire to derive the input switching network for logic simulation directly from mask data motivated the design and construction of a program for connectivity analysis and node extraction by Baker [8], [10]. The input mask specification is transformed to a series of bit maps on a "lambda-grid." Explicit representation of rectangles is suppressed but local adjacency relations between squares (currently 2 to 3 microns on the side) on this grid are used to infer connectivity. Although it is computationally expensive (on a conventional computer) to obtain this grid from layout (rectangle-based) representations, a three-square window, passed over the rasterized lambda-grid representation, suffices to extract the required topology. At the same time other information, such as transistor types and sizes, are obtained. In addition to the use of this network representation for logic simulation, many checks can be made to ascertain well-formedness. Perhaps the most useful check is the determination that an internal node can be both pulled up and pulled down, and hence capable of doing logical work. Multiple pass transistors in series can be detected. Experience has shown that this program is invaluable for checking layout well-formedness, and it holds an important place among the group of heavily used design programs at M.I.T.

Another type of artwork analysis required for correct design is design-rule checking. Two programs for this purpose have been written. Following the development of his topological node extractor program, Baker extended the use of a scanning window over a lambda-grid of the layout to provide design rule checking [8], [10]. For the NMOS Si-gate (single-poly, single-metal) process currently used at M.I.T., a 4×4 square window suffices. Once again, creation of the lambda-grid raster is expensive on a conventional computer, but the possibility of special hardware for both rasterization and window scanning is currently being investigated. In the second design rule checking program, the input mask data is represented as a set of rectangles, and a set of primitive operations is defined. The primitives have been chosen so that they can be combined to test for a wide variety of design rules, sufficient to satisfy the needs of any process currently contemplated. A macro language definition facility has been provided to enable the circuit designer to build his own custom set of design rules. This approach was developed at M.I.T. Lincoln Laboratory by Giovinazzo [11], and adapted to M.I.T. campus needs by Seiler [12].

It is recognized that artwork analysis tasks will be increasingly expensive computationally as the number of mask artifacts rises. At M.I.T., three approaches are being taken to lessen the impact of this cost. First, the objects represented on the masks have usually been restricted to orthogonal rectangles and wires lying parallel to the X or Y axes. Some of the artwork-analysis tools do not support angled rectangles or wires and in addition some do not support objects whose vertices are not located on an integer lambda-grid. This restriction is not universally accepted, since it is recognized that for some designs, angles can result in substantial space savings. Secondly, the complexity of these algorithms is under study, and substantial gains in performance are expected from sorting and partitioning of data objects. Some of these techniques are already in use in the rectangle-based design rule checker. Lastly, it is judged that substantial speed-ups can be achieved by use of special hardware architectures for these tasks.

In the discussion above, several programs for artwork analysis have been described. The generation of mask artwork is an equally strong activity. In general, designers at M.I.T. have not chosen to produce a full layout by hand, to then be digitized into machine-readable form. Instead, the emphasis has been on the development of textual and graphic representations, and the typical designer, while he may choose to sketch out small cells of his design, interacts directly with a computer-based layout program, either through a keyboard on an alphanumeric terminal or with the aid of graphic input devices. The notion of generating an entire chip design before digitization is avoided. The goal is to benefit from the computational services of the design system, either for generating cells or analyzing them, at the earliest possible stage, so as to insure well-formedness for all the representation domains at each step of the architectural hierarchy. Another important aspect of this approach is the equivalence of textual and graphic representations. In those layout systems that provide for both textual and graphic input, output representations can also be obtained in both forms. Experience has shown that there are big differences in human factors preference for textual and graphic forms. While there is a tendency for users to prefer graphics (particularly color), some designers prefer to deal strictly with text input, and issues of representing local cell detail along with global routing paths present many difficulties, still under study, for graphic systems.

It is, of course, possible to design in a low level interchange language, such as CIF [13]. At M.I.T., however, there has been a desire to embed layout languages in full-scale programming languages, so as to automatically obtain extensible functionality. Two such approaches currently exist. In AIDS, an interactive layout language embedded in APL, Penfield has provided a broad array of constructs to facilitate rapid and correct layout [14]. The designer provides input in text form, and output can be obtained as CIF code or a four-color plot. A LISP-based system, DPL [15], [16], developed by Sussman and Batali, provides for the specification of parameterized prototypes. Input is by text, and output is CIF or graphic display. Both of these programs are undergoing continual evolution, in order to provide new data types and representations, such as stretchable cells.

There is a strong tendency at M.I.T. to use programs to generate artwork. Designs are often compiled from higher level representations, and in the case of frequently used canonical cells, such as PLA's and register files, programs have been written for the direct synthesis of mask artwork. Glasser and Penfield have written a PLA generator [17] utilizing AIDS. This capability is being added to other layout systems. Other current work includes automatic optimization (for minimal area) for both PLA's and finite state machines (FSM's).

Since routing is an aspect of layout that is heavily influenced by complexity issues, the M.I.T. approach has been to attack fundamental routing limitations in terms of space bounds. Rivest and LaPaugh have focused on the problems of layout for orthogonal rectangles with interconnect on all four sides. Initial results have given bounds for the channel area around isolated rectangles [18], and the techniques used for this study are being extended to assess the complexity characterization of a complete inter-rectangle routing strategy. On a utility basis, some routing programs have been written for cell interconnect and busses, but they are to be viewed in the context of an interactive design style that contemplates human modification of the router output in ways that are currently not understood for algorithmic representation.

Shrobe has built a LISP-based system, DAEDALUS for interactive, graphic-based design making use of the notion of prototype cell and the ability to alter cell instances in the presence of layout stipulations that would otherwise cause design-rule violations. This program utilizes techniques for constraint propagation, so that a correct layout is maintained as the design evolves. In addition, flexible windowing techniques have been utilized to aid the routing process so that the designer can see local detail for terminations, but block-level boundaries for routing path choices. Since DAEDALUS and DPL manipulate the same database format, known as LISPIC, a design can be done partially in DPL and partially in DAEDALUS. Many of the DAEDALUS capabilities are available in DPL also.

It should be emphasized that the M.I.T. approach to design requires a sophisticated computing environment. A wide variety of hardware and software has been utilized, including at least six programming languages and four different computers. At present these facilities are only loosely connected, with internal networks where possible, and taking advantages of CIF as an interchange language where applicable. At this time there is no intent to unify all of these programs in one system, since they collectively represent several design styles with extensive overlap. Instead, clusters of these programs are expected to evolve into distinct systems representing different approaches to the design problem. While this multiplicity may be regarded as a luxury, it does allow comparison of divergent techniques.

IV. Two Radically New Design Styles

In Section III, a view of design-automation activities at M.I.T. was presented. Programs to generate and manipulate representations in several design domains were described. Each of these programs can help support many different design styles. In this section we discuss two design styles emerging at M.I.T. that appear to be radically new.

The first of these is based on concepts in artificial intelligence. The ultimate goal of this work is to understand, from an AI point of view, the human design process. Part of this goal is accomplished by writing computer programs that act more or less as a human design engineer would when confronted with the same circumstances. Since no rigorous theories seem to apply in this area, the work envisioned has a high heuristic content.

Rather than model the human design process in general, Sussman has chosen to do so in a particular domain, namely that of VLSI design. This is an excellent choice, since VLSI design is neither too simple nor too complicated; its artifacts in the various domains discussed in this paper are readily represented in a computer; and the results of design programs can be tested by several techniques including syntactic checking, analysis, simulation, and ultimately manufacture and testing. The applicability of VLSI to design of computer systems assures that there will be adequate opportunity for Sussman and his colleagues to observe and participate in actual VLSI designs at M.I.T.

The LISP-based tools described in Section III are the foundation on which the model of a VLSI designer will ultimately be built. When this is done, there will be two results of major importance. One is a model of the human design process, which hopefully will shed light on design in other domains as well as VLSI. The other is the computer program which embodies the model. This program (or system of programs), by being able to do roughly what a human designer would do in similar situations, can serve as an intelligent design assistant, that is, as an extremely helpful CAD tool. This approach to design automation is so different from anything in use currently that it must be regarded as radically new.

The other new approach is based on the observation that VLSI design is difficult because of the sheer complexity of the systems being designed. The discipline that has dealt most directly, and most effectively, with complexity is computer science. Thus those techniques that have been found successful in computer programming are believed to hold promise for design of IC's. The simplest way (although perhaps not the only way) to absorb those techniques is to regard the specification of an IC design as equivalent to writing a computer program. In fact, the idea is to write a program, in a suitable language, that will denote the design—to document the design, a program listing is obtained; to change the design, the program is edited; and to create the data necessary to produce the design in mask form, the program is executed. Naturally the program in question can call others, and the architecture of a design can be reflected in this subroutine call structure. What is needed for this approach to succeed is a high-level IC design language and a set of underlying programs which execute statements in that language. As far as the geometric layout aspects of VLSI design are concerned, both AIDS and DPL satisfy that goal. Neither one at present extends to the other important domains of description. Both AIDS and DPL, being embedded in general-purpose interactive languages (APL and LISP respectively) can be regarded as special-purpose extensions of their host languages, and, therefore, both may be programmed in.

This programming-based style is appealing at M.I.T., because many of those involved in VLSI designs are computer scientists who feel quite at ease writing programs; indeed it would be difficult to envision many of them doing a design any other way. This type of style is so different from the styles in current use in industry that the term "radically new" is appropriate.

V. Summary

The M.I.T. VLSI research program contains several projects in design automation. Many of the design aids in use at M.I.T. are capable of supporting several design styles. Two emerging design styles are so different from what is currently done in industry that they are termed "radically new." The design aids can best be understood in the context of a design space with various domains of representation, each capable of expressing the architecture of a design as a hierarchy of modules.

References

[1] G. S. Miranker, "The Use of Conflict in the Translation and Optimization of Hardware Description Languages," Ph.D. dissertation, Dep. Electrical Engineering and Computer Science, M.I.T., May 1979.

[2] G. L. Steele, Jr., "The Definition and Implementation of a Computer Programming Language Based on Constraints," Ph.D. dissertation, Dep. Electrical Engineering and Computer Science, M.I.T., Aug. 1980.

[3] G. J. Sussman and R. M. Stallman, "Heuristic techniques in computer-aided circuit analysis," *IEEE Trans. Circuits Syst.*, vol. CAS-22, pp. 857–865, Nov. 1975.

[4] J. De Kleer and G. J. Sussman, "Propagation of constraints applied to circuit synthesis," *Int. J. Circuit Theory Appl.*, vol. 8, pp. 127–144, Apr. 1980.

[5] R. E. Bryant, Ph.D. dissertation, Dep. Electrical Engineering and Computer Science, M.I.T., in preparation.

[6] ___, "An algorithm for MOS logic simulation," *Lambda*, Fourth Quart., 1980, vol. I, no. 3, pp. 46–53.

[7] ___, "MOSSIM: A logic-level simulator for MOS LSI: User's manual," unpublished, July 1980.

[8] C. Baker and C. Terman, "Tools for verifying integrated circuit designs," *Lambda*, Fourth Quart., 1980, vol. I, no. 3, pp. 22–30.

[9] C. Terman, "Simulation Tools for VLSI Design," Ph.D. dissertation, Dep. Electrical Engineering and Computer Science, M.I.T., in preparation.

[10] C. M. Baker, "Artwork Analysis Tools for VLSI Circuits, M.S. thesis, Dep. Electrical Engineering and Computer Science, M.I.T., June 1980.

[11] A. J. Giovinazzo, "A mask design rule checking system," in *Proc. IEEE Int. Conf. on Circuits Computers ICCC80*, pp. 932–936, Oct. 1–3, 1980.

[12] L. Seiler, "Introduction to the design rule checker," unpublished, Mar. 1980.

[13] C. Mead and L. Conway, *Introduction to VLSI Systems.* Reading, MA: Addison-Wesley, 1980. ch. 4.

[14] P. Penfield, Jr., *AIDS-79 User's Manual*, unpublished, Mar. 1980.

[15] J. Batali and A. Hartheimer, "The design procedure language manual," unpublished, Sept. 1980.

[16] G. J. Sussman, J. Holloway, and T. F. Knight, Jr., "Design aids for digital integrated systems, an artificial intelligence approach," in *Proc. IEEE Int. Conf. on Circuits Computers ICCC80*, pp. 612–615, Oct. 1–3, 1980.

[17] L. A. Glasser and P. Penfield, Jr., "An interactive PLA generator as an archetype for a new VLSI design methodology," in *Proc. IEEE Int. Conf. on Circuits Computers ICCC80*, pp. 608–611, Oct. 1–3, 1980.

[18] A. S. LaPaugh, "A polynomial time algorithm for optimal routing around a rectangle," in *Proc. IEEE 21st Symp. Foundations of Computer Science*, pp. 282–293, Oct. 13–15, 1980.

Stanford Overview in VLSI Research

ROBERT W. DUTTON, SENIOR MEMBER, IEEE

Invited Paper

***Abstract*—The Stanford University environment provides an unique combination of systems applications coupled with a state-of-the-art IC fabrication facility. Tools for VLSI design are being developed in concert with the evolution of chip designs—the largest design to date contains more than 10 000 transistors. Design methodologies are discussed as well as specific tools and approaches for IC synthesis, layout, and analysis. Recent results of a database experiment are presented. In addition, examples of specific system architecture experiments are considered.**

I. Introduction

The VLSI efforts at Stanford University have evolved from a diversity of research projects with one common theme—the effective engineering[1] of "order" in the face of complexity. The four groups, represented in this paper, each come from a distinct physical location as well as intellectual temperament at Stanford. The structure of the sections which follow is aimed to cut across the activities in the different groups. The organization of the paper follows a topical coverage of: 1) a definition of the groups, 2) experiments in design methodologies, 3) synthesis, 4) layout, 5) analysis, 6) databases, and 7) system architectures and the man–machine interface. Before launching into the discussion of the VLSI research at Stanford, it is useful to describe the environment.

Fig. 1 shows a two level diagram of the research groups and their physical locations. The figure also shows the present computer resources, the ETHERNET [1] communications, or location of the IC Lab (as well as a possible site for a new VLSI Lab). While the research units are separated physically, the computer NETWORK provides a robust interconnection for communication, distributed processing and resource sharing. For example, all groups use the DEC 20 as a mail system. The VLSI Lab will house both a new fabrication facility as well as teams of designers drawn from the systems laboratories and Computer Science Department.

The Computer Science (CSD) and Electrical Engineering (EE) Departments at Stanford are distinct and yet share the faculty and courses taught in the Computer Systems Laboratory (CSL). Within the CSD there are three groups involved in VLSI: the theory group (Ullman), the database

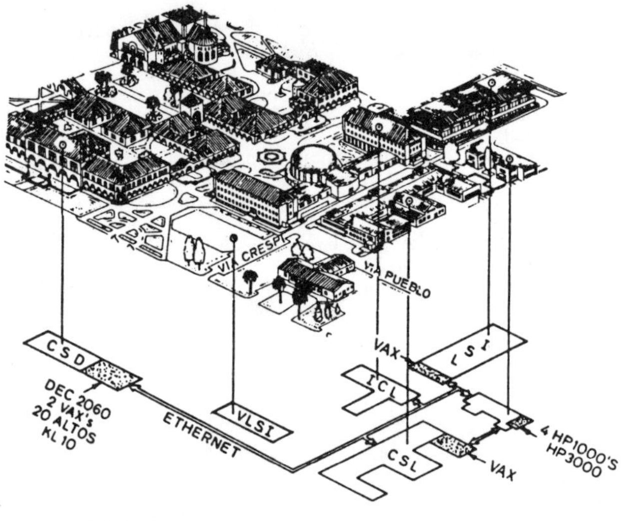

Fig. 1. Two-level view of Stanford. (a) The birds-eye map of buildings. (b) An organizational view of Laboratory and Department entities as well as computer resources.

group (Wiederhold), and the heuristic programming group (Brown,Feigenbaum). Each group is known for its contributions in fields other than VLSI. The present thrust within these CSD groups is to apply their research techniques and tools to the complexity problems of VLSI.

The CSL addresses itself to the more classical hardware/software issues of the computer area. Four groups have explicit interests and activities in VLSI: the testing group (McCluskey), the system software group (Baskett, Hennessy), the systems architecture group (Baskett, Clark), and the layout and specification group (vanCleemput).

The Information Systems Laboratory (ISL) rapidly has become involved in VLSI both in developing algorithms for information processing via hardware (Morf, Kailath) as well as in addressing the IC design process itself as an application of information theory. Areas such as cryptography, distributed networks, and image processing all have growing projects related to IC's. The groups involved specifically in IC design address both theoretical (El Gamal) and applied (Newkirk, Mathews) design.

The ICL is involved in building custom IC's for specific system applications—for example, rehabilitation systems for the blind and medical ultrasonics for imaging. Technology development generally has been considered in a system context. Recently the laboratory has developed an NMOS process specifically for VLSI applications and completely processed educational course project chips on a fast-turn-

Manuscript received November 25, 1980; revised January 22, 1981. This work was supported in part by the Army Research Office under Contract DAAG29-80-K-0046, the Defense Advanced Research Projects Agency under Contracts MDA903-79-C-0680 and MDA903-80-C-0107, and by the National Science Foundation under Contract NSF-MCS-79-04864.

The author is with the Center for Integrated Systems, Stanford University, Stanford, CA 94305.

[1]The choice of the word implies no particular disciplinary bias.

Reprinted from *IEEE Trans. Circuits Syst.*, vol. CAS-28, no. 7, pp. 654–666, July 1981.

TABLE I
ORGANIZATION VIEW OF PAPER AS WELL AS OVERVIEW OF DEPARTMENTAL AND LABORATORY INTERACTIONS INVOLVING VLSI RESEARCH

		Electrical Engineering Department			Computer Science Department
		ICL	ISL	CSL	
II	Methodologies			SPRINT (PCB) (vanCleemput)	Heuristics (Brown, Feigenbaum)
				← SCALD (ECL Wirewrap) → (McWilliams, Baskett)	
		← VLSI Education (3 offerings of Introduction to VLSI and advanced courses/seminars) →			
			← Testing → (Mathews, Newkirk)	(McCluskey)	
III	Synthesis	Chip Fabrication (Plummer, Meindl)		Microcoded PLA's (Hennessy, Clark) PLA's from DDL (vanCleemput)	Regular Expressions (Ullman)
IV	Layout		Statistical (El Gamal, Mathews, Newkirk)	Hierarchical Place and Route (vanCleemput)	Optimal Routing (Ullman)
V	Analysis	Process and Device Simulation (Dutton)		Functional Simulation & Verification (vanCleemput)	
VI	Data Base				Design and Evaluation (Wiederhold)

around basis.[2] Within the ICL the activities include research in the areas of systems (Meindl), device technology (Plummer), and computer-aided design (Dutton).

Given this brief organizational introduction to the Stanford environment, the following subsections will address specific project activities. It will become clear from the discussion that projects cut across the traditional disciplines mentioned above. In addition, a few topics have efforts that may appear to overlap. At this point we feel this overlap is reasonable since the respective groups each bring new perspectives to the problems. Two areas where this multifaceted approach is apparent include wire routing and chip implementation languages. It is too early in the research to discuss the overall interface of tools into a system. All groups are driven by their own specific design examples. The most frequent occurrence of program interfacing at this point results from the needs of chip designers to overcome limitations imposed by a particular tool or to cross-check their results. A good example of this cross-checking is taking place for the synthesis of PLA's. The major considerations for interfacing at present involve communications. In particular, the ETHERNET is being used as an experimental hardware communications link. In addition there are growing language needs to communicate between tools in describing both structural and functional characteristics of design. Table I gives a matrix-listing of projects and groups involved in the topics presented in the following sections. As each topic is discussed, the table may be a useful reference point to keep perspective on the overall activity.

[2]During the Winter Quarter 1980 the Stanford IC Lab assembled its own set of multiproject chips—a total of 16 designs and four chip reference frames—and completed fabrication of the projects within a period of three weeks. The ring-oscillators on the test chips operated at a pair delay of 2.5–3 ns. Statistics concerning results of the individual projects are available [2].

II. EXPERIMENTS IN DESIGN METHODOLOGIES

Design methodologies are most often technology-specific, and prototype system capabilities developed at Stanford reflect this fact quite clearly. The evolution of system designs toward the increasing use of IC's as modules has been one important trend. The second trend involves the increased complexity of IC's, which now embody major system functions on single chips. In fact, many of the system concepts and design methodologies are becoming relevant in the context of chip design. Two examples of mature non-IC design systems developed at Stanford are discussed below, as well as the evolution of new IC-specific systems. The motivation for discussing the non-IC technology design tools within the context of the VLSI theme is two-fold: 1) many lessons have been learned that stimulate our further thinking, and 2) the subsequent discussion of specific tool development and database experiments draws directly on understanding the SPRINT and SCALD systems.

A. SPRINT

This system for the layout of printed circuit boards and of hybrid circuits was developed over the period 1975–1979 [3]. The primary purpose of the system was to provide a test vehicle for the development of new algorithms to support new technologies. Derivatives of this system are currently in use in industry both for the layout of printed circuits and for bipolar gate arrays. An interesting concept that evolved during the SPRINT project was that of a central design file, a local database for printed circuit design. The key feature of this design file was to have rapid access to all the data needed in the design process. Where it was first thought that the inefficiency of some industrial PC design systems was due to the use of a commercial available database management system (DBMS), it has now become clear that a judicious selection of the data base schema is the most critical issue. The SPRINT system consists of a dozen different programs for automatic component placement, automatic conductor routing, automatic gate assignment, via elimination, interactive component placement, interactive routing, interactive artwork editing, and artwork production.

B. SCALD

The SCALD system began as a project in a computer architecture course at Stanford [4]. The outcome was the initial design of a 15 MIPS machine with more than 5000 ECL gates which was completed in less than 2-1/2 man years of effort. The features of the system that made this design efficiency possible included: 1) the use of an hierarchical drawing system for both documentation and creation of design files, 2) the macroexpansion of the design to create the full hardware implementation, and 3) the implementation of automatic wire-wrap output to create the actual artifacts of the design. One key aspect of the design methodology was the fact that all engineering change orders (ECO's) were instantiated via the macroexpansion capability, so that many design errors were avoided. In fact, the total number of errors and corrections for the initial design of the MARK I processor, now being built at Lawrence Livermore National Laboratory, was substantially smaller than for commercial designs of systems of comparable complexity [5]. Following the development of SCALD I, and based on the design experience of the MARK I, a timing verification capability was developed [6]. In contrast to logic simulation, the timing verification checks consistency of specified timing, independent of actual logic patterns. This involves the initial specification of windows of assumed activity and inactivity, as created during the input of system drawings. The evolution of the SCALD system involves a redesign of the system to account for IC-specific features, including two-dimensional constraints both for layout and for the resulting path-lengths for signal delays [7].

C. VLSI Education

The offering of a course based on the Mead–Conway text [8] has provided an excellent basis for testing IC-specific design methodologies. Moreover, the course has necessitated the implementation of a set of design tools and their trial by fire. As discussed earlier, one unique aspect of the design course at Stanford is the internal fabrication capability, which is being used to process project chips. In future course offerings, as well as for research group activities, quick-turn-around fabrication will be provided as needed. The experimental nature of the course and its limited scope precludes overly ambitious designs. On the other hand, the trend of increasing project sizes given a fixed design time is a very positive reflection on the evolution of the tools toward VLSI.

For the initial offerings of the course, the students have had only primitive layout tools at their disposal. They have used a succession of low-level layout languages, entering text to describe their designs and plotting them to perform checks. These language processors produce CalTech Intermediate Form (CIF) as output, the proper form for designs to be included in the Multi-Project Chip sets [9].

The most recent of these simple layout languages is CLL (Chip Layout Language). Similar to CIF in expressive power, it is intended as a simple description language rather than an interchange format. It includes named symbols, precisely defined wires and more convenient transformations, while restricting the designer to a 1/2-lambda grid and right angles only. (An embedding of CLL in Pascal, which also included a simple routing scheme, yielded no further improvements in design speed or correctness.)

A more powerful layout language is currently being developed, a major purpose of which is to provide a test vehicle for a statistical router. This router is based on a two-dimensional statistical layout model that predicts a chip's wiring requirements and final size before routing is begun [10]. These estimates are used to adjust the initial placement provided by the designer, sharply reducing the number of times routing itself must actually be done. Since the typical CLL designer spends about 30 percent of his effort describing and debugging interconnections [2], the router will contribute markedly to his productivity.

Owing to the obstacles to testing—building an environment for a chip and furnishing necessary laboratory equipment—comparatively few class designers test their projects in any detail. The current tool to ease testing is the MINIMAL tester. This tester is little more than a 40-pin socket interfaced via standard RS232 protocols between a terminal and the computer; it provides means for driving and sensing voltages on the pins, using as few parts as possible. Despite its low speed (less than 1000 drive/sense operations per second), it is capable of functionally testing most simple designs, since storage times approaching one second are common. Prototyping kits for the MINIMAL tester have been delivered to CalTech, CMU, UCLA, USC-ISI, Washington University, St. Louis, MO, and Duke University.

A recently completed, LSI-11-based tester can functionally test most chips that the MINIMAL tester cannot. The LSI-11 (a single-board microcomputer manufactured by Digital Equipment Corp.) serves as a buffer between the main computer and two MINIMAL testers. These testers are driven through a direct memory access interface to the LSI-11, yielding about 150000 drive/sense/synchronize operations per second. This tester accommodates up to 80-pin chips.

These tools and others currently being developed will be evaluated in the context of the design course over the next year. The sheer number of designers—130 during the 1979–1980 academic year—taxes every aspect of this design environment and permits both thorough testing and evaluation of each tool.

D. On-Chip Testing

Testing is being considered from the perspective of analysis, hardware, and software support. Issues concerning design for testability have evolved over the past several years [11] and certainly much more work is needed in this area. One specific set of projects underway is aimed at understanding faults. These efforts include:

1) Work in the area of design of watchdog processors and associated software routines. The term watchdog processor means a custom piece of hardware with the specific function of monitoring activities and faults of a specific digital block. At present the Stanford PASCAL compiler is being modified so that it emits code which contains messages to be sent to a watchdog processor. Ways to compress these messages so as to limit the required memory space and bandwidth are one area of interest.

2) Work is in progress aimed at determining chip architectures which will allow the chips to be easily tested—probably capable of completely autonomous self-test.

3) A test chip has been designed in order to collect data on temporary failures. The plan is to produce a number of these chips which are designed to be self-checking and to monitor their error outputs in order to identify and characterize the failures that occur.

4) Chip designs are being investigated with the objective of implementing in silicon some of the algorithms which are currently used in software for providing fault-tolerance.

E. Heuristics of Design

The knowledge-based VLSI design project is directed toward the study and the application of Artificial Intelligence concepts and techniques in the context of very large scale integrated circuit design. The project has as its goals:

1) The identification and articulation of the information used in the design of integrated circuits.

2) The development of methods for representing and reasoning with this information.

3) The development of a knowledge based, expert system for assisting in the design of integrated circuits.

The project is a part of Stanford's Heuristic Programming Project [12]. The initial task of our VLSI project is to develop an experimental, graphics editor which will assist designers in creating IC mask layouts from a component type description of a circuit. The editor will be used to investigate techniques for overcoming the limitations of current IC design systems. For example, most current IC design systems operate in a "breadth first" manner. They require a designer to fully specify the overall circuit at a given level of detail, for example, logic level, before moving on to a more detailed level of specification, for example, schematic level. Moreover, except for some rigidly structured design systems such as master slice or standard cell systems, most systems provide only very limited assistance to a designer in transforming one level of an IC design to a more detailed level or in maintaining the relationships between the various levels of circuit specification.

The editor is based on three central concepts:

1) Mixed-level design notations.

2) Mixed, multilevel design specification.

3) Knowledge-based, expert system design aids.

The editor will utilize a mixed level design notation which will enable a designer to suppress details until they are needed in the design process. The editor will also permit a designer to suppress detail by specifying the circuit components and interconnections at various, mixed levels of abstraction, with either the designer or, as possible, the editor supplying additional detail as it is needed. The editor will automatically maintain the relationships between the more abstract and the more detailed circuit specifications. We plan to use this mixed-level specification capability of the editor to explore a least commitment approach to IC design. This approach avoids making premature design decisions by delaying decisions until they are necessary.

In addition to developing notation and methodology, we are developing knowledge-based design aids which utilize encoded factual and judgmental (heuristic) knowledge to integrate design constraints, help transform abstract circuit descriptions into more detailed specifications, and justify design decisions. The editor will contain several design aids to help a designer in supplying circuit details. The primary design aid will be an expert system (i.e., system performance at a level comparable to that of human experts) to generate the geometric details of a mask layout from a more abstract, topological layout of a circuit. This expert system will be driven by a knowledge base which includes layout heuristics as well as technology design rules. Other design aids in the editor will be used to perform topological routing and to maintain circuit connectivity during design manipulation.

The initial editor will be a useful design aid for student-level NMOS designs. Moreover, it will serve as a vehicle for acquiring layout knowledge and as a testbed for experimentation with various design methodologies.

III. Synthesis

The synthesis of hardware based on high-level language descriptions is a very attractive means to cope with the complexity. Although it is clear that the layout is only one detail in the design process, the aspects of synthesis specific to layout are treated in the next section. The synthesis examples cited in this section address the broader aspects of synthesis given a particular architectural set of objectives. In dealing with microcoded machines it is necessary to understand and simulate the operation of the code as well as the hardware. A synthesis environment for design and documentation of microcoded PLA machines is described, including facilities for both simulation and automatic layout [13]. In a second synthesis example, the use of regular expressions is considered as an alternative for simplifying the problems of going from specification to layout [14].

A. Microcoded Machines–SLIM

Simulation Language for Implementing Microcode (SLIM) is a language and processing system for defining microcode to be implemented using PLA's. This is particularly suitable in the VLSI environment as the high-level microcode specification can be debugged in software and finally directly compiled into hardware. The microcoded control for a chip is written in SLIM as a high-level language program.

A SLIM program consists of the control specification and a functional environment description. The control specification is a program written in a form that can be easily translated into a finite state machine. The environment is a description of the functional units which the controller interacts with. It is described in a standard programming language and used to simulate the microprogram. When a SLIM program is processed, the designer may specify either PLA generation (silicon compilation) or the generation of a functional simulation program. The simulation program utilizes the environment description to describe the functional aspects of the chip.

Since the end product of a SLIM compilation is a finite state machine implemented with PLA techniques, a method of incorporating details about the implementation is required. This specification should include mappings between functional specifications in the environment and actual PLA outputs, as well as timing specifications that force outputs to occur either earlier or later than they occur in the program. Including these details separately allows a more functional orientation in the microcode description.

The relationship between the microcode specification of the control program and the PLA is defined by the input signals, the output signals, and the mapping between environment procedures and output signals. PLA signals are defined by means of input and output signal declarations; their relationship to the environment routines is established through signal definitions. Since a procedure or function in the environment can logically correspond to one or more signals, a method of defining the mapping between environment functions/procedures and signals is needed. Allowing this capability has two significant effects: it allows the microcode description to be functionally oriented, and it significantly decreases the amount of code needed to describe PLA implementation of the microcode. The mapping between environment procedures and signals that are generated in the PLA is given in the definition section of an environment procedure.

Simulation of a SLIM program is done by the user creating a Pascal program which embodies the semantics of the microcode. Two types of simulation output can be produced: 1) functional simulation with state traces and outputs of additional information as dictated by the environment procedures, and 2) functional simulation with input/output signal tracing.

PLA generation is done from a SLIM description in two parts. The first part analyzes the microcode structure and creates product term lists for each output. The effect of signal definition and pipelining is integrated before making these lists. The PLA layout is then done by a separate program that reads the signal descriptions and the product term lists. This structure allows the insertion of programs that optimize the finite state machine.

The use of SLIM in actual chip design has proved to be very valuable. It has been heavily used in the Geometry Engine [15], Image Memory Processor [16] and by several other smaller projects. Without SLIM the larger projects would not have been feasible. The use of SLIM has also made incremental development and design alteration possible, as well as contributing substantially to the correctness of the final design.

B. PLA Synthesis from DDL

DDL-P is a Pascal implementation of a large subset of Digital Design Language (DDL), a register transfer level hardware description language first proposed by Duley at University of Wisconsin [17]. This system includes a compiler and interactive two-value simulator for the design of synchronous digital systems. DDL-P includes facilities for declaring synchronous or asynchronous registers, data paths, and combinational or sequential networks. The control of this hardware is described with a finite state machine notation.

While the analysis of an existing design is an understood task, synthesis is a very difficult task to be automated. As an attempt in that direction, a system that automatically generates PLA's for the control circuit of a digital machine from a DDL-P (register transfer language) description, has been developed. The system identifies the control part in the description and translates it into Boolean equations which may be mapped into a PLA. This is a very convenient tool for the design of a finite state machine. The control circuit of any digital system for which a state diagram can be drawn can be designed easily using this system. This synthesis procedure assumes that the data paths of the system are designed manually. From a practical perspective, this is a reasonable requirement and as such this synthesis procedure nicely complements the Caltech structured layout style, by automatically designing the control PLA. This system was used successfully to synthesize the control circuitry of Intel 8008 and 8080 microprocessors without difficulty. Table II shows results of PLA minimization for several digital system building blocks. The number of inputs and outputs, the number of product terms (before and after minimization) and CPU times using the DEC 20 system are shown. One extension to the PLA minimization system currently under development is a PLA partitioning system. A preliminary result shows that the 8080 control PLA (see Table II) can be partitioned into 16 smaller PLA's thereby reducing the total area by a factor of 5.

C. Synthesis Using Regular Expressions

An effort is under way to compile very high-level descriptions of circuits into efficient layouts. Ideally we would

TABLE II
RESULTS OF PLA MINIMIZATION

Name	Source	# of in	# of out	# Prod before	# Prod after	CPU time DEC 20 (sec)
Intel 8008	DDL-P	31	46	168	68	19
Intel 8080	DDL-P	65	66	181	129	43
PDP 11	DDL-P	102	61	1651	348	156
μ proc PLA	manual input	23	40	273	165	390
4 x 4 multiplier	manual input	8	8	225	129	92

like to translate functional descriptions of circuits into logic for those circuits. We are also concerned with applying the theory of algorithms to develop efficient algorithms or heuristics for a variety of layout problems that are exposed as we build the compiler. Initial work has involved translating regular expressions into circuits that recognize their languages. Recall that regular expressions are built from symbols representing inputs, with the operators + (union), * (any number of, or closure), and juxtaposition (concatenation) [18]. Thus the expression $((a+b)(c+d))^*$ represents the set of all sequences of a's, b's, c's, and d's, such that every even position is a c or d, and every odd position is an a or b, the total length being even. We can envision a circuit that would signal at those times when the sequence of input signals received since startup satisfied that property, and that would not signal otherwise.

Regular expressions are parsed into binary trees, and then a crude layout for the circuit is constructed. Each node of the tree is translated into a rectangular box whose function it is to recognize the strings denoted by the subexpression dominated by the node of the parsed tree. Boxes are wired together and placed by a straightforward scheme that will become progressively more sophisticated in the future.

Two natural problems have been spawned by this project. One key problem is to produce small layouts for boxes that recognize single abstract symbols of the regular expression; these symbols may be represented by logical combinations of input wires. We are examining the automatic design of good layouts for small circuits that are specified by logic equations and conditions on the boundary. The general technique uses a "bin packing" algorithm on the minimal set of contacts deduced to be needed in the circuit.

Another problem is to wire together rectangles with wires running between specified ports on the borders of the rectangles. Often the connecting wires will be in the same order on the edges of both rectangles. In the case that the relative vertical position of the rectangles is fixed, we can only optimize the horizontal distance. If neither is fixed, we must find the relative vertical position that allows minimum horizontal distance. We have been looking at the optimization problems that are suggested by the above conditions and have discovered efficient algorithms that produce optimal wiring using any finite set of legal angles

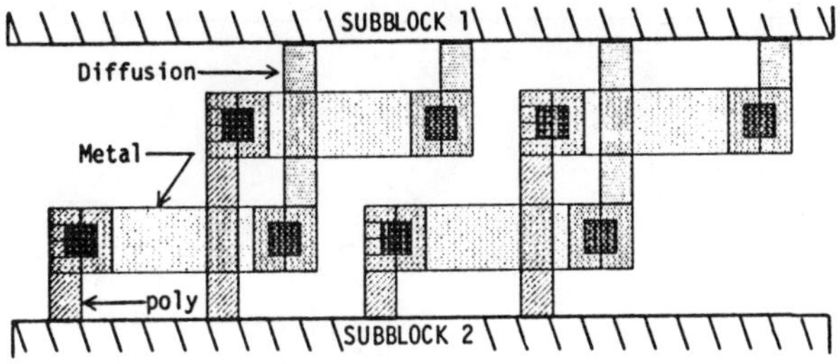

Fig. 2. Multilayer connections for ports misaligned by one unit.

for wire, in the following cases: 1) no crossovers are allowed, vertical position fixed or variable, and 2) crossovers allowed, vertical position fixed.

In the first case, assuming only horizontal and vertical wires are permitted, the Caltech paper shows an example of optimal, no crossover wiring. That diagram was generated by a "river router," which apparently uses the algorithm we proved optimal. Fig. 2 shows an example of an optimal wiring with crossovers. Depending on physical dimensions, that wiring could use less space than the river router would use.

There are a number of unsolved cases, including the case where the order of ports on one face is variable. There are also special situations where the distance between ports is sufficiently small that layer changes become a problem. In that case, less use, if any, can be made of crossing wires, and optimal wiring appears to require solving some combinatorial problems that do not come up in situations where there is adequate distance between ports.

IV. LAYOUT STRATEGIES

A. Hierarchical Placement and Routing

The SICLOPS system is a joint effort between SANDIA Laboratories and Stanford University [19]. The intention of the project is to layout a hierarchically specified integrated circuit in a completely automatic way. The layout problem encountered in this environment consists of finding optimal placement of rectangular blocks of arbitrary size and shape. The criterion used in this placement problem is to minimize the total area of the integrated circuit chip. In addition to a placement algorithm, routing algorithms were developed for solving the layout problem with 100 percent completion. The routing algorithm consists of three distinct steps: initial topological routing, topological routing improvement, and track/channel assignment. One of the problems that surfaced during the research into hierarchical layout was that due to differences in shape of the various blocks, layout efficiency could be seriously affected. Several approaches were studied: limiting the dimensions of the blocks in some way or redesigning all the blocks such that a tight fit would be obtained. The second possibility led to a chip planning tool developed by Slutz [20]. Fig. 3 shows the results of an example where an entire PDP11 CPU chip (more than 10 000 transistors) was placed and routed in one-half hour of DEC 20 CPU time. Fig. 3(a) shows the chip plan with various functional blocks

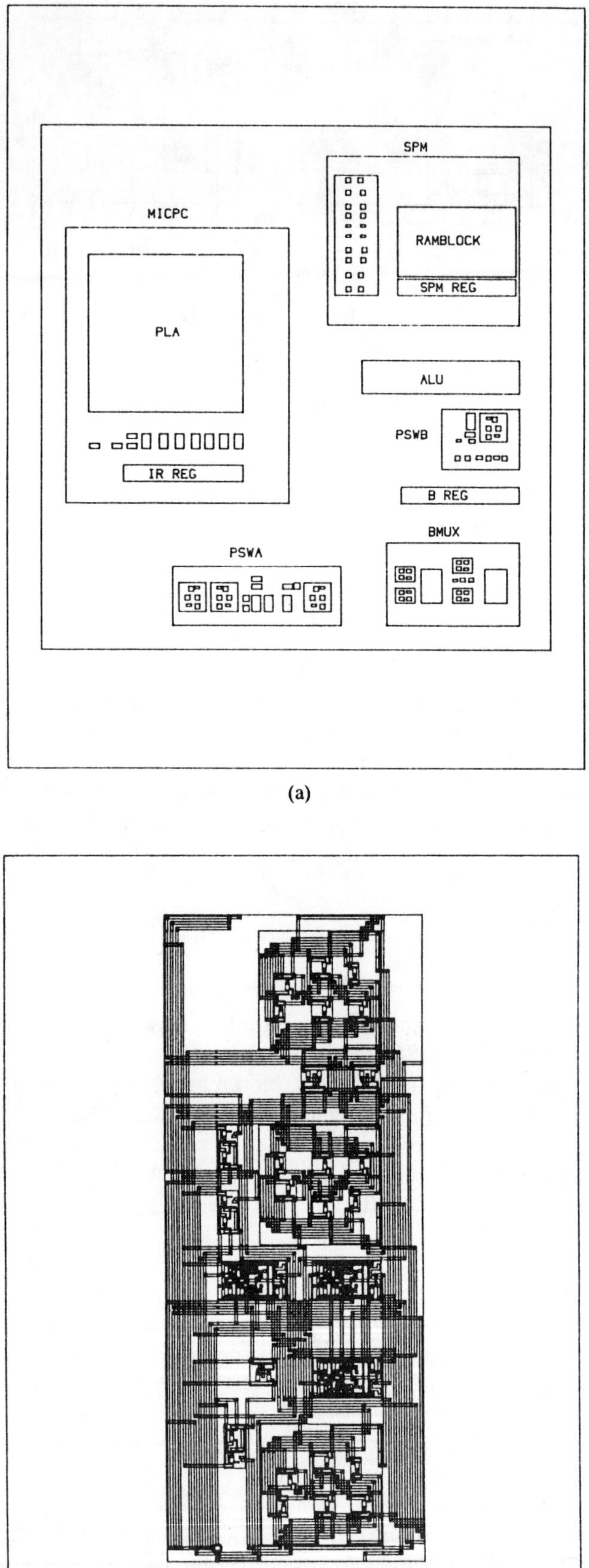

(a)

(b)

Fig. 3. The automatic layout of PDP 11 CPU. (a) The chip plan with subblocks indicated. (b) The detailed routing of the PSWA block.

labeled. The details of the routing for the program status word section of the chip (PSWA) are shown in Fig. 3(b). Work is currently in progress on the automatic repartitioning of the hierarchically specified logic design in such a way that the blocks satisfy certain area and shape factor constraints.

B. Routing Based on Statistical Modeling

Statistical techniques are very effective in providing useful solutions to computationally complex problems such as: large systems of particles, noise modeling, and computer performance evaluation. Recent research indicates that these techniques can be utilized for layout of VLSI systems.

In a recent publication [10] it was shown that for master slice integrated circuits a simple statistical model for interconnections could be used to estimate routing channel widths and consequently to predict chip area. For integrated circuits composed of unequal size functional blocks a similar model has been developed [21]. Given an initial placement of the blocks (which is usually done by the designer) and an estimate of average interconnection length and pin density (which could be done by sample routing), the model is used to estimate chip area and routing channel widths. These estimates are then used to start the routing process or to discard the initial placement. The routing is done as follows: 1) based on the channel-width estimates a modified placement is obtained using optimization techniques, 2) interconnections are assigned to the channels in a manner analogous to the estimation procedure, 3) power and ground are routed using the algorithm reported elsewhere [22], 4) a channel router is used to assign tracks, and 5) minor placement reorganization with repeated use of item 4) is used to complete the routing. The objectives of this router include: 1) to obtain 100-percent routing completion in acceptable time, 2) to approximate the routing density achieved by a layout expert, and 3) to allow a high degree of flexibility to the designer. Statistical measures of routing (and placement) efficiency which could be used to further direct the layout process are topics of further study.

V. Analysis

There are new requirements involved in VLSI design because of both the shrinking size of the devices and the explosive growth in the number of functional blocks on an IC. At the microscopic level the gate length of scaled MOS devices poses technological as well as electrical limits on performance. In order to be able to fabricate one-million devices with a predictable set of characteristics and statistics on a chip, simulation at the device and technology levels has become essential. At the functional level it is also of increasing importance to simulate the top-down performance as well as to verify the correctness of subsequent hardware implementations which can have a variety of topologies and detailed realizations. In this section the research activities involving analysis of device-technology and functional simulation/verification are discussed.

A. Device Technology

Circuit simulation has become an essential tool for the IC designer, since on-chip parasitics and performance of integrated devices cannot be breadboarded. Within the last few years it has become apparent that the modeling and analytic understanding of silicon processing technology must be used both for design and control of scaled-down devices. The reason for this new thrust in modeling comes from the stringent requirements placed on dimensional control—both across the wafer surface and into the bulk of the silicon—as minimum masking dimensions approach the control limits of the processing technology. Stanford has pioneered the development of simulators to model fabrication processes and hence to predict the limits and control of scaled technologies [23]. Moreover, there is an ongoing effort to understand the underlying physics and incorporate this understanding and data directly into the simulation capabilities.

Beyond the efforts to understand and develop models for IC technology, there is a need to couple the bottom level technology factors to all higher levels of design in the most efficient way. Research efforts are in progress to develop two-dimensional device modeling capabilities that couple with technology factors and directly produce the electrical parameters relevant to circuit design. Results to date have produced a general two-dimensional Poisson solver with direct input compatibility with the SUPREM program [24]. A key application of this analysis capability is the extraction of electrical parameters for process and circuit design. The first example of this use has been the study of capacitance components in the Stanford NMOS process. Using both two-dimensional process and device analysis, the sensitivities of source–drain sidewall capacitance have been modeled as a function of both implantation dose and line width parameters [25]. During the next year the overall set of design rules currently used for VLSI design with NMOS will be characterized using the two-dimensional process and device analysis tools. The objective is to determine process control limits and to understand the circuit constraints imposed by the design rules. In addition to the analysis aspects of process modeling, the design aspects will be treated as follows. The coupling of technology-specific details to circuit design and design rule checking is at present a tedious and handcrafted process. Although design rule checking (DRC) and topology extraction are currently available via straightforward algorithms, these approaches to date have not included the necessary technology specific details to either make the methods accurate or efficient. It is the specific objective of the process-modeling research to make all process and device specific data directly available to the other parts of the hierarchical design process.

B. Simulation and Verification Using High-Level Language Descriptions

The intention of the ADLIB development project was to create a single language for the representation of behavioral information at all levels of abstraction [26]. Most description languages in existence today are limited to a rather narrow level of abstraction, often the register-transfer level. ADLIB is an extension of Pascal with facilities to represent hardware primitives and operations. The language has been used for a wide variety of descriptions, varying from high level such a description of packet switching in the ARPA network to the description of the timing behavior of an MOS transistor. Fig. 4 is a simple circuit including a J–K flip-flop and other logic. The SDL description is shown in Table III and the ADLIB subprocess for the flip-flop appears in Table IV.

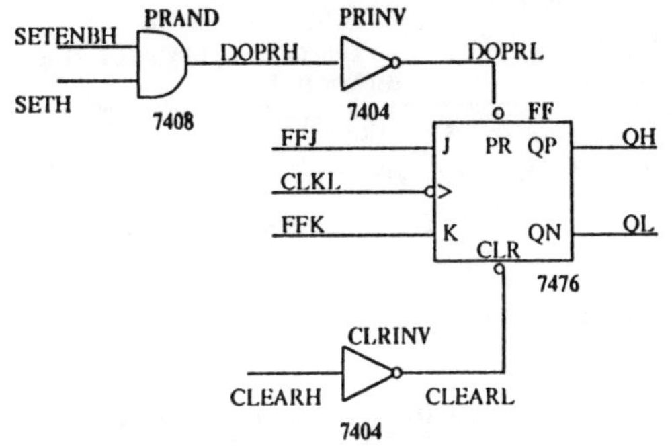

Fig. 4. A logic circuit example for ADLIB/SDL descriptions (see Tables IV and V).

TABLE III
SDL DESCRIPTION OF CONNECTIVITY AND COMPONENTS FOR THE CIRCUIT SHOWN IN FIG. 4

```
                (* SDL ENCODING OF CIRCUIT *)

NAME:    GATE EXAMPLE;

7408:    PRAND(PD = "20.0E-9");
7404:    PRINV(PD = "10.0E-9"), CLRINV;
7476:    FF;

END;

NETSEGMENT;

SETENBH  =   PRAND.A;
SETH     =   PRAND.B;
DOPRH    =   FROM(PRAND.Y), TO (PRINV.I);
DOPRL    =   PRINV.O, FF.PR;
CLKL     =   FF.CLK;
CLEARH   =   CLRINV.I;
CLEARL   =   CLRINV.O, FF.CLR;
QH       =   FROM (FF.QP);
QL       =   FF.QN;

ENDNETS;

ENDC;
```

SABLE is a multilevel hierarchical simulation program. The input to the SABLE simulator consists of a block diagram described in SDL, and of the behavior descriptions for each of the blocks in ADLIB [26]. SABLE is a multilevel simulator in several ways: it allows the mixture of logic primitives such as AND, OR with functional blocks such as microprocessors; it also allows the use of multiple values for the signals during the course of a single simulation run (2-valued mixed with 9-valued simulation).

The goal of the verification project is to show how one may formally prove the equivalence of two ADLIB/SDL hardware descriptions. In a hierarchical hardware design, one of these descriptions will represent the specification for

TABLE IV
ADLIB DESCRIPTION OF FUNCTIONAL INPUT FOR THE CIRCUIT SHOWN IN FIG. 4

```
COMPTYPE    JKFF "7476" (PD: REAL);
DEFAULT     PD = 25.0E-9;
INWARD      PR, CLR, CLK, J, K:  BOOLNET;
OUTWARD     QP, QN: BOOLNET;
VAR NEXTQ : BOOLEAN;
SUBPROCESS
PRESET    : UPON NOT (PR AND CLR) CHECK PR, CLR DO BEGIN
            ASSIGN NOT PR TO QP DELAY PD;
            ASSIGN NOT CLR TO QN DELAY PD;
            NEXT Q: = NOT PR
            END;
VAL-CHECK-1: UPON PR AND CLR AND CLK CHECK J, K DO
            WRITELN  (OUTPUT, '**CHANGING J, K WHILE CLOCK', 'HIGH');
VAL-CHECK-2: UPON NOT (PR OR CLR) CHECK PR, CLR DO
            WRITELN (OUTPUT, '**PR, CLR BOTH LOW IN 7476');
BEGIN       (* JKFF * )
            NEXTQ := FALSE;
WAITFOR DELAY 0.0;
WHILE TRUE DO BEGIN
   WAITFOR PR AND CLR AND NOT CLK CHECK CLK;
   NEXTQ := (J AND NOT NEXTQ) OR (NOT K AND NEXTQ);
   ASSIGN NEXTQ TO QP DELAY PD;
   ASSIGN NOT NEXTQ TO QN DELAY PD
   END
END; (* JKFF 7476 *)
```

a system, while the other represents a proposed design.

In the ADLIB/SDL verification system, a hardware designer will supply a simulation relation specifying the desired correspondence between two ADLIB/SDL descriptions. Both descriptions will then be symbolically simulated. The results of two simulations should be consistent with the simulation relation. This approach, while similar to that used at IBM and elsewhere, is complicated by the presence of different levels of abstraction and asynchronous timing in ADLIB.

A modified hierarchical design methodology will allow the verification system to accommodate different levels of abstraction. The new methodology recognizes that refinements in the levels of abstraction occur in two ways: The level of structural complexity increases in a top-down design, while progressively less compact, less abstract data representations may be used. The verification system will use event-driven symbolic simulation to support asynchronous timing.

VI. DATABASES

Database technology has dealt with both the management of large volumes of data and the problems caused by concurrency of data access. Many existing database systems support interarea communications functions; instances are found of systems dealing with inventories, where production and sales issues interface, or patient management, where individual care and global health care concerns come together. Databases are used as well to resolve the multidisciplinary design issues in aircraft design and to manage the problems of engineering changes in computing systems [27]. Given this background, the database group foresees that databases may provide tools to serve the VLSI designers; however, the tool is complex and will not be effective unless well understood and adapted to the demands of this application.

TABLE V
HIERARCHICAL STRUCTURE OF THE PDP-11 DESCRIBED IN SDL

Level	Components
cpu	PDP-11
reg	BLEG, PROCOUNT, BUSCOUNT, RAM16x16, ALU, STATUS, AMUX
rf1	PRIOARB, TIMER
rf2	TCFF2, RAM16, 40181, 40182
ff	TCFF, LFF16, DFF4, BUSD16, MUX16, RAM
gate	DFF, LFF, XOR, BUSD, MUX
trans	TRANSP, INV, NAND, 3NAND, 4NAND, NOR, 3NOR, 4NOR
bottom	TN, TP, PLA-SMALL, PLA-MED, PLA-LARGE

The approach being used in this work includes the use of commercial database systems to assess the general suitability of database approaches to VLSI design data. One objective is to determine whether commercial database systems, used knowledgeably, can perform adequately, or if they do not, where the bottlenecks are. The current set of experiments use a network system, DEC's DBMS-20. This is a system based on the published CODASYL [28] database definition, which has a strong orientation to business applications, such as inventory management. The database designer can, through schema specification, select which of the logically appropriate linkages should be implemented [29]. A successor experiment planned in this area is to investigate use of a relational database system, RIMS, which has strong automatic query optimization capabilities [30].

An evaluation was made of DEC's DBMS-20 performance on data retrieval times. Specialized design files of circuit information were used as the control to the experiment. The user begins with his design written in a Structural Design Language (SDL) [31]), a straightforward hierarchical description language. The SDL description is input to the SDL compiler, which then produces a dump file containing the logical description of each component described in the design. This design methodology is strictly hierarchical: the logical description of one component is described in terms of smaller and lower level components. For example, gates are described in terms of transistors, flip-flops are described in terms of gates, registers are described in terms of flip-flops, etc. Each description level is given an unique name in the design file. Table V shows the hierarchical structure of the PDP-11 processor, as described in SDL [32]. The top level component is the PDP-11 itself, with component type named to be "cpu." Lower level components describe the gradual breakdown into smaller conceptual pieces such as flip-flops, gates, transistors, etc. The dump file from the compiler is loaded into the SPRINT database [3], the specialized control example, via a "hardwired" schema.

Initially the CODASYL schema was modeled very closely to the SPRINT schema, and due to SPRINT's hierarchical structure, CODASYL network capabilities were not utilized in the first iteration. Later the schema was modified to take advantage of network structures. This provided linked

TABLE VI
RESULTS COMPARING A CUSTOM DATABASE (SPRINT) WITH A DBMS-20 IMPLEMENTATION FOR TWO EXAMPLES

		SPRINT	DBMS-20	records read	words read
ALU	I/O time	21 s	33 s	1514	7754
	CPU time	30	45		
PDP-11	I/O time	66 s	115 s	5925	28501
	CPU time	120	190		

access to the library file of components. A loader program was written to load the same dump file produced by the SDL compiler into the DBMS-20 database. Comparison of loading times showed the SPRINT database to be faster as expected, but DBMS-20 times were quite acceptable. The initial loading of the PDP-11 example took SPRINT about one minute and about four minutes for DBMS-20.

For the actual test of the retrieval times, a macroexpander program was used. This program reads the logical description of a high level component from the database. The user then specifies the levels to be expanded. Thus if the user specified the program to expand all levels down to the transistor, then the resulting output will be the logical description of the original component described totally in terms of transistors. This description may not be very useful to the designer, but could be the input to a simulator program. Also, in the VLSI environment, this could be the first step in producing the layout diagram.

The macroexpander program needs to do a tremendous amount of random read operations, especially if the component that is being expanded is large and described at a high level. Table VI gives the results of expanding ALU and PDP-11.

These results show less than a factor of two degradation in performance of the DBMS-20. This is an acceptable trade-off for the increased flexibility and generality of CODASYL, and dispute the original theories that there would be at least an order of magnitude difference in performance.

After demonstrating acceptable times for both reading and writing into the database using DBMS-20, several experiments were performed. The first attempt was to flag necessary upper level structures when changes were made to lower level components. The method of flagging used is best characterized as height-first.

It must be noted that the number of detailed elements, if a complex chip were to be described in detail at the lower levels, exceeds the storage and manipulation capability of quite large machines. It is still desirable to provide access to lower levels. This dichotomy is handled by having a mix of instantiated and noninstantiated elements. Instantiated elements represent special, designer specified cases; non-instantiated elements are generated when requested by automatic invocation of generation routines. For more information on the experiments see [33].

The significant aspect of using a commercial database system is that it allowed implementation of both initial designs and modifications relatively quickly and easily. For example, implementation of flagging changes to upper level structures took one student two weeks and only a moderate amount of effort. This ability to experiment is important to obtain the knowledge needed for designing useful database systems for VLSI design.

VII. SYSTEMS ARCHITECTURE

VLSI presents the challenge of integrating complete systems on a single piece of silicon. However, mapping existing architectures into silicon as fabrication technology allows it does not always yield the expected performance. For example, because wiring propagation delays in MOS technologies are frequently greater than gate delays. Architectures that minimize wiring lengths will often perform better.

The Computer Systems Laboratory (CSL) system architecture group has general-purpose and special purpose microprocessor projects underway. The general-purpose projects is to build a "standard" 32-bit microprocessor that Directly Executes a suitable Language (DEL), such as Mesa or Pascal. The special purpose architecture projects focus on two custom NMOS chips for high-performance graphics, described below. All of these projects push the limits of VLSI and provide natural forcing functions for design tool development.

Graphics is the "window" for both IC and digital system design. Yet the hardware has been both a performance and perceptual bottleneck. Moreover, the software that supports the graphics is a painful reflection of both the design styles and general system constraints (i.e., database, software interface, and associated hardware components). The projects reported below reflect two evolutionary projects: 1) hardware experiments involving smart/fast terminals for color, and 2) software for schematic-type input specification.

A. Graphics Hardware

The Stanford University Network (SUN) Graphics Terminal represents a recent effort in hardware development to both address VLSI designer needs as well as to exploit custom graphics hardware chips. The SUN terminal provides quality graphics including high resolution, multiwindowing capabilities and high speed at a moderate cost [34]. Two related projects involving custom IC design for graphics applications are the Geometry Engine and the Image Memory Processor.

The geometry engine [15] is a four-component vector function unit that can perform the most common repetitive geometric computations of graphics. It is microprogrammed to do floating-point matrix transformations of 4-component vectors, clipping of lines, polygons and characters to the user's floating-point viewing-space window, and scaling to the integer-space viewpoint of the destination screen. The structure of the system is described elsewhere [15].

The performance limitations of conventional image memories are in the time required to do the arithmetic of scan-conversion, i.e., determining which pixels to alter, and in the memory bandwidth. Normal organizations have one I/O port to the host processor and one output port to the CRT refresh controller. The host is often a general-purpose processor with the image memory part of its normal memory address space. Thus scan-conversion is done by the host. The process is usually limited by the computation rate of the processor in combination with the bandwidth of the memory.

The Smart Image Memory System [35] scan-converts polygons, lines and characters using a custom NMOS Image Memory Processor (IMP), which has been designed, simulated, laid out, and is presently being fabricated in the Stanford Fast-Turn-Around facility. Multiple copies of the IMP in combination with either 16K or 64K memory chips will yield a low-cost Smart Image Memory System that should provide three improvements over existing systems: 1) vector-drawing rates comparable to those of random-deflection CRT's, 2) previously unattainable, arbitrary-font character drawing rates, and 3) polygon drawing rates comparable to those of very expensive, dedicated systems. This Image System combined with the Geometry Engine System described above should result in a very powerful personal-computer graphics system as well as needed custom hardware to improve the efficiency of the SUN terminals.

B. *Graphics Software*

DRAW is an interactive schematic editing and evaluation program, developed at Hewlett-Packard and written in PASCAL. It uses the HP 2648 raster-scan graphics terminal with a digitizing tablet. Stanford-generated modifications of DRAW now allow for hierarchical design, with user-definable components nested at any level [36]. Cells may include programs, which are executed when the cell is used. These cells may examine the parameters and connections of the particular instantiation of a cell, and make decisions about the topology and component values. Outputs include a component list, without hierarchical information and a dictionary which preserves the hierarchical information. As an example, it is possible to describe an N-bit ALU by a combination of graphical information and of attached textual procedures. This is in contrast with most of today's schematic representation systems, where one has to draw separate representation for a 4-bit and for a 9-bit ALU. Current research concerns a system that attempts to preserve the advantages of fully hierarchical design for artwork. This allows design rule checking, logic verification and component value computation to be done on a cell-by-cell basis, with no expansion of the artwork. An editor with appropriate constructs can do this with no loss of area and the only drawback is the possible proliferation of cell types. This problem can be avoided by allowing programmatic modifications to the cells. Uses include automatic layout of common constructs, adaptable cells for gate arrays, and automatic designs to designated performance criteria.

VIII. Conclusions

VLSI research at Stanford now cuts across three laboratories in the Electrical Engineering Department as well as the Computer Science Department. A breadth of internal applications exist for experiments in VLSI at Stanford. A state-of-the-art fabrication laboratory is used to build the IC's designed by system architects. The experience at Stanford in design tools includes a breadth of technologies—SPRINT for PCB's and SCALD for ECL-wirewrap. Present methodologies for VLSI grow out of this past experience as well as the new thrust of structured NMOS design. Chip testing tools and experiments in this area are of major interest in the project environment. Within the computer science activities novel methodologies are being considered using heuristics. Synthesis tools are focused primarily on system subblocks for VLSI. Simulation and synthesis tools for microcoded PLA's as well as implementations based on register-transfer language descriptions have been discussed. An approach for circuit synthesis using regular expressions as a means to translate high level language descriptions is being pursued. Layout aids involve activities in language description tools as well as experiments in: 1) routing based on statistical modeling, and 2) hierarchical placement and routing. Analysis tools have been developed to simulate both high level functional behavior as well as circuit and device performance at the technology level. It is clear that these extremes of the "top" and "bottom" ends of the design spectrum have a profound impact on intermediate level choices of logic circuits. Recent database experiments show very promising results using commercial database implementations with appropriate schema. At a systems architecture level there are a variety of projects involving: the design of special purpose computers and IC design stations with special emphasis on graphics—hardware, software, and network capabilities. These design projects—more than any other single factor—motivate the development, testing and refinement of design tools at Stanford.

Acknowledgment

This paper represents the collective work and writing of many people. The contributions by topic and the associated authors are indicated in Table I. In addition the author would like to express his thanks to each individually for their cooperation and enthusiasm in making this manuscript a reality. From the Computer Science Department he acknowledges the contributors are Anne Beetem, Harold Brown, Jeff Ullman, and Gio Wiederhold. Portions of the paper concerning the Information Systems Lab (EE) were prepared by Abbas El Gamal, Robert Mathews, and John Newkirk. Authors from the Computer Systems Lab (EE) include Forrest Baskett, James Clark, John Hennessy, and Ed McCluskey. A special thanks goes to Willem vanCleemput for his many contributions including suggestions concerning overall organization. There are a number of research sponsors to be thanked. The author would like to thank the Army Research Office for its support of efforts involving structured design based on hierarchical concepts. He is pleased to acknowledge special interest and encouragement of Dr. Robert Kahn of DARPA.

References

[1] R. Metcalfe and D. Boggs, "Ethernet: Distributed packet switching for local computer networks," *Communications of the ACM*, vol. 19, no. 7, July 1976.

[2] R. Mathews and J. A. Newkirk, Summary of Student Design-Project Results, 1980.

[3] W. M. Vancleemput, T. C. Bennett, J. A. Hupp, and K. R. Stevens, "SPRINT: An interactive system for printed circuit board design," in *Proc. IEEE Computer Society Int. Computer Software and Applications Conf.*, (Chicago, IL), pp. 113–119, Nov. 1977.

[4] T. M. McWilliams and L. C. Widdoes, Jr., "SCALD: Structured computer-aided logic design," in *Proc. 15th Design Automation Conf.*, (Las Vegas, NV), pp. 271–277, June 1978.

[5] S1 Project Staff, "Advanced digital computer technology base development for Navy applications: The S1 project," Lawrence Livermore National Lab. Rep. UCID 18038, Sept. 1978.

[6] T. M. McWilliams, "Verification of timing constraints on large digital systems," in *Proc. 17th Design Automation Conf.*, (Minneapolis, MN), pp. 139–147, June 1980.

[7] M. Marshall, "VLSI pushes super-CAD techniques," *Electron.*, pp. 73–80, July 1980.

[8] C. Mead and L. Conway, *Introduction to VLSI systems*. Reading, MA: Addison-Wesley, 1980.

[9] *LAMBDA Magazine,* Fourth Quarter 1980. (Publication of Redwood Systems Group, P.O. Box 50503, Palo Alto, CA, 94303).

[10] A. El Gamal, "Two-dimensional stochastic model for interconnection in masterslice integrated circuits," *IEEE Trans. Circuits Syst.*, to be published.

[11] E. J. McCluskey and S. Bozorgui-Nesbat, "Design for autonomous test," in *1980 Testing Conf.*, (Philadelphia, PA), Nov. 11–13, 1980.

[12] P. E. Friedland, "Knowledge-based experiment design in molecular genetics," Doctoral dissertation, Stanford Univ., Stanford, CA, 1979.
R. K. Lindsay, B. G. Buchanan, E. A. Feigenbaum, and J. Lederberg, *Applications of Artificial Intelligence for Chemical Inference: The DENDRAL Project*. New York: McGraw-Hill, 1980.
E. H. Shortliffe, *Computer-Based Medical Consultations: MYCIN*. New York: Elsevier/North Holland, 1976.

[13] J. L. Hennessy, "A language for microcode description and simulation in VLSI," TR #193, Computer Systems Lab, Stanford Univ., Stanford, CA, 1980.

[14] R. W. Floyd, J. D. Ullman, "The compilation of regular expressions into integrated circuits," to be published.

[15] J. H. Clark, "A VLSI geometry processor for graphics," *IEEE Comput. Mag.*, July 1980.

[16] J. H. Clark and M. R. Hannah, "Distributed processing in a high performance kSmart image memory," *LAMBDA Magazine*, fourth quarter, 1980.

[17] D. L. Dietmeyer and J. R. Duley, "Register transfer languages and their translation," in *Digital System Design Automation*, (ed. M. Breuer), Woodland Hills, CA: Computer Science, Press, pp. 117–218, 1975.

[18] J. E. Hopcroft and J. D. Ullman, *Introduction to Automaton Theory, Languages and Computation*. Reading, MA: Addison-Wesley, 1979.

[19] B. T. Preas, "Placement and routing algorithms for hierarchical integrated circuit layout," Ph.D. dissertation, Stanford Univ., Stanford, CA, Aug. 1979.

[20] E. A. Slutz, "Shape determination and placement algorithms for hierarchical integrated circuit layout," Ph.D. dissertation, Stanford Univ., Stanford, CA, Aug. 1980.

[21] A. El Gamal and Z. Syed, "A stochastic model for interconnections in custom integrated circuits," *IEEE Trans. Circuits Syst.*, to be published.

[22] Z. Syed and A. El Gamal, "On the routing of power and ground networks," in preparation.

[23] D. A. Antoniadis and R. W. Dutton, "Models for computer simulation of complete IC fabrication process," *IEEE J. Solid-State Circuits*, vol. SC-14, pp. 412–422, Apr. 1979.

[24] J. A. Greenfield and R. W. Dutton, "Nonplanar VLSI device analysis using the solution of Poisson's Equation," *IEEE Trans. Electron Devices*, vol. ED-27, pp. 1520–1532, Aug. 1980.

[25] D. J. Chin, M. R. Kump, H. G. Lee, and R. W. Dutton, "Process design using coupled 2D process and device simulators," in *Int. Electron Devices Meeting*, Paper 9.3, p. 223–226, Tech. Dig., Dec. 8–10, 1980.

[26] D. D. Hill, "Language and environment for multilevel simulation," Ph.D. dissertation, Stanford Univ., Stanford, CA, Mar. 1980.

[27] Felix P. Mallmann, "The management of engineering changes using the primus system," in *17th Design Automation Conf. Proc.*, pp. 348–366, 1980.

[28] CODASYL Data Base Task Group, Apr. 1971 Rep. ACM, New York, 1971.

[29] G. Wiederhold, R. El-Masri, "The structural model for database design," in *Proc. Int. Conf. Entity-Relationship Approach to Systems Analysis Design*," pp. 247–267, Dec. 1979.

[30] J. Simpson, *et al*, User's Manual for the RIMS/FPDB System; Science Applications Inc., Palo Alto, CA, Mar. 1980.

[31] W. vanCleemput, "SDL: A structural design language for computer-aided design of digital systems," Stanford CSL Tech. Rep. #136, 1979.

[32] E. Slutz, SDL Description of DEC PDP-11, Stanford Univ. Stanford, CA, CSL, CIS, 1979.

[33] Gio Wiederhold, Anne Beetem, and Garrett Short, "A database approach to communication in VLSI design," Stanford CSD Tech. Rep. #826, 1980.

[34] A. Bechtolsheim and F. Baskett, "High performance raseter graphics for microcomputer systems," in *ACM Computer Graphics Conf. Proc.*, vol. 14, no. 3, p. 43–47, July 1980.

[35] J. H. Clark and M. R. Hannah, "Distributed processing in a high-performance smart image memory," pp. 40–46, vol. 1, no. 3, Lambda, fourth quarter, 1980.

[36] L. K. Scheffer, "A methodology for improved verification of VLSI designs without loss of area," *CalTech Conf.*, Jan 19–21, 1981.

Design Aids for VLSI: The Berkeley Perspective

ARTHUR RICHARD NEWTON, MEMBER, IEEE, DONALD O. PEDERSON, FELLOW, IEEE, ALBERTO L. SANGIOVANNI-VINCENTELLI, MEMBER, IEEE, AND CARLO H. SÉQUIN, MEMBER, IEEE

Invited Paper

Abstract—**Computer aids for integrated circuit (IC) design have been an integral part of the Berkeley IC program since its inception in the early 1960's. While our initial work concentrated on the development of simulation programs, such as SPICE2 and SPLICE, we are currently developing an integrated design system for VLSI circuits. The system is based on 1) the use of dedicated design stations, connected among themselves and to peripheral equipment via a high-speed computer network, and 2) a portable operating system framework for the integration of a set of design programs. This paper describes our ongoing research in the area of layout design and verification, simulation, synthesis, and optimization for VLSI design.**

I. Introduction

THE DEVELOPMENT and use of computer aids for Integrated Circuit (IC) design at Berkeley has rapidly expanded in the recent past. A core group of faculty which have, over the years, been interested in IC technology and design (Brodersen, Gray, Hodges, Meyer, Neureuther, Newton, Oldham, and Pederson) have been joined by faculty and research students from the Systems area (Kuh, Polak, Sangiovanni-Vincentelli) and Computer Science (Ousterhout, Patterson, Séquin). This large group has the common goal of developing an integrated set of design-aids for Very Large Scale Integration (VLSI) embedded in a flexible and adaptable VLSI design system.

Computer aids for IC design and evaluation have been an integral part of our IC program at Berkeley since its inception in the early 1960's. In our first projects, either special purpose programs were written to analyze our devices and circuits, or existing programs were adapted to suit our needs. In particular, the nonlinear performance of the integrated circuits, involving both electronic and temperature effects, made circuit simulation essential. These efforts lead to the SPICE [1] and SLIC [2] circuit simulation programs at Berkeley and the TIME program [3] at Motorola. The SPICE2 program [4], [5] has become widely used within the IC industry worldwide as well as by a great many universities. Since all of our programs are issued in the public domain and each group who receives a copy is at liberty to reissue the program, it is virtually impossible to know how many copies of SPICE programs are actually being used; however, we estimate there are over 1000 copies of SPICE2 now in use.

Manuscript received December 1, 1980; revised February 9, 1981.
The authors are with the Department of Electrical Engineering and Computer Sciences, University of California, Berkeley, CA 94720.

In our IC program, each research student is responsible for the entire design and fabrication of a circuit and rapidly becomes an expert at the layout process. As designs have become larger and more complex, we have followed industrial practice and introduced computer aids such as digitization and interactive correction of layouts, leading to computer-controlled pattern generation. As new design-aids are developed at Berkeley, they are used extensively in the classroom as well as by our own research students before being released to other universities and the electronics industry.

The following section provides an overview of the integrated design system under development in our laboratory. In particular, the hardware and software requirements of our approach to an integrated design station are presented, and the particular design methods we propose to support for VLSI circuits are described. In the remainder of the paper, our research in the area of computer aids for IC design, layout, simulation, verification, and optimization, is described. While some recent results are presented, the main intent of this paper is to describe work in progress and work planned for the near future, so as to indicate the direction of research at Berkeley in the area of design aids for VLSI.

II. The Design Environment

Throughout the IC industry it is now clear that an integrated set of computer design aids, coupled with an unified approach to data management, is essential for VLSI design. With the decreasing cost of hardware, a dedicated special-purpose *design station* can be provided for VLSI designers on a per-designer basis. It is anticipated that this design station would have the computational power of a high-performance minicomputer of today, coupled to a high-resolution color graphics display, and interfaced to other design stations and common resources via a high-speed computer network. Thus a *design system* for VLSI circuits would consist of a number of design stations coupled together, each with access to other computers for mass storage and high-speed computation as well as to common peripherals such as line-printer and color graphics

Reprinted from *IEEE Trans. Circuits Syst.*, vol. CAS-28, no. 7, pp. 666–681, July 1981.

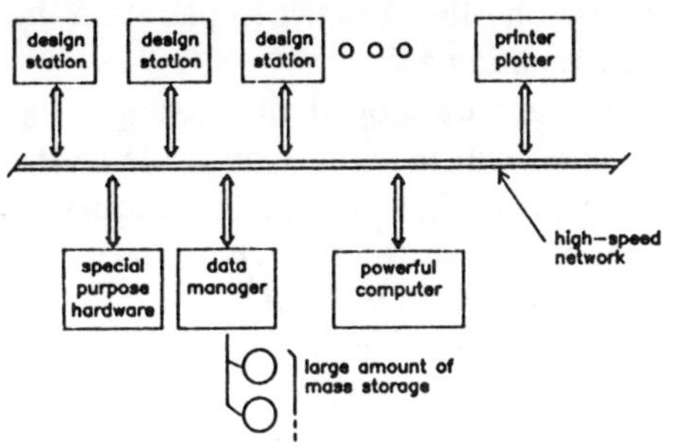

Fig. 1. Topology of a design system for VLSI using dedicated design stations and a high-speed network.

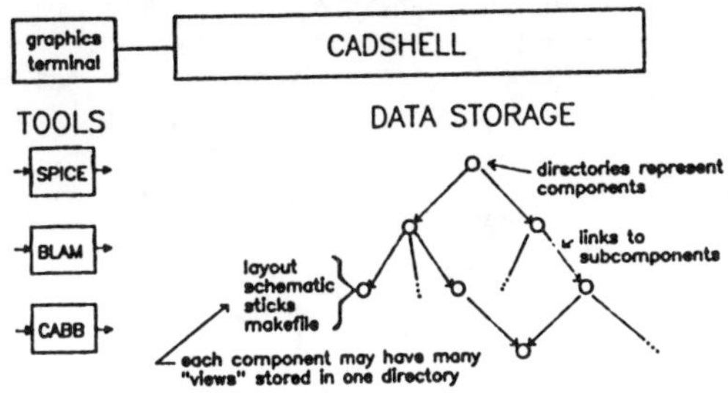

Fig. 2. Software architecture of a design system for custom VLSI circuit design.

hard copy devices. A possible topology for such a system is shown in Fig. 1.

It is anticipated that each design station will use a number of processors suited to specific tasks, such as a graphics processor and a communications processor. Special purpose hardware may be added to the design station or attached to the network to speed up certain frequently performed tasks. For example, a *sparse matrix* processor may be used to speed up circuit simulation, while another dedicated processor on the network may be used to create rasters of IC layout for a hardcopy device. At Berkeley, we are developing a design system for VLSI based on the use of single-user design stations of this type.

Unfortunately, cost-effective hardware suitable for such a design station does not exist today. When such hardware does become available, it is not clear that the operating system provided by the hardware manufacturers will meet the needs of the design system. Hence, we must develop the software for our design station on an existing computer system in such a way that it can be transferred to the new hardware as it becomes available. There are two solutions to this problem. The first approach is to develop the design software in such a way that it can be transported between operating systems with only minor modification. This solution may not allow the design software to take advantage of low-level operating system utilities and hence may result in an inefficient implementation. With recent advances in computer operating system technology, it has become feasible to transport an entire operating system from one hardware configuration to another. Hence, a second approach to this problem is to develop the design software taking full advantage of the features provided by a particular operating system and then to transport the entire operating system, with the embedded design software, from one set of hardware to another. We are exploring the latter approach.

In the past, design software has often been tailored to a specific design style or *design method*. To employ a different approach would then typically require a complete rewrite of the software in the design system. At this time an optimal design methodology for VLSI is not known. In fact, it is clear that different design approaches are necessary for different applications. A design style suitable for low volume custom circuits would not suit a high volume, general purpose component designer. At Berkeley we are developing a software framework which supports the design programs in such a way that the system can adapt to a variety of different design methods without the need to rewrite common components used in the differing approaches. The use of compact software modules, or *tools*, which communicate via a common data interface is a major characteristic of our approach to the problem of VLSI design.

A. The Design System

As mentioned above, we are developing our design system within the framework of a flexible operating system and intend to transport the operating system, with the embedded design software, to suitable new hardware as it becomes available. We have chosen the UNIX[1] operating system [6], [7] as our initial vehicle for this development since it already provides many of the features needed in our planned design system. Initial experiments with UNIX include the use of its hierarchical file system for representing the hierarchical aspects of a chip design as well as the use of the UNIX *shell* concept [8] to assist the designer in interacting with the design software.

Fig. 2 illustrates the general software architecture of a design system we are currently investigating [9]. In this figure, the graph on the right represents the hierarchical file system used to store the design data. Each node in this graph represents a component, or *module*, of the circuit under development and is implemented in the file system as a *directory* of files on the computer. Directed branches indicate references, or *links*, to submodules. At each directory node many files may exist, each one containing information relating to a specific "view" of that module. For example, there may be files containing mask layout descriptions, a schematic diagram, a simulator input file, and perhaps an English description of the intended function of the module. Since a file system is used, the number of files permitted at each node is potentially unlimited. Restructuring the design simply involves changing the link structure. The design software is responsible for maintaining the integrity of the various files for each component. The design-aid programs, or *tools*, interact with this simple database under the control of a supervisory program. This supervisor program is responsible for maintaining the relationship between representations and keeping the descrip-

[1]UNIX is a trademark of Bell Laboratories.

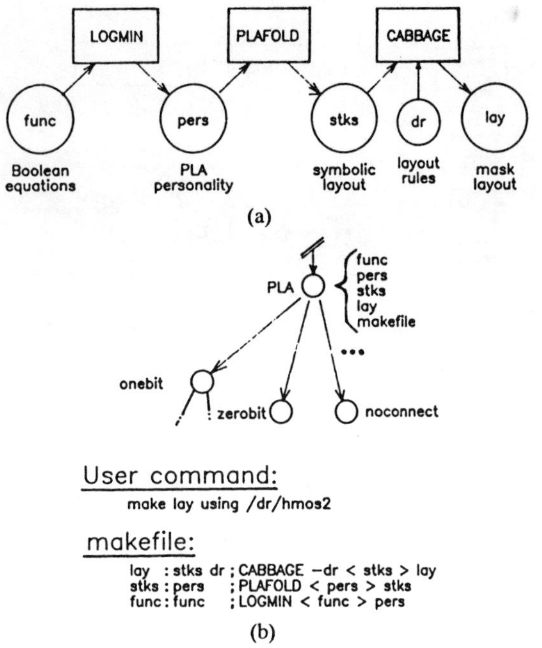

Fig. 3. (a) Block diagram of PLA generation system. (b) Possible data structure for PLA and corresponding *makefile* example.

tions "up-to-date" after a change has been made. The program MAKECHIP (inspired by the MAKE program [10]), operates by maintaining a *makefile* at each node in the database. The makefile describes the relationships (dependencies) between the various data descriptions which may exist at that node, as well as the rules by which a target description may be obtained from the information on which it depends. For illustration, the block diagram of a PLA generator is shown in Fig. 3(a), with a fragment of the database structure for a PLA and its corresponding makefile shown in Fig. 3(b). Three design-aid programs convert the functional description of the combinational logic to be implemented by the PLA into the detailed, final layout. The first program, LOGMIN, performs two-level logic minimization. This process is followed by some topological optimization performed by PLAFOLD. Finally, a symbolic layout of the PLA and an associated set of layout rules are used to generate the mask layout of the circuit via the CABBAGE compaction program. Each line in the makefile represents a dependency. For example, the first line states that the mask layout (lay) depends on the symbolic layout (stks) and the layout rules (dr). It also states the the mask layout can be generated from the symbolic layout by running the CABBAGE program, as illustrated in Fig. 3(a), where stks is input and lay is the resulting output. The designer can invoke the MAKECHIP program and request that the layout be updated using layout rules /dr/hmos2, as illustrated in Fig. 3(b). The MAKECHIP program then begins by noting the date and time at which each file referenced in the makefile was last modified. If the layout is dated more recently than any description on which it depends, MAKECHIP will not take any action. Otherwise, the MAKECHIP program will invoke, in the required order, all the programs necessary to update the layout description. In this way, the MAKECHIP utility can be used by the designer to maintain the integrity of the data representing a design.

As new tools are developed or existing programs are improved, changes will inevitably be made to the underlying structure of the design system. To simplify the interaction of the designer with the underlying tools, a high-level interface, called the CADSHELL, will be provided, as illustrated in Fig. 2. This interface provides a buffer between the user and the design system. A simple user command in this shell may result in many processes being executed on the underlying system. As new tools are added to the system or technology changes are incorporated, major syntax changes can be absorbed by reconfiguring the CADSHELL interface, thus minimizing the impact of those changes on the user of the design system. Although the CADSHELL has not been implemented at this time, we have emulated a number of its features using UNIX commands, and this system provides an effective framework for the development of the design software.

B. Design Methodology

In the previous sections we have described a framework for the design system. It is necessary to implement a particular design approach, or *design method*, in this framework to produce a VLSI circuit. Of course, the choice of a particular design method depends heavily on the constraints on the end product of the design: the importance of yield, design time, and so on. One particular area that will benefit a great deal from our computer aids is that of custom system design up to "medium volume" levels. System designers in this area have become used to relying on a "catalog of parts," such as TTL and MECL in components or polycells (e.g., [11], [12]) and gate arrays (e.g., [13]) in custom IC's. Each component in the design is generally invariant and, therefore, inflexible. That is, a TTL package cannot grow new pins to accommodate a new application and the user must choose another, more suitable part. Also, in most polycell systems the output drive capability of a cell is not automatically modified by the design system as a function of the attached load capacitance. As a result, the designer must interact with the system at a relatively low level when, in many situations, it would be possible for the computer to make these decisions relating to the electrical or physical design.

To obtain error-free design of VLSI circuits, it is necessary to engage the computer much more directly in the synthesis process. This leads to the concept of *parameterized* circuit modules, or *macros* [14], where each module may be described by a catalog entry and an associated set of parameter values. The parameters are used by the designer and/or the design system to specify the desired physical and electrical characteristics of the module. For example, the register stack of Fig. 4 is characterized at the functional level by the number of bits per word (NBITS) and by the number of words (NWORDS). The presence of SET and RESET lines, the position of the connection points on the periphery of the module, and the requested aspect ratio may be other parameters of this module [15]. A computer program may then *generate* the layout of a

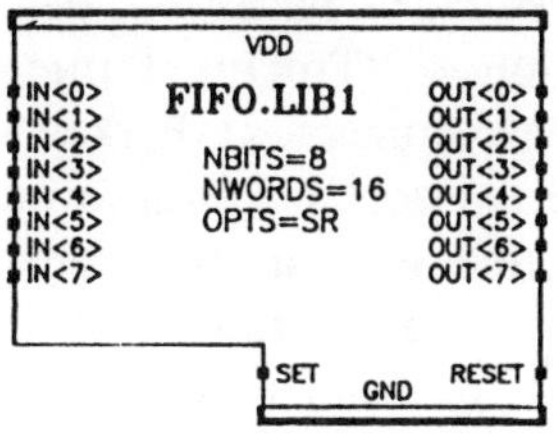

Fig. 4. Register stack as an example of a parameterized function module.

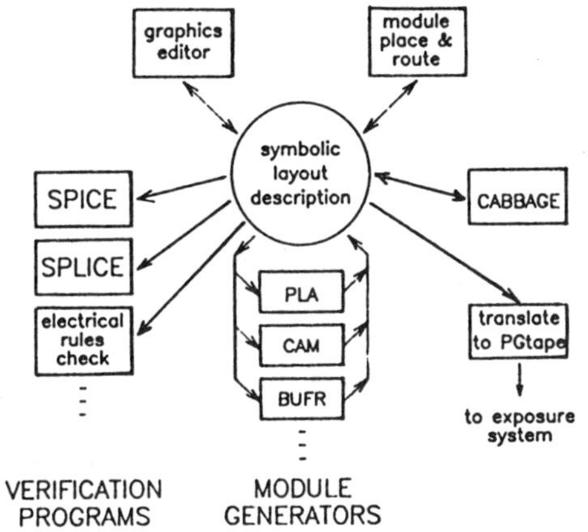

Fig. 5. Block diagram illustrating the use of module generators and verification tools.

particular instance of the register array whose electrical and physical characteristics are determined by the parameters with which the program was called and the design equations included in the module generator. The program which generates the module may return useful information to the designer or to the design system, such as bounding-box dimensions of the resulting layout or estimates of speed and power requirements for performance evaluation. Some functional characteristics of a cell may also be included as parametric information. For example, the logic equations to be implemented by a PLA may be passed to the PLA generation program.

Programs which perform these electrical and physical design functions are referred to here as *module generators*. We are developing such generators for our design system, and the block diagram of Fig. 5 illustrates how these module generators may be used. The designer interacts with the system using a graphics editor and a catalog of parameterized cells. Underlying each cell is a module generator which synthesizes layouts for cell instances in a symbolic layout language [16]. High-level design aids may include a two-dimensional placement and routing program for the modules used by the designer (e.g., [17], [18]).

Verification is an important aspect of IC design to ensure that the resulting circuit will satisfy all the functional requirements. Not even module generators can be relied upon to produce error-free designs for all combinations of values of their input parameters. For this reason, it is essential that design verification programs be developed which can handle VLSI circuits. It will be mandatory to exploit the structure (regularity, hierarchy, etc.) of the design in order to obtain acceptable performance from such tools.

III. Layout

As the complexity of integrated circuits increases, the generation, management, and analysis of integrated circuit mask information becomes of critical importance in ensuring error-free designs. Sophisticated graphic editors are required to capture hand-designed components of the circuit, and design-aids for cell placement and routing are necessary to assist the designer. The graphic editors developed at Berkeley include BEGE [19], CAESAR [20], and KIC [21]. Using these editors we can manipulate mask level data or symbolic layout data as described in the next section.

Interactive design-aids for layout synthesis can be divided into two categories. *Compaction* techniques reduce the area occupied by a layout by adjusting the size of noncritical components, such as interconnections, while ensuring that all the electrical and physical layout rules are satisfied. An initial layout topology is provided by the designer or by another design program. Compaction programs, such as the CABBAGE program described later, work to optimize the layout area while maintaining essentially the same topology. Local, or "peephole," optimizations, such as the introduction of bends, or jogs, in interconnections and the rotation of cells or devices, may also be used.

Placement and routing programs relax some of the constraints above by attempting to alter the topology of the circuit by rearranging the relative placement of cells in order to improve performance. Placement and routing programs have been used for some time with gate array [23] and polycell [11] designs. New techniques are being developed for the placement and routing of arbitrary rectagons in an hierarchical manner [17], [18]. A key component of these programs is the router. At Berkeley we have recently developed a new, high performance channel routing algorithm which we intend to use in new programs under development for module placement and routing.

Unfortunately, most hand layouts contain errors and it is often impossible to prove that even a layout synthesis program will generate correct layouts in all case. For this reason, *analysis* programs are required to verify that both physical and electrical layout rules are satisfied. Physical rules, such as minimum spacing between components on the same or different masks, can be performed by set operations on the data files containing the mask data (rectangles, polygons, etc.) [24]. Unfortunately, the time complexity of algorithms which perform these operations can lead to very long run times for a physical rules check of a VLSI circuit. In order to substantially reduce the run time of these checking programs, we are developing a physical rules check program which exploits the hierarchy of a circuit as well as its regularity within arrays of similar cells. Often the electrical characteristics of the circuit, in particular circuit connectivity, must also be considered in order to permit proper interpretation of geometry, such as complicated contacts or transistors. We are developing programs for electrical rules checking, in conjunction with our physical rules checking program.

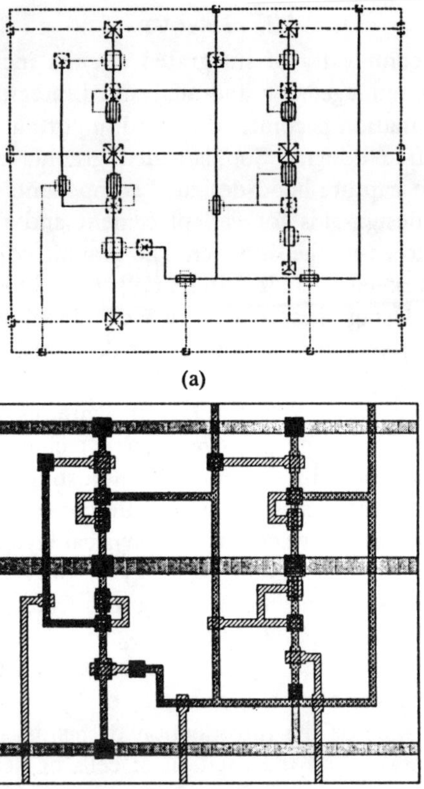

(a)

(b)

Fig. 6. (a) A CABBAGE-generated black-and-white plot of stick diagram for a toggle flip-flop. (b) Corresponding uncompacted layout representation.

A. Symbolic Layout

In most design systems by the time a design has been converted to mask data format, almost all the information relating to circuit connectivity has been discarded. To perform electrical rules checking, a program must first reconstruct this information by detailed analysis of the mask data. If the design description were available in an hierarchical format, where components such as transistors and modules such as PLA's are expressed as unique cells with explicitly specified interconnections between them, then many verification tasks could be simplified greatly. Symbolic layouts, in the form of block diagrams, or *stick diagrams* [25], provide this information. A graphical view of a stick diagram, as used in the CABBAGE system, and its corresponding mask level layout, is shown in Fig. 6.

Not only can symbolic layout be used to aid the verification of the circuit but, by separating cells and interconnections, computer programs can be used to optimize the area utilization of the circuit by modifying only the noncritical interconnections; these techniques are employed in CABBAGE. Our design system will use a symbolic layout language as its primary medium for communication between tools.

While the various tools used in the framework of IC design may require a variety of different internal data structures for efficient manipulation and perusal of the information, the input and output formats of these programs need not be equally varied. Just as CIF2.0 [26] provides an interchange format among a large number of input, output, and storage devices that deal with geometrical mask information, a STructured Interchange Format (STIF) is under development at Berkeley, ultimately to serve as a standardized portable data format that captures all relevant information about an IC design at the symbolic layout level as well as the geometrical mask level.

After soliciting input from many researchers at major universities and in industry, a first version, STIF1 [16], was implemented during the second half of 1980 [27] to gain experience with the implementation and usage of such a language and with the development of a cohesive set of tools working from such a format. STIF1 primarily addresses issues related to layout geometry, logical and physical interconnectivity, and hierarchical representation and encapsulation of cells. Future extensions of the language will permit the inclusion of models for simulation with each cell, information on history and status of the cell, and possibly even relevant test patterns.

The intent of the designer is preserved as much as possible in a STIF description. In particular the hierarchical structure, the iterative nature of regular arrays, logical connectivity, and models specifying the behavior of each particular cell are explicitly retained.

STIF1 also contains geometric primitives (rectangles, polygons, etc.), so that the details of an actual mask layout can be specified unambiguously. For compactness of the description and to provide proper structuring, groups of geometrical items can be lumped together into cells and these cells can be called individually or as whole arrays. To provide an efficient means for describing ROM's or PLA's, a suffix notation is used specifying placement of different versions of a cell throughout an array.

Cells at any level can be encapsulated in protective frames. Such frames must not overlap with other frames or any other feature in the layout. These frames may be defined on each mask layer and may consist of a set of disjoint polygons on any single layer. To permit interconnection to these protected cells, the frames have formal interconnection points. These framed cells then are the basis for circuit abstractions. The interior of such a cell can be suitably simulated and modeled, searched for layout or electrical rule violations, and verified for consistency among all representations. For the construction of larger system, an abstract representation of this cell can then be used with confidence, without the need for repeating verification or detailed simulation every time the cell is used.

A simple syntax and a terse and compact format have been chosen for STIF to reduce both the complexity of the parsing process as well as storage space and file transmission time. The language is entirely descriptive rather than procedural, thus avoiding many difficulties with its implementation and providing a cleaner framework for future verification tools.

B. Compaction

The CABBAGE program [22], [28] implements a set of computationally efficient algorithms which convert a loose, symbolic, topological layout plan of an integrated circuit cell into a compact and correct geometrical layout. In fact,

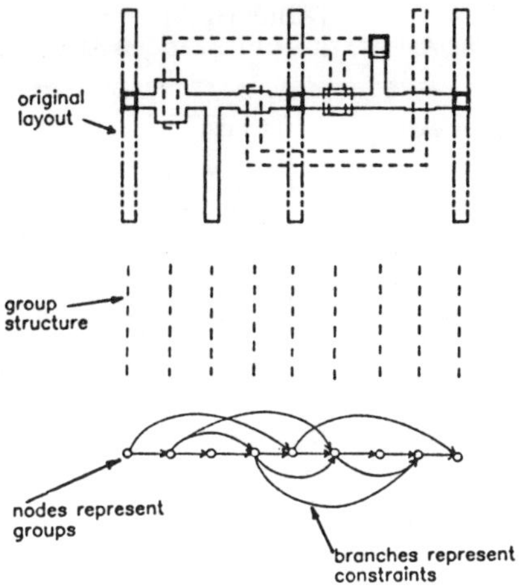

Fig. 7. Horizontal graph structure of a layout. Note that *groups* may contain many geometrical items.

the program produces a "correctly spaced" symbolic layout description from which the final artwork may be generated by replacing the layout symbols with their equivalent mask geometries.

There are two common approaches to the use of symbolic layout techniques. In the *fixed grid* (or coarse grid) method [29]–[31], the layout is drawn on equally spaced grids. The grid spacing is determined by a set of dominant layout rules in such a way that it is impossible for user input to violate spacing rules. A conversion program may then be used to replace each symbol in the user input with its equivalent mask layout description. The compactness of the resulting layout depends on the quality of the user input (spaces in the layout left by the user are not removed) and the coarseness of the grid.

The second type of symbolic layout method avoids these difficulties by performing geometric modifications, such as compaction and expansion of elements, as well as limited topological changes. This approach is called the *relative grid* method, since the loose layout plan indicates only the relative placement and interconnection of element symbols [32]–[35].

CABBAGE, our compaction program, implements the relative grid approach. Its compaction algorithm is based on a critical path analysis method. Two graphs are constructed to represent the layout and the constraints between adjacent groups of geometry. One graph represents x-axis constraints, the other graph y-axis constraints. A simple layout, its group structure, and the equivalent graph representation are shown in Fig. 7. The branches of the graph represent the minimum and maximum spacing limits between adjacent groups of geometry. A *most constraining path* is determined through each graph and the layout is reduced to the smallest possible correct size that can be achieved by independent x-axis and y-axis movements. Noncritical elements are placed midway between constraints to improve yield. Force-directed heuristics are used to introduce jogs into the layout where they might help the compaction process.

CABBAGE provides a fast, interactive tool for the compaction of cells. Fig. 8(a) shows the correctly spaced symbolic layout of a cell after compaction, as seen by the user of the system, and Fig. 8(b) shows the corresponding layout plot of the cell. While checking for layout errors may be necessary in complex cell designs, techniques which not only *detect* rule violations but also *correct* the layout if possible are clearly preferred. This is another major advantage of the CABBAGE approach.

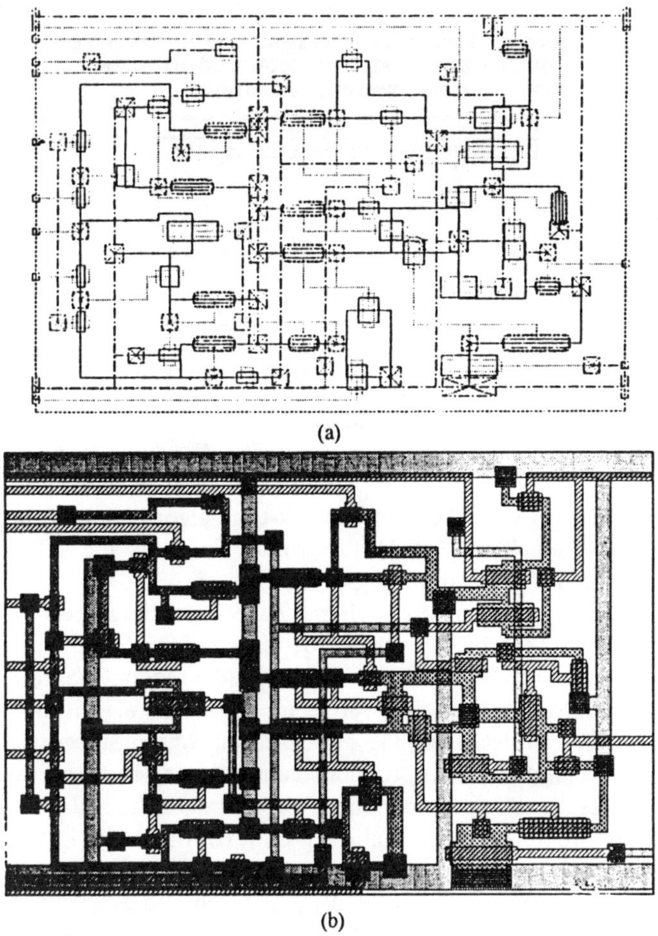

Fig. 8. (a) A correctly spaced stick diagram of a latch-driver cell. (b) Cell layout after compaction by the CABBAGE program.

C. Channel Routing

A channel router is an important building block of a system of design aids for the hierarchical layout of integrated circuits. Channel routing has been used successfully in the layout of regular structures such as gate arrays and polycells as well as in the layout of random logic.

A channel router solves the problem of routing a specified net list between two rows of terminals across a two-layer channel. Nets are routed with horizontal segments in one layer and vertical segments on the other. One objective of channel routing is the minimization of the number of tracks needed to route the nets. Unfortunately solving the resulting optimization problem exactly, perhaps with a branch-and-bound method as in [36], may take an enormous amount of computing time for problems of practical size, say more than 50 nets and 300 pins. Therefore, heuristic algorithms have been proposed to generate optimum or near optimum solutions in a reasonable amount of

TABLE I
COMPARISON OF HORIZONTAL TRACKS AND CPU TIME REQUIRED FOR OPTIMAL SOLUTION (OPT), LEFT-EDGE ALGORITHM (LEFT E), AND THE TWO NEW ALGORITHMS

Problem	Solution			
name (#nets)	opt	left e	alg.1(cpu)	alg.2(cpu)
ex.1 (21)	12	14	12 (0.05)	12 (0.07)
ex.2 (30)	15	18	15 (0.07)	15 (0.43)
ex.3b (47)	17	20	17 (0.17)	17 (0.57)
ex.3c (54)	18	19	18 (0.18)	18 (0.72)
ex.4b (57)	17	23	17 (0.23)	17 (0.77)
ex.5 (62)	20	22	20 (0.22)	20 (0.88)
dif.ex (72)	26	39	30 (0.40)	28 (1.60)

TABLE II
RESULTS OF TRACK REQUIREMENTS AND CPU TIME FOR THE "DIFFICULT" EXAMPLE, AS COMPARED WITH LTX

name (#nets)	density	LTX	alg.1(cpu)	alg.2(cpu)
dif.ex (72)	19	21	21 (1.00)	20 (2.10)

computing time. The so called left-edge algorithm is commonly used [36], [37], [11] in routers such as the LTX program [11] developed at Bell Labs.

Recently, two new algorithms have been developed at Berkeley [38] which produced better solutions for all test cases than other published routers and used only a few seconds on a VAX 11/780 computer. The algorithms, described in detail in [38], are based on a graph-theoretic interpretation of the optimum channel routing problem. A few results of the applications of the algorithms to test cases are reported in Tables I and II. In Table I, the examples 1 through 5 were taken from [36] and the "difficult example" was provided by Deutsch and Schweikert of Bell Labs. Here no dogleg (splitting of horizontal segments of nets) is allowed. In the columns of the table, the number of horizontal tracks used by the optimum solutions,the left-edge algorithm, Algorithm 1 and Algorithm 2 are listed. Note that in 6 out of 7 cases the new algorithms were able to obtain the optimum solution. In Table 2, doglegs are allowed. In the LTX system more than 50 doglegs were created where Algorithms 1 and 2 created only 19 and 18 doglegs.

IV. SIMULATION

With the exception of highly constrained design methods, it is not normally possible for the designer nor the design software to produce a guaranteed error-free design with known performance under all operating conditions. This is particularly true if optimal performance is desired over a wide range of operating conditions. Simulation has proven effective for the evaluation of specific designs and has been applied from the analysis of fundamental component and processing technology through to the analysis of high-level abstractions of system architectures.

As we move into the era of VLSI, it is clear that to simply extend existing simulation techniques to circuits containing over 100 000 devices is not practical and hence new algorithms and simulator architectures must be developed. One underlying theme present in all our work in this area is an attempt to exploit the *structure* of the circuit and its component parts wherever possible. Both the hierarchy of a design and its use of repeated substructures can be exploited during simulation. In certain design approaches, the use of simplified models for high-level circuit components can also provide considerable reduction in computational effort.

In the following sections we describe our ongoing work in the area of simulation for VLSI from process and device analysis through to the analysis of the logical performance of circuits. The final section describes our work in the development of models for simulation.

TABLE III
AREAS FOR THE APPLICATION OF IC PROCESS ANALYSIS TOOLS

IC PROCESS APPLICATION AREAS	
LITHOGRAPHY Optical E-Beam X-Ray Ion-Beam	**ETCHING** Wet Ion Milling Plasma Etching Ion-Assisted Etching
THERMAL PROCESSING Oxidation Doping Ion Implantation Annealing	**DEPOSITION** Evaporation Sputtering Chemical Vapor Deposition Electroplating Epitaxial Growth

A. Process and Device Simulation

As IC fabrication technology moves toward finer linewidths the topographical features and doping profiles of devices are becoming an increasing concern. Not only is greater linewidth control required for high resolution but it must be accomplished in a context where features in the third dimension are as large as the linewidth itself. When process models can be established, simulation algorithms can often be used to explore process effects by simulating time evolution of the line-edge profile.

It is potentially possible to find both physical models and algorithms to simulate most IC fabrication processes. Table III contains a variety of examples. They can be divided into four major areas: lithography, etching, deposition, and thermal processes. Simulation models have been developed for most of these processes over the last ten years and two user-oriented CAD programs which have emerged are the SUPREM program [39], developed at Stanford for the simulation of thermal processes such as diffusion, implantation, and oxidation, and the SAMPLE program [40], [41], developed at Berkeley. SAMPLE permits the simulation of individual lithography, etching, and deposition processes in a user-oriented environment.

In the device analysis area, program TWIST [42], [43] has been developed for the two-dimensional static analysis of MOS transistors. This program has been used to study weak-inversion and weak-injection punchthrough phenomena in short-channel MOS devices. It employs a nonuniform mesh and successive over-relaxation. TWIST runs interactively on a 16-bit minicomputer and provides three-dimensional plots on a graphic terminal.

B. Circuit Simulation

Circuit simulators provide precise electrical waveform information about the circuit under analysis. The majority of circuit simulators in use today contain models for a wide range of active devices and hence are largely independent of technology. For this reason, these programs must employ general algorithms for the solution of the set of coupled, nonlinear, ordinary differential equations which describe the integrated circuit and so cannot exploit the characteristics of a particular technology.

The SPICE circuit simulation programs were initially designed for the cost-effective analysis of circuits with a few hundred devices or less. Because of the need to verify the performance of larger circuits, many SPICE users, despite the cost, have successfully simulated circuits containing thousands of transistors. On an IBM 370/168 computer, the average cost of a SPICE analysis is 4 ms/device/clock/timepoint. For a 10 000 device circuit, with 3 clocks and for an analysis of 10 μs at 1 ns steps, the computation time would be in excess of 15 computer days! Nevertheless, the success of circuit simulation in design evaluation has been such that designers wish to continue to simulate large circuits at the level of accuracy provided by this type of program.

We have been able to improve the execution speed of SPICE by the use of table look-up models for nonlinear circuit elements as well as by exploiting the hierarchical structure of the circuits and their repeated use of the same cell. By applying node tearing techniques [44] to the interface between cells in the circuit, inactive cells may be bypassed during the equation solution phase. However, it is anticipated that these techniques alone will provide less than an order of magnitude speed improvement. This is not sufficient improvement in performance to permit cost effective device-level analysis of VLSI circuits. We are currently looking toward the interaction between hardware and software to provide further performance improvement.

Our first experiments in this direction involved the development of the SPUDS program [45], a subset of SPICE2 implemented on an HP1000 minicomputer. SPUDS has illustrated the potential savings that can be achieved with the use of special-purpose hardware instructions suited to the operation of a simulator. In particular, special sparse matrix instructions implemented in microcode have helped to reduce the equation solution time of SPUDS by an average of 20 times for nonlinear time-domain transient analysis when compared to a direct implementation of the same SPICE2 subset running on the same computer.

In our efforts to exploit circuit structure during simulation, it is clear that parallel processing can provide significant performance improvements. We are designing a new version of the SPICE2 program for use with vector-oriented machines such as the CRAY1 and the Floating Point Systems processors [46]. With these architectures, it is possible to perform parallel block decomposition and elimination of structurally similar sparse submatrices as well as the parallel evaluation of similar device model equations.

One of the major bottlenecks associated with the use of the above architectures is the "gather/scatter" operation [47] associated with the formulation of data for parallel processing. We are investigating the use of multiprocessor networks, in particular the perfect shuffle connection [48], as an approach to the implementation of circuit simulation which may reduce the impact of this problem.

C. Timing and Mixed-Mode Simulation

If simulation algorithms are tailored to specific technologies or applications substantial savings in computations can be achieved. Many components of digital MOS or I^2L circuits may be considered unilateral in nature. This characteristic, as well as the facts that these families are saturating so that accumulated voltage errors are lost at the extremes of signal swing, and that large digital circuits are relatively inactive at the gate level, are exploited in Timing Simulation. Timing simulators [49]–[51] can increase simulation speed by up to two orders of magnitude while maintaining acceptable waveform accuracy. These savings are achieved by using node decoupling techniques in conjunction with simplified table look-up models for nonlinear devices. Where a library of cells is used during the design, simplified *macromodels* of the cells may be used to reduce computation further. Our timing simulator MOTIS-C [50] has been used successfully for the analysis of integrated circuits containing over 5000 transistors.

The speed of timing simulation is still insufficient for the analysis of an entire VLSI circuit. Gate level logic simulators [52]–[54] and Register Transfer Level (RTL) simulators [55], [56] may be used to increase analysis speed further, provided accurate waveform analysis is not required. However, some parts of a circuit may require the waveform accuracy of circuit simulation while in other parts of the circuit, a more efficient gate level logic analysis is sufficient. In this situation, *mixed-mode* simulation can be used and our mixed-mode simulator, SPLICE [57], [58], combines circuit, timing, and logic analyses for MOS integrated circuits. Other mixed-mode simulators include DIANA [59], ADLIB [60], and a program developed at Bell Labs [61].

Fig. 9 illustrates the application of mixed-mode simulation to a 256-bit RAM circuit. During the design of the decoder, a detailed waveform analysis is necessary to predict performances accurately. The waveforms obtained from SPLICE for the timing analysis of the decoder are shown in Fig. 9(a). Once the operation of the decoder has been verified, a more simple logic-level model may be derived and included in the program, as shown in Fig. 9(b). Here, the essential timing characteristics have been abstracted. This simple representation of the decoder may then be used in an overall analysis of the RAM. The waveforms in Fig. 9(c) illustrate the simple one/zero logic characteristics of the decoder outputs, while signals at the sense amplifiers are obtained from a detailed electrical analysis.

A number of new techniques have been used in the SPLICE program to improve both its speed and its accuracy. In particular, the use of selective-trace or event-driven analysis to exploit latency for electrical analysis can improve execution speed by up to an order of magnitude for large circuits. New, compact table look-up models are also used to improve speed at relatively low-storage penalties [62]. Another critical area in mixed-mode analysis is the interface between different analysis modes. By introducing combined "Norton" states in logic analysis mode, where a notion of source impedence is introduced [62] rather than a simple low-high-unknown analysis of the logic gates, the

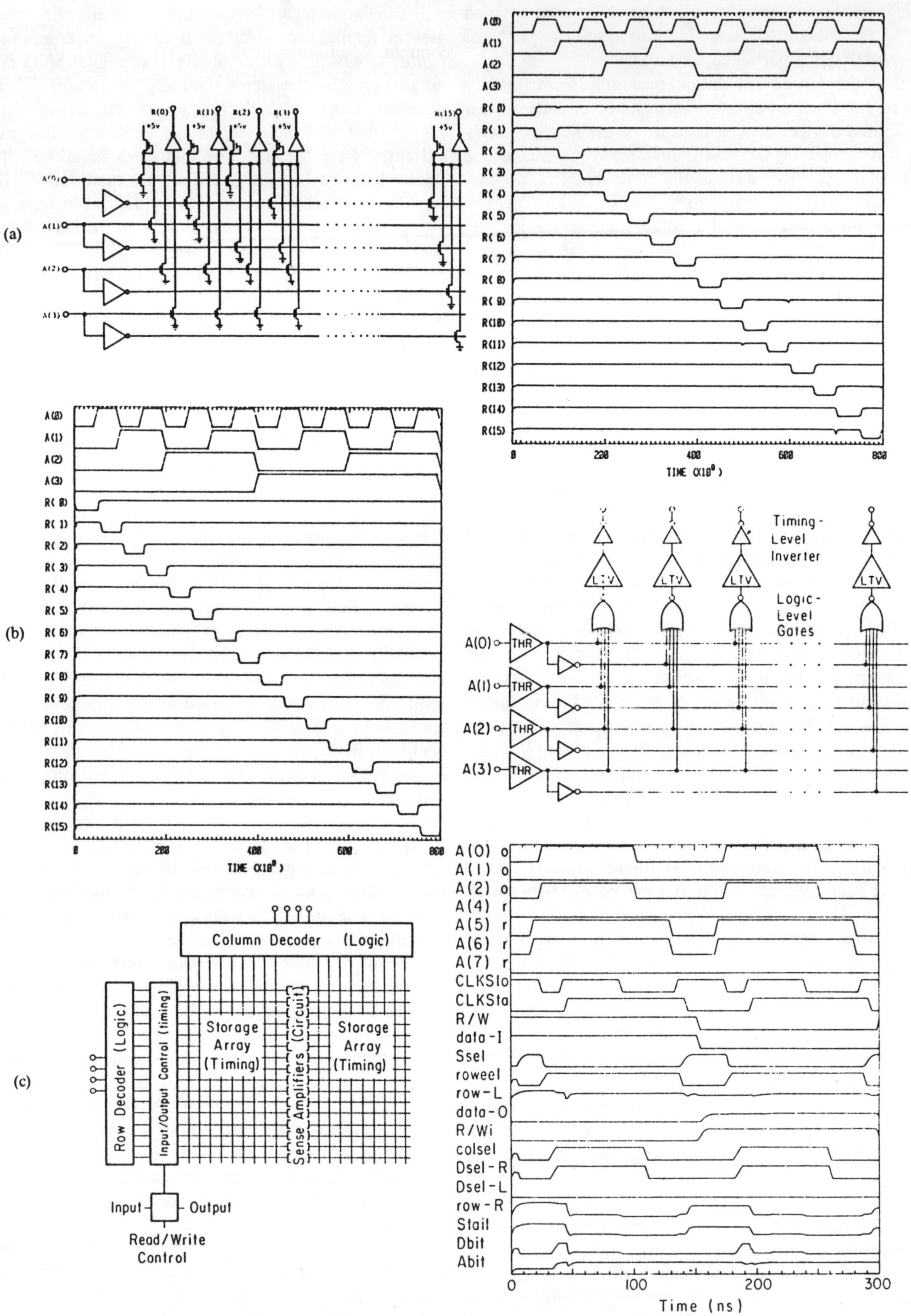

Fig. 9. (a) Timing analysis of row decoder. 36 CPU seconds on CDC6400 computer. (b) Mixed logic and timing analysis of row decoder. 3 CPU seconds on CDC6400. (c) Block diagram of RAM circuit and SPLICE waveforms for write/read operation.

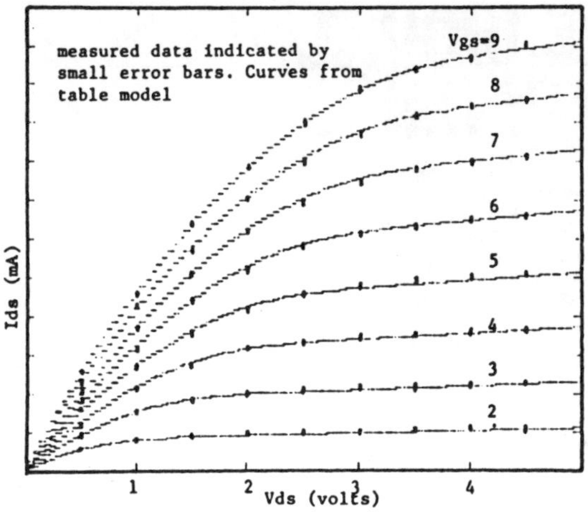

Fig. 10. Table model and measured data for MOSFET: $l_{eff}=1.75\ \mu$, $w=50\ \mu$, $t_{ox}=650$ A, junction depth$=0.5\ \mu$.

transformation between electrical and logic analyses can be achieved without undue loss of accuracy. The use of these combined states also enhances greatly the logic analysis of MOS structures by simplifying the modeling of bidirectional circuit elements and of dominant and nondominant gates [63].

A major problem in timing analysis has been its inability to deal effectively with tightly coupled feedback circuits. In particular, MOS transmission gates and floating (nongrounded) capacitors have caused severe execution speed reductions and inaccurate results. We have recently developed new algorithms for the analysis of floating capacitors which have reduced these problems significantly [64], [65]. We are implementing these algorithms in the SPLICE program to investigate their performance for the analysis of large circuits.

D. Modeling

In conjunction with our work in the area of simulation, electronic device modeling has been a constant thread of activity in our integrated circuit efforts. Accurate models are essential for all of the simulation techniques described above and without such models, the results of the simulation may not reflect the performance of the circuit under analysis.

Our recent work has focussed on the development of empirical models for MOS transistors which predict the characteristics of the devices accurately without requiring large amounts of computer time. With the decreasing cost of computer memory, we are exploring techniques for combining one-dimensional table look-ups with simple arithmetic operations to improve execution speed. Fig. 10 shows a comparison of measured data for a short-channel MOS transistor with the table-generated characteristic. Four tables, each of 100 entries, are used in this model. Initial implementation of this model in SPICE2 has resulted in an order of magnitude reduction in model evaluation time for transient analysis compared to an analytic implementation of comparable accuracy. We are currently working to extend these techniques to the modeling of channel charge.

When a group of transistors is used to implement a common function, such as a cell or logic gate, it is often possible to exploit the known structure of the circuit and use a simplified representation which maintains the essential characteristics of the cell at reduced computational expense. Such a reduced representation is called a *macromodel* [66]–[68]. Our work in this area is centered around the use of model tuning techniques which help the designer to correlate the performance of the macromodel with the detailed description it replaces.

As the minimum feature size on an integrated circuit is reduced, while the size of the circuit itself increases, the properties of the interconnect become increasingly important to the overall performance of the circuit. Signal delays and signal degradation can determine circuit operation. For this reason, explicit models for interconnect are necessary for accurate simulation, and we are incorporating such models in our simulators. The parameters of such models will be provided interactively by the designer or directly by design programs.

V. Module Design

A large fraction of all digital integrated circuitry is composed of some of the same generic register-transfer-level modules: ROM's, PLA's, finite state machines (FSM's), RAM's, caches, register files, ALU's, barrel shifters, etc. While individual instances of these modules often are of different sizes and functions, their general construction is often the same. It thus pays to invest some effort to formalize the procedures according to which these modules are constructed, so that the computer can then aid their design. To accommodate a wide range of applications, the construction rules are formulated in a parameterized manner.

The generation of these modules can be performed at different levels. For certain types of modules which typically form large arrays, the individual building blocks and the assembly rules should be stated in such a way that a compact and optimized layout is directly generated; this approach typically would be appropriate for memory arrays, ROM's and PLA's. Some initial work in this area has been described by IBM [14] and MIT [69] and a data path design system for ALU's has been developed at Caltech [15].

Less regular modules, or those which depend on a large number of parameters, would involve rather complicated expressions and conditional statements if the construction of an optimized layout were included directly in the module generator program. In these cases the result of the generator program should be a loose topological representation, i.e., a stick diagram, and the transformation to a viable layout should be left to a separate compaction program such as CABBAGE. Preliminary programs for PLA's and FSM's are operational and are being used by students building class project chips. These programs are described below. The feasibility of more complicated generators for circuits such as a cache memory module and a contents-addressable memory (CAM) is under investigation.

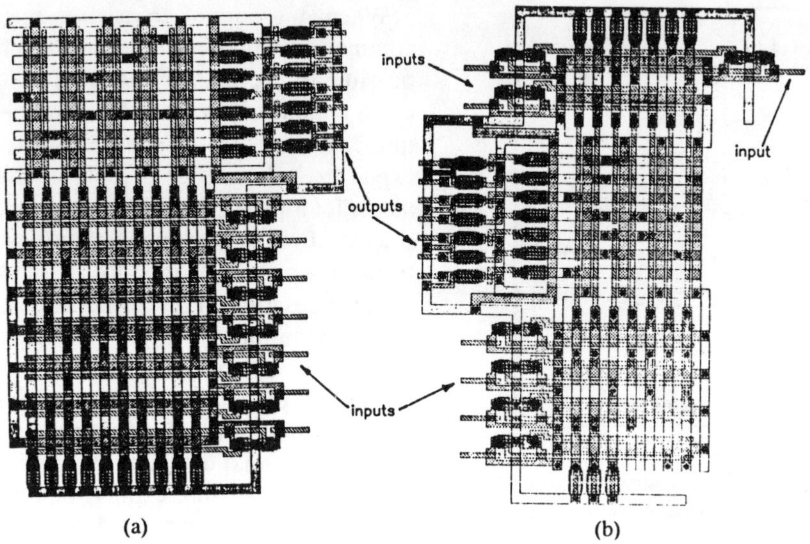

Fig. 11. (a) Programmable logic array before folding. (b) Programmable logic array after folding.

The design of a PLA implementation of a multiple-output combinational logic function (sometimes called a "naked" PLA when input and output registers are not considered) may be divided into three major steps: functional design, topological design, and physical design. These steps are illustrated in Fig. 5. The functional design process involves translating the Boolean equation specification of the multiple-output logic function into a minimal, sum-of-products or product-of-sums form. Both logic partitioning [70] and two-level logic minimization, e.g., [71], [72], have been used to produce optimal functional-level designs.

Once the functional representation of the logic best suited to PLA implementation has been determined, the next task is to map the logic into a topological representation of the final PLA structure. While the physical implementation of the PLA depends on technological and performance issues, e.g., [73], [74], [25], some topological optimization can be performed which may reduce the "sparsity" of the PLA and hence improve area utilization and performance. One such topological optimization is the *folding* process [75]. In conjunction with colleagues at IBM, we have developed algorithms which use a graph-theoretic representation of the folding problem and employ heuristic techniques to perform optimal PLA folding [76], [77]. The layout of a small PLA before and after folding is shown in Fig. 11.

The symbolic layout of the PLA generated by this folding algorithm must then be designed to meet electrical and technological performance requirements. This final step involves such tasks as sizing of busses and drivers as a function of PLA complexity [69] as well as meeting layout rule requirements.

We have developed a set of programs which accepts Boolean equations and generates an optimal PLA layout that implements these logic functions. The PLA generation system is illustrated in Fig. 12. The Boolean equation input is first parsed and converted to two-level, AND–OR form by EQNTOTT. This description is then functionally minimized using the PRESTO program, developed by D. Brown, at Tektronix [72]. At this point, MKPLA [78] may be used to generate a CIF2.0 layout of the unfolded PLA. MKPLA is written in Pascal and performs both electrical and physical design of the PLA. The BLAM [79] program performs the folding, PLAID [79] builds a symbolic layout of the PLA, including input and output buffers, and CABBAGE adjusts the layout to ensure all rules are satisfied. CABTOCIF [80] can then be used to generate a CIF2.0 layout of the PLA. The PLA of Fig. 11(b) was generated by this system.

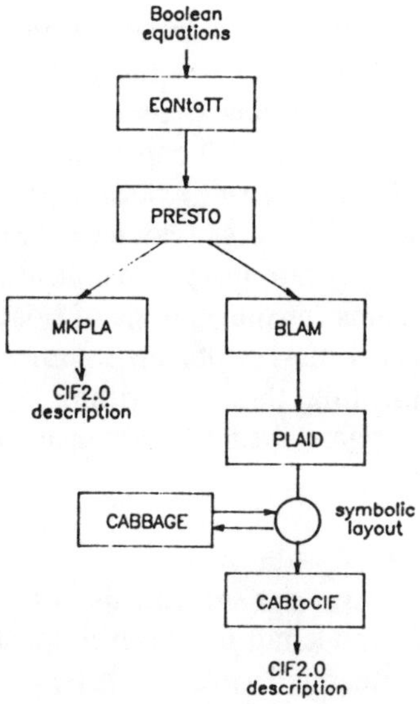

Fig. 12. Detailed block diagram of PLA generation system illustrating the programs used.

We are in the process of implementing a number of industrial and university designs using this system to measure its performance and extend the algorithms to include other practical constraints on PLA design.

Another program, MKFSM [78], can be used in conjunction with MKPLA to add feedback registers to the PLA and generate a finite state machine implementation. Input to the program consists of explicit state assignments and transitions. We are currently developing heuristic tools for optimal state machine design. Future work will also be directed towards algorithms for logic partitioning and decomposition of state machines. While the PLA was the first candidate of our module generator project, module generators for other commonly used building blocks are under development in our laboratory.

VI. Parametric Optimization

Since we consider performance and cost important aspects of most designs, we are continually seeking optimal solutions in both design and verification. As described in the previous sections, combinatorial optimization can be used in layout, such as compaction and routing, and module generation, such as logic minimization and PLA folding. In this section, we describe a software system for interactive parametric optimization recently developed at Berkeley [81].

Given an initial design of a particular system, parametric optimization consists of adjusting parameter values to satisfy a set of given specifications which are not met by the initial configuration and/or to optimize a performance function. In general, a designer uses a cut-and-try procedure with the help of simulation programs to perform a heuristic parametric optimization. Unfortunately, pure heuristics are generally inefficient in searching a multidimensional parameter space, and they cannot be relied upon to produce an optimal configuration. An optimization system can be indispensable in relieving the designer from the drudgery of search in the parameter space and to allow him to concentrate on the conceptual aspects of this task.

Since optimization techniques are fairly expensive and their cost grows more than linearly with the dimension of the parameter space, their use in VLSI design is most important for the design of basic cells, such as logic gates, PLA cells, or amplifiers. In order to obtain good performances for a VLSI system containing the replication of several basic cells, each cell must be designed optimally. Optimization techniques can also be used in model parameter fitting, or tuning. In particular, when VLSI systems simulation is performed at different levels of abstraction, the models at each level must be adjusted to minimize the loss of accuracy in the simulation at higher levels.

Despite considerable research activity in computer optimization of electronic circuits (see [82], [83] for a review of such efforts), optimization techniques have not been used as widely as might be expected. The basic reasons for this situation are the inadequacy of the optimization algorithms used, the difficulty of coupling optimization packages to simulation programs, and the lack of familiarity with these tools.

In the past, the optimization algorithms used in circuit design (in general, penalty function with conjugate gradients as a subroutine) had weak convergence properties and were too primitive for the task at hand. For example, until recently, optimization algorithms were not able to solve design problems involving yield maximization. The coupling of optimization algorithms with existing circuit simulators has been difficult due to the lack of features such as sensitivity computation in the time domain, necessary for the most efficient algorithms. Moreover, the use of optimization algorithms required a deep understanding of rather sophisticated mathematical concepts which were beyond the background of most circuit designers.

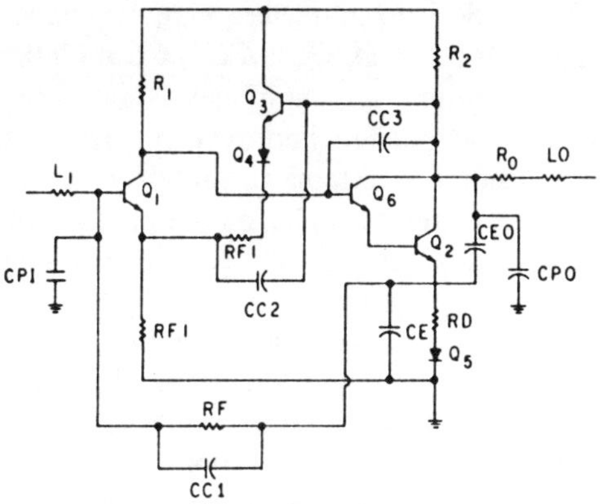

Fig. 13. Schematic diagram of the wideband amplifier circuit.

The INTEROPTDYN (INTERactive OPTimization of DYNamic systems) package is a prototype software developed at Berkeley to cope at least partially with these problems. The optimization algorithm implemented in this package, the Gonzaga–Polak–Trahan phase I–phase II method of feasible directions [85], has *guaranteed* convergence properties and can solve problems with objective function and complex inequality constraints. The latter may even involve performance specifications expressed as inequalities to be satisfied for all the values of parameters such as frequency, temperature, and/or time ranging continuously over an interval. The algorithm has several parameters which control its performances and which can be interactively adjusted.

The system is based on INTRAC-C, the interpreter of an interactive language evolved from INTRAC, an elementary interactive language [86]. A basic feature of the system is that existing Fortran subroutines may be included in a relatively simple, painless way. This feature is fundamental when the system has to be interfaced with existing simulation programs such as SPICE2 for function and derivative computation.

SPICE2 is being upgraded with sensitivity computations in the frequency and time domain. An experimental version based on SPICE2 has been tested on a relevant design problem: the design of a monolithic wideband amplifier shown in Fig. 13. The amplifier as manually designed had a computer predicted $f_{-3\text{ dB}} = 780$ MHz. The objective was to show that optimization could obtain a significant improvement over an excellent manual design. The objective function to be maximized was the bandwidth of the amplifier, subject to constraints on the input and output impedance over the entire band of the amplifier, set to be

50 Ω±5 Ω. The design parameters were the element values of RF, RF1, R0, R2, RD, CC1, CC2, CC3, CE0, and CE. After the optimization, an amplifier was obtained with $f_{-3\,\mathrm{dB}}=1.2$ GHz, a 54-percent improvement over the manual design. This result was obtained with 54 s of CPU time on a VAX 11-780 computer. The only optimization system with similar characteristics in existence today is the APLSTAP system [84].

At present, we are working on a new system called DELIGHT [87], based on a RATFOR-like programming language, called RATTLE, which replaces INTRAC and removes some fundamental limitations of INTRAC as an algorithm development and command/execution environment.

VII. Summary

Since we began our work in the area of design-aids for integrated circuits in the early 1960's, our primary motivation has been to support our own integrated circuit design efforts at Berkeley, both for research and instruction. With the advent of VLSI, the circuits under development in our own laboratory will require extensive use of computer-aids, both for design and verification. It is no longer sufficient to provide designers with a loosely related set of programs, but rather an integrated *design system* is required. We are developing a design system based on the use of a high-performance design station on a per-designer basis, connected to other design stations and additional hardware and software support via a high speed network. We are developing our software in a UNIX-based environment and anticipate transporting the operating system and design software to the design station when it becomes available. The UNIX environment provides an excellent framework in which to embed our design programs.

We are continuing to develop programs for design verification while also working in the areas of layout synthesis and optimization. Simulators SPICE, SPLICE, and SAMPLE are under continuous development as new algorithms and models are developed. The use of a symbolic layout approach to design and design verification is a key aspect of our research program. The symbolic layout language is being used by both our verification and our design programs. While our recent work in the area of design has focussed on the optimal implementation of combinational logic in the form of PLA's, we plan to extend this work to optimal implementation of synchronous machines using multiple PLA modules, as well as to the synthesis of other important building blocks for VLSI. An important component of the system will be a computer program for the placement and routing of the generated modules.

We are working closely with our many industrial affiliates, in particular Hewlett Packard, Tektronix, and the Digital Equipment Corporation, in the development of this design system and the associated design and verification programs which it will support.

Acknowledgment

Our work in the area of design-aids for VLSI is supported by many industrial and government sponsors. In particular, the Army Research Office, Hewlett Packard, National Science Foundation, and Tektronix have provided continued support over the years. We also acknowledge the support of Bell Laboratories, Digital Equipment Corporation, Harris Semiconductor, International Business Machines, Intel, Signetics, and in particular the Defense Advanced Research Projects Agency.

References

[1] L. W. Nagel and D. O. Pederson, "Simulation program with integrated circuit emphasis," in *Proc. 16th Midwest Symp. Circ. Theory*, (Waterloo, Canada), Apr. 1973.
[2] T. E. Idleman, F. S. Jenkins, W. J. McCalla, and D. O. Pederson, "SLIC—A simulator for linear integrated circuits," *IEEE J. Solid-State Circuits*, vol. SC-6, pp. 188–204, Aug. 1971.
[3] F. S. Jenkins and S. P. Fan, "TIME—A nonlinear DC and time-domain circuit simulation program," *IEEE J. Solid-State Circuits*, vol. SC-6, pp. 188–192, Aug. 1971.
[4] L. W. Nagel, "SPICE2: A computer program to simulate semiconductor circuits," ERL Memo ERL-M520, Univ. of California, Berkeley, May 1975.
[5] E. Cohen, "Program reference for SPICE2," ERL Memo. ERL-M592, Electronics Res. Lab. Univ. of California, Berkeley, June 1976.
[6] *Bell Syst. Tech. J.*, vol. 57, no. 6, Jul.–Aug., 1978, contains a collection of papers describing UNIX.
[7] "UNIX programmer's manual," 7th edition, Virtual VAX-11 Version, Dep. of Electrical Engineering and Computer Sciences, Univ. of California, Berkeley, Dec. 1979.
[8] S. R. Bourne, "UNIX time-sharing system: The UNIX shell," *Bell. Syst. Tech. J.*, vol. 57, no. 6, pp. 1971–1990, Jul.–Aug. 1978.
[9] A. R. Newton, "The VLSI design challenge of the 80's," in *Proc. 17th Design Automation Conf.*, pp. 343–344, June 1980.
[10] S. I. Feldman, "Make—A program for maintaining computer programs," UNIX Programmer's Manual, 7th edition, vol. 2A, Jan. 1979.
[11] G. Persky, D. N. Deutsch, and D. G. Schweikert, "LTX—A minicomputer-based system for automatic LSI layout," *J. Design Automation and Fault-Tolerant Computing*, vol. 1, no. 3, pp. 217–255, May 1977.
[12] A. Feller, "Automatic layout of low-cost quick-turnaround random-logic custom LSI devices," in *Proc. 13th Design Automation Conf.*, pp. 79–85, June 1976.
[13] R. J. Blumberg and S. Brenner, "A 1500-gate, random-logic, large-scale integrated (LSI) masterslice," *IEEE J. Solid-State Circuits*, vol. SC-14, pp. 818–822, Oct. 1979.
[14] J. W. Jones, "Array logic macros," IBM J. Res. Develop., pp. 98–109, Mar. 1975.
[15] D. Johansen, "Bristle blocks: A silicon compiler," in *Proc. 16th Design Automation Conf.*, pp. 310–313, June 1979.
[16] C. H. Séquin, "Standard interchange formats for integrated circuit design," NATO Advanced Study Institute on Design Methodologies for Very Large Scale Integrated Circuits, Lovain-la-Neuve, Belgium, July 8, 1980.
[17] B. T. Preas and C. W. Gwyn, "General hierarchical automatic layout of custom VLSI circuit masks," *J. Design Automation and Fault-Tolerant Computing*, vol. 3, no. 1, pp. 41–58, Jan. 1979.
[18] U. Lauther, "A min-cut placement algorithm for general cell assemblies based on a graph representation," in *Proc. 16th Design Automation Conf.*, pp. 1–10, June 1979.
[19] M. A. Cutter, "BEGE: A berkeley graphics editor," MS Rep., Dep. of Electrical Engineering and Computer Sciences, Univ. of California, Berkeley, June 1980.
[20] J. Ousterhout, "CAESAR: An interactive editor for VLSI circuits," Internal Memo., Dep. of Electrical Engineering and Computer Sciences, Univ. of California, Berkeley, Jan. 1981.
[21] K. H. Keller, "KIC: A graphics editor for integrated circuits," MS Rep., Dep. of Electrical Engineering and Computer Sciences, Univ. of California, Berkeley, June 1981.
[22] M. Y. Hsueh and D. O. Pederson, "Computer-aided layout of LSI circuit building blocks," in *Proc. IEEE Int. Symp. on Circ. and Syst.*, pp. 474–477, (Tokyo, Japan), 1979.
[23] P. M. Carmody, *et al.*, "An interactive graphics system for custom physical design," in *1979 IEEE Int. Solid-State Circ. Conf.*, pp. 246–247, Feb. 1979.
[24] A number of physical layout rule checking programs are commercially available. The program distributed by the NCA company is an example.
[25] C. Mead and L. Conway, *Introduction to VLSI Systems*. New York: Addison-Wesley, 1979.
[26] R. W. Hon and C. H. Séquin, "A guide to LSI implementation," Xerox Palo Alto Research Center, Jan. 1980.
[27] J. Kohn, "STIFPLOT: A STIF based IC plotting tool," MS Rep.,

Dep. of Electrical Engineering and Computer Sciences, Univ. of California, Berkeley, Sept. 1980.

[28] M. Y. Hsueh, "Symbolic layout and compaction of integrated circuits," ERL Memo. UCB/ERL M79/80, Univ. of California, Berkeley, Dec. 1979.

[29] D. Gibson and S. Nance, "SLIC—Symbolic layout of integrated circuits," in *Proc. 13th Design Automation Conf.*, pp. 434–440, June 1976.

[30] R. P. Larson, "Versatile mask generation technique for custom microelectronic devices," in *Proc. 15th Design Automation Conf.*, pp. 193–198, June 1978.

[31] K. Hardage, "ASAP: Advanced symbolic artwork preparation," presented at the Conf. on Advanced Research in Integrated Circuits, MIT, Cambridge, MA, Jan. 1980.

[32] Y. E. Cho, A. J. Korenjak, and D. E. Stockton, "FLOSS: An approach to automated layout for high-volume designs," in *Proc. 14th Design Automation Conf.*, pp. 138–141, June 1977.

[33] A. E. Dunlop, "SLIP: Symbolic layout of integrated circuits with compaction," *Computer-Aided Design*, vol. 10, no. 6, pp. 387–391, Nov. 1978.

[34] J. D. Williams, "STICKS—A graphical compiler for high-level LSI design," in AFIPS Conf. Proc., vol. 47, pp. 289–295, June 1978.

[35] A. D. Ivannikov and P. P. Sipchuk, "Computer-aided design of MOS integrated circuit layout," in *Proc. IEE—CADMECCS*, (Brighton, England), pp. 47–51, July 1979.

[36] B. W. Kernigham, D. G. Schweikert, and G. Persky, "An optimum channel routing algorithm for polycell layouts of integrated circuits," in *Proc. 10th Design Automation Workshop*, pp. 50–59, 1973.

[37] G. Persky, D. N. Deutsch, and D. G. Schweikert, "LTX—A system for the directed automation design of LSI circuits," in *Proc. 13th Design Automation Conf.*, 1976.

[38] T. Yoshimura and E. S. Kuh, "Efficient algorithms for channel routing," *IEEE Trans. Circuits Syst.*, to be published.

[39] R. W. Dutton, D. A. Divekar, A. G. Gonzales, S. E. Hansen, and D. A. Antoniadis, "Correlation of fabrication process and electrical device parameter variations," *IEEE J. Solid-State Circuits*, vol. SC-12, pp. 349–355, Aug. 1977.

[40] W. G. Oldham, S. N. Nandgaodnkar, A. R. Neureuther, and M. O'Toole, "A general simulator for VLSI lithography and etching processes: Part I—Application to projection lithography," *IEEE Trans. Electron Devices*, vol. ED-26, pp. 717–723, Apr. 1979.

[41] W. G. Oldham, A. R. Neureuther, C. Sung, J. L. Reynolds, and S. N. Nandgaonkar, "A general simulator for VLSI lithography and etching processes: Part II—Application to deposition and etching," *IEEE Trans. Electron Devices*, vol. ED-27, pp. 1455–1459, Aug. 1980.

[42] S. Liu, B. Hoefflinger, and D. O. Pederson, "Interactive two-dimensional design of barrier-controlled MOS transistors," *IEEE Trans. Electron Devices*, vol. ED-27, pp. 1550–1558, Aug. 1980.

[43] S. Liu, "A unified CAD model for MOSFETS," Ph.D. dissertation, Dep. of Electrical Engineering and Computer Sciences, Univ. of California, Berkeley, Apr. 1981.

[44] A. L. Sangiovanni-Vincentelli, L-K. Chen, and L. O. Chua, "A new tearing approach—Node-tearing nodal analysis," in *Proc. IEEE Int. Symp. Circ. and Syst.*, pp. 143–147, Apr. 1977.

[45] E. Cohen, "Performance limits of a dedicated CAD system," in *Proc. IEEE Int. Symp. Circuits Syst.*, pp. 708–711, Apr. 1980.

[46] A. Vladimirescu and D. O. Pederson, "A computer program for the large-scale analysis of solid-state integrated circuits," in *Proc. IEEE Int. Symp. Circuits syst.*, (Chicago, IL), Apr. 1981.

[47] D. A. Calahan and W. G. Ames, "Vector processors: Models and applications," *IEEE Trans. Circuits Syst.*, vol. CAS-26, pp. 715–725, Sept. 1979.

[48] H. S. Stone, "Parallel processing with the perfect shuffle," *IEEE Trans. Comput.*, pp. 153–161, Feb. 1971.

[49] B. R. Chawla, H. K. Gummel and P. Kozak, "MOTIS—An MOS timing simulator," *IEEE Trans. Circuits Syst.*, vol. CAS-22, pp. 901–909, Dec. 1975.

[50] S. P. Fan, M. Y. Hsueh, A. R. Newton, and D. O. Pederson, "MOTIS-C: A new circuit simulator for MOS LSI circuits," in *Proc. IEEE Int. Symp. Circuits Syst.*, pp. 700–703, Apr. 1977.

[51] G. R. Boyle, "Simulation of integrated injection logic," ERL Memo ERL-M78/13, Univ. of California, Berkeley, Mar. 1978.

[52] S. A. Szygenda and E. W. Thompson, "Digital logic simulation in a time-based, table-driven environment. Part 1. Design Verification," *IEEE Comput.*, pp. 24–36, Mar. 1975.

[53] M. A. Breuer, "General survey of design automation of digital computers," *Proc. IEEE*, vol. 54, Dec. 1966.

[54] P. Wilcox and A. Rombeck, "F/LOGIC—An interactive fault and logic simulator for digital circuits," in *Proc. 13th Design Automation Conf.*, pp. 68–73, 1976.

[55] M. Barbacci, "The ISPL language," Dep. of Computer Science, Carnegie-Mellon Univ., Pittsburgh, PA, 1977.

[56] A number of references may be found in *IEEE Comput.*, vol. 7, no. 12, Dec. 1974.

[57] A. R. Newton, "The simulation of large-scale integrated circuits," ERL Memo ERL-M78/52, Univ. of California, Berkeley, July 1978.

[58] A. R. Newton, "Techniques for the simulation of large-scale integrated circuits," *IEEE Trans. Circuits Syst.*, vol. CAS-26, pp. 741–749, Sept. 1979.

[59] G. Arnout and H. De Man, "The use of threshold functions and boolean-controlled network elements for macromodelling of LSI circuits," *IEEE J. Solid-State Circuits*, vol. SC-13, pp. 326–332, June 1978.

[60] D. D. Hill and W. M. Van Cleemput, "SABLE: Multi-level simulation for hierarchical design," in *Proc. IEEE Int. Symp. on Circuits and Syst.*, pp. 431–434 (Houston, TX), Apr. 1980.

[61] V. D. Agrawal, *et al.*, "The mixed mode simulator," in *Proc. 17th Design Automation Conf.*, pp. 618–625, June 1980.

[62] A. R. Newton, "Timing, logic, and mixed mode simulation for large MOS integrated circuits," NATO Advanced Study Institute on Computer Design Aids for VLSI Circuits, Sogesta-Urbino, Italy, July 1980.

[63] LOGIS is a commercially available logic simulator written by F. Jenkins and available on the ISD timeshare system.

[64] A. R. Newton, "The analysis of floating capacitors for timing simulation," in *Proc. 13th Asilomar Conf. on Circuits, Systems and Computers* (Asilomar, CA), Nov. 1979.

[65] G. De Micheli, A. R. Newton, and A. L. Sangiovanni-Vincentelli, "New algorithms for timing analysis of large circuits," in *Proc. IEEE Int. Symp. on Circuits and Syst.*, pp. 439–444, (Houston, TX), Apr. 1980.

[66] E. B. Kosemchak, "Computer aided analysis of digital integrated circuits by macromodeling," Ph.D. dissertation, Columbia University, NY, 1971.

[67] N. B. Rabbat, "Macromodeling and transient simulation of large integrated digital systems," Ph.D. dissertation, The Queen's University of Belfast, U.K., 1971.

[68] M. Y. Hsueh, A. R. Newton, and D. O. Pederson, "The development of macromodels for MOS timing simulators," in *Proc. IEEE Symp. on Circuits and Systems*, pp. 345–349 (New York), May 1978.

[69] L. A. Glasser, "An Interactive PLA Generator as an Archetype for a new VLSI design methodology," in *Proc. IEEE Int. Conf. on Circuits and Computers*, pp. 608–611, Oct. 1980.

[70] H. Fleisher and L. I. Maissel, "An introduction to array logic," IBM J. Res. Develop. pp. 98–109, Mar. 1975.

[71] S. J. Hong, R. G. Cain, and D. L. Ostapko, "MINI: A heuristic approach for logic minimization," IBM J. Res. Develop., vol. 18, no. 5, pp. 443–458, Sept. 1974.

[72] The PRESTO program was developed by A. Svoboda and D. Brown, Tektronix, Inc.

[73] Signetics Bipolar and MOS Memory Data Manual, pp. 156–188, 1979.

[74] P. W. Cook, *et al.*, "1 μm MOSFET technology: Part 3—Logic circuit design methodology and applications," *IEEE Trans. Electron Devices*, vol. ED-26, pp. 333–345, Apr. 1979.

[75] R. A. Wood, "A high density programmable logic array chip," *IEEE Trans. Comput.*, vol. C-28, pp. 602–608, Sept. 1979.

[76] G. D. Hachtel, A. L. Sangiovanni-Vincentelli, and A. R. Newton, "Some results in optimal PLA folding," in *IEEE Int. Conf. Circuits and Comp.*, pp. 1023–1028, Oct. 1980.

[77] D. D. Hachtel, A. L. Sangiovanni-Vincentelli, and A. R. Newton, "Some results in optimal PLA folding," to be published.

[78] H. Landman, "Automatic layout of optimized PLA structures," MS Rep. Dep. of Electrical Engineering and Computer Sciences, Univ. of California, Berkeley, in preparation.

[79] M. Hofmann, "A method for topological compaction of programmed logic arrays," MS. Rep. Dep. of Electrical Engineering and Computer Sciences, Univ. of California, Berkeley, Sept. 1980.

[80] M. W. Bales, "The CABTOCIF program," CADMAN on Berkeley UNIX System, June 1980.

[81] M. A. Bhatti, T. Essebo, W. Nye, K. S. Pister, E. Polak, A. L. Sangiovanni-Vincentelli, and A. L. Tits, "A software system for optimization-based interactive computer-aided design," Memo. UCB/ERL M80/14, Univ. of California, Berkeley, Apr. 1980.

[82] R. K. Brayton and R. Spence, *Sensitivity and Optimization*. The Netherlands: Elsevier, 1980.

[83] J. W. Bandler and M. R. Rizk, "Optimization of electrical circuits," Mathematical Programming Study 11, pp. 1–64, 1979.

[84] G. D. Hachtel, T. R. Scott and R. P. Zug, "An interactive linear programming approach to model parameter fitting and worst-case circuit design," *IEEE Trans. Circuits Syst.*, vol. CAS-27, pp. 871–882, Oct. 1980.

[85] C. Gonzaga, E. Polak, and R. Trahan, "An improved algorithm for a class of optimization problems with functional inequality constraints," Memo. UCB/ERL M78/56, Univ. of California, Berkeley, 1978.

[86] J. Wieslander and H. Elmqvist, "INTRAC, A communication module for interactive programs," Dep. of Automatic Control Memo. LUFTD2/(TRFT-3149)/1-060/1978, Lund Institute of Technology, Lund, Sweden, 1978.

[87] W. Nye, E. Polak, A. L. Sangiovanni-Vincentelli, and A. L. Tits, "DELIGHT: An optimization-based computer-aided design system," invited paper, in *Proc. 1981 Int. Symp. on Circuits and Systems*, (Chicago, IL), Apr. 1981, to be published.

Design Aids for VLSI: A Perspective Revisited

Berkeley's goal in its IC/CAD research is to forge an advanced, flexible design environment. Results are now coming in from a Unix-based, 1980 VLSI design perspective.

A. R. Newton, D. O. Pederson, and A. L. Sangiovanni-Vincentelli,
University of California, Berkeley

The Integrated Circuit/Computer-Aided Design Research Program in the Electrical Engineering and Computer Sciences Department and the Electronics Research Laboratory at Berkeley has been ongoing for more than 15 years. The IC/CAD Program is broad, involving over 20 faculty members who have at least one project in the CAD area and more than 90 graduate students actively involved in a CAD research project. The program is also deep. In the simulation area alone there are over 36 research projects, which include theoretical analysis of the properties of algorithms, the design of computer architectures best suited to these algorithms, the design of algorithms best suited to a particular computer architecture, the construction of programs to verify our techniques and to support our own IC design activities, and the circuit and technology modeling research necessary to make such programs useful.

In this overview of our research activities it is impossible to describe all of this ongoing work. For that reason, we focus on some of the projects with which the authors and their research students are involved.

In 1980, we described a perspective for VLSI design.[1] Many of our recent projects have shown the results of the perspective. In that paper, we introduced the concept of a flexible, file-based data management framework for the support of a variety of custom and semicustom design styles. This framework was well suited to the 32-bit Unix workstation-based computing environment we envisioned. The Hawk/Squid system, which embodies these concepts and which forms the basis of our design system, has been in use at Berkeley for two years. The Squid data manager[2] supports many representations of the design data, including mask layout, schematic diagrams, documentation, and symbolic layout. The Python spacing program was incorporated with Squid for generating efficient symbolic layouts.[3]

In that paper, we also introduced the concept of module generators for use in a macrocell-based design system. This approach now forms the basis of many commercial silicon compiler offerings today. At Berkeley, we initially focused our attention on the programmable logic array as a target design style for a module generator. In particular, emphasis was placed on the efficiency (performance, silicon area, etc.) of the final layout. Our PLA design tools have since been used extensively.

For many years, we have been concerned with tools for design verification, circuit simulation in particular. In recent years, we have continued our work in this area by developing relaxation-based[4,5] and mixed-level[6,7] simulators, as well as investigating the relationship between algorithms and computer architectures.[8,9] Recent work in this area has shown great promise for the implementation of relaxation-based algorithms on advanced multiprocessors.[10] The coupling of simulation with advanced optimization algorithms has also led to an effective circuit design optimization tool.[11]

Along the way we have gained a great deal of experience both in the tool area and in the area of design styles and approaches to complexity management. It is still our plan to forge an advanced and flexible IC

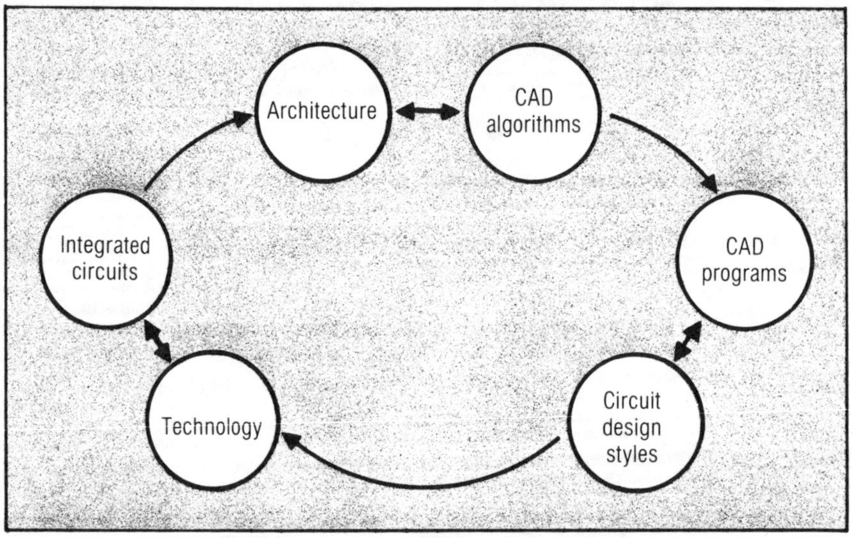

Figure 1. The components of an integrated IC/CAD environment.

Reprinted from *IEEE Design and Test of Computers*, pp. 106–115, April 1985.

design environment, coupling design methods with state-of-the-art, IC-oriented design and verification tools. A key factor in our work is the close coupling between design styles (floor-planning techniques, layout design styles, etc.) and the algorithms and computer programs needed to support them. The effect is synergistic—changes in either area can have dramatic effects on the overall efficiency and performance of the design. Another major aspect of our research is the relationship between computer architecture and the CAD algorithms best suited to such architectures. Substantial performance improvements can be achieved by designing both the architecture and the algorithm to match a specific CAD task. These relationships are summarized in Figure 1. In particular, the relationship between design styles and CAD algorithms, the theoretical properties of these algorithms, the relationships between computer architectures and algorithms, the relationship between graphics and procedure, approaches to efficient synthesis at many levels, and the relationship between algorithmic and rule-based approaches are all important aspects of our work.

As in our earlier work, these projects are coupled closely with other ongoing CAD, IC design, and technology development projects at Berkeley, in particular, the design of general-purpose microcomputers,[12] signal-processing design, and the design of special-purpose CAD multiprocessors.[13] In the following overview, some of our research projects are described briefly and a number of the new research problem areas in which we are working are presented.

Design and verification

Circuit simulation. Since the mid 1960's, circuit simulation has been an important tool for both the design and verification of integrated circuits. Companies spend many millions of dollars per year in computer costs (a number of companies run over 60,000 simulations a month), and the increasing size of circuits results in longer and longer runs. For a number of years, designers and managers have attempted to reduce the need for large, expensive simulation runs by developing new, more conservative design styles. By characterizing the small circuit cells very carefully, and by using efficient timing verification, performance errors can be reduced. While such an approach does work where high circuit performance is not a requirement, for high-performance circuits, such as state-of-the-art microprocessors or RAMs, circuit simulation of large blocks continues to be a necessity.

Spice. The majority of circuit simulators today calculate circuit transient response by

UC Berkeley IC/CAD research areas

The basis for a significant portion of the IC/CAD research at Berkeley was described in a design perspective published in 1980. Research on each of the CAD realms was undertaken concurrently so that the issue of IC design could be approached as a whole. This summarizes the seven areas described fully in the accompanying reports.

Computer-aided design environment. VLSI design requires a flexible design environment that can adapt to new design styles and new technologies. System-level, logic-level, circuit-level and CAD algorithm designers must be supported in a unified system. The Hawk/Squid system at Berkeley supports hierarchical synthesis and verification tools to accomplish some of this.

Design and verification. Because of the extreme complexity of VLSI design, IC engineers may use conservative design techniques to reduce the need for large, expensive computer simulations. These techniques do not address the needs of high-performance ICs. The Spice3 program, a faster version of Spice2, has been developed to tackle part of this problem. Waveform relaxation and iterative timing analysis also show promise; but these methods may miss some design errors. Rule-based expert systems can find many of these elusive errors by critiquing VLSI circuits at the transistor level.

Special-purpose hardware. VLSI design requires new, multiprocessor architectures with significant performance improvements over conventional computers. New algorithms are needed for these multiprocessors to take full advantage of their potential; both waveform relaxation and iterative timing analysis are well suited to such machines.

Automatic design process. If the bulk of today's custom IC designs are implemented using regular structures, human designers are still needed to implement much of the IC's control logic. While it is often only a small part of a chip, the control logic requires the most work to create—and has proven difficult to automate. Using the "fast and simple" approach leads to inefficient implementation.

Floor-plan layout. The macrocell design style has emerged as the new approach to VLSI design—but it leaves open the problem of how to best assign physical parameters so that both area usage and performance are optimized. Simulated annealing reduces the likelihood of a local solution overriding the discovery of a better overall solution, but requires a great deal of computer time.

Testing. Manual test generation is almost impossible for VLSI, while the runtimes for computer test generation are often prohibitive. One solution may be test generation strategies for macrocell-based design styles and self-test architecture for each macrocell created by module generators.

Parametric optimization. Changing IC parameters forces the designer to run time-consuming circuit simulations to test his changes. The Delight_Spice system frees the designer from selecting the parameter values, but requires long runtimes and an understanding of the algorithms involved. The development of a cost model to relate price, demand, and value of some IC product specifications is under way for such a optimization system.

Other major projects at Berkeley relating to IC design include the Magic project, carried out by John Ousterhout and his students, and the work of Randall Katz, both of which are described in part in the February *Design & Test*; the placement and routing work of Ernest Kuhl and his students; the graphics and symbolic layout work of Carlo Sequin and his students; and the photolithographic and device analysis work of Andrew Neureuther and William Oldham. Other faculty members active in the IC/CAD area are Brian Barsky, Robert Brodersen, Leon Chua, Paul Hilfinger, David Hodges, Chen-Ming Hu, Robert Karp, Ping Ko, David Messerschmitt, Robert Meyer, Richard Muller, Michael Stonebraker, and Clarke Thompson.

numerically integrating a complete set of simultaneous nonlinear differential equations. This method is general, but the computational complexity grows rapidly as the number of devices increases. The Spice2 circuit simulator[14] can take several hours on a Vax 11/780 to simulate circuits with only a few hundred devices. While the circuit simulation tools in use today continue to serve us well, they were designed in the early 1970's for the simulation of circuits with a few hundred transistors at most and for use in a punched-card, batch environment. By recoding Spice2 in the C programming language and by updating its algorithms to improve convergence for MOS and modern bipolar circuits, our Spice3 version runs significantly faster than Spice2. Spice3 provides a more flexible testbed for working with both new algorithms and with new device models and technologies. We plan to incorporate a number of new features in Spice3 necessary for the design of high-performance circuits, in particular, gallium arsenide transistor models and microwave circuit components.

As well as working with Spice, we are investigating a number of personal computing environments and program development systems for CAD tool development. We have implemented a time-domain transient circuit simulator (Bias) in a number of programming languages and for a number of computers (Rong).[15] Comparisons indicate that a Lisp-based environment not only permits rapid program development but also provides a unified coupling to a symbolic processing system (Karti).[16] Many of the preprocessing and circuit auditing tasks necessary to reduce the circuit simulation burden are best implemented in such a symbolic processing system.

Relax and Splice. Faster simulation methods have been developed recently that attempt to exploit not only spatial sparsity but also the temporal sparsity and the unidirectional nature of MOS digital circuits. We have developed two simulators based on these new relaxation algorithms: Relax, based on waveform relaxation[4] and Splice, a mixed-level simulator which uses iterated timing analysis, or ITA,[5] for the circuit level. On a uniprocessor, these pro-

Making a CAD framework

In recent years, the notion of procedural circuit design[1-3] and the rule-based expert system technology have emerged as key components in the design process. These techniques, coupled with the ever-moving boundary between entire systems, printed-circuit boards, and chips, have broadened the requirements for an integrated design system.

What is needed to support this work is a flexible design and programming environment that allows a variety of approaches to design to coexist and permits system-level, logic-level, and circuit-level designers, as well as CAD algorithm developers, to work together in a single, unified environment. The keys to such a system are common levels of abstraction and standard interfaces among them, as well as a powerful set of synthesis and verification tools which form the basis on which further research is carried out.

Our research on integrated data management systems for IC design began in 1980. The model for this system[4] was a flexible data manager (Squid) using a file-system-based approach and a window-based graphics editor (Hawk). These two tools, in conjunction with a connectivity-oriented data structure, provide a framework for trying new design methods and for incorporating new tools, such as verification or synthesis tools.[5]

Procedural interface. Both Hawk and Squid provide a procedural interface to the CAD system designer and use dynamic linking and generator concepts for efficient data management. A key component of the system is its graphical data structure, which supports mask layout, schematic, and symbolic layout data as well as textual representations of the design. This data structure evolved from our work on intermediate forms, STIF[6] and SLL[7] in particular. The semantic model of the Squid data structure is very similar to that used in the EDIF interchange format.[8] The data structure uses the protection frame and terminal mechanisms to provide a powerful hierarchical data abstraction. The protection frames are generated automatically using a set of mask operations.[5,9] This abstraction technique has made the incorporation of both hierarchical synthesis (e.g. Python, Panda,[10] and YACR[11]) and verification tools (e.g., Lyra[12]) straightforward. With such a system, the tool developer can focus on the algorithmic aspects of the problem without loss of efficiency in the final layout.

Over the past year, the Hawk/Squid system has been used for the design of the CMOS version of the Soar[13] chip. The chip uses over 35,000 MOSFETs and is designed in a 3-micron P-Well CMOS process. During that design project, all PLAs were synthesized automatically using EQNTOTT,[14] Espresso-IIC,[15,16] and Panda, with much of the routing performed interactively using "cable routers," river routers, and the YACR channel router. Interactive layout-rule checking was performed using Lyra. These tools were incorporated into the Hawk/Squid framework on a demand-driven basis by the CMOS Soar designers.[17]

The use of procedure, rather than just graphics (or data), for describing IC designs was pioneered at Caltech[1] and applied later at MIT in the DPL project.[2] These and other efforts have inspired a large number of projects at university and industrial sites for in-house use. The term silicon compiler is often associated with procedural design* and a number of new companies are advertising silicon compilers. Unfortunately, the majority of systems offered so far do *not* offer the user a procedural design capability. Rather, they can be characterized as cell-based systems where the circuit building blocks are parameterized cells. These cells can be assembled in a variety of ways, depending on the design style the compiler is using.

Early limited success. True procedural design, where the IC designer can write programs which, when executed, produce layout, is still of key importance to the productivity of the custom IC designer. However, the early procedural design systems have had limited success for a number of reasons:

- The relationship between graphics and procedure was not exploited sufficiently. Graphics and procedure are generally treated as disjoint descriptions of the design. The graphical mask layout was the result of the procedure, rather than an active part of it.
- Verifying the correctness of a procedurally generated design is generally performed at the mask layout level. At that point, no correspondence between the active procedure which created an object and its geometrical layout is maintained. As a result, debugging the design is very difficult.

*The term was first used at Caltech to describe the early procedural design work there.

grams show from one to two orders of magnitude speedup over conventional circuit simulators for time-domain transient analysis. A detailed review of this work is presented by Newton and Sangiovanni-Vincentelli.[17] We are continuing our work with relaxation-based approaches. In particular, we are studying the convergence properties and timestep control strategies necessary to make these programs as rugged and reliable as conventional circuit simulators. We are also developing relaxation-based algorithms for use on multiprocessors as described in the next section.

In the design of state-of-the-art integrated circuits there are many errors that techniques like timing analysis, switch-level simulation, layout rule checking, or even circuit simulation cannot find. Of course, by requiring that strict design rules be followed, the probability of such an error can be reduced substantially. However, the errors still can (and often do) occur. Static interactions among blocks and sneak feedback paths, dc errors, or capacitive feedthrough errors are examples of such problems. Even detailed circuit simulation cannot find some errors unless the correct input stimuli are known a priori. Many of these errors are found by experienced designers during a design-review process. If such an error is missed by the designer, it can be very difficult to find in the final design.

Rule-based critique. Rubicc[18] is a rule-based expert system written to perform a design critique of VLSI circuits at the cell, or transistor, level. The goal of this project was to explore the feasibility of using expert system technology to provide meaningful feedback to circuit designers on the quality of their designs in a manner similar to a critique provided by an experienced designer. The use of a rule base facilitates the encapsulation of the knowledge base about the design of circuits in a given technology and allows for incremental additions to the knowledge base.

Rubicc accepts a Spice-like netlist of a circuit and produces a critique summary, listing errors and suggestions for circuit improvements. In its original form, Rubicc contains 110 rules (both forward and backward chaining). It performs meaningful cri-

- Often the designs produced by the early systems did not achieve high density or high performance, nor did they meet power requirements.

In our present work, we are addressing all of these issues and plan to incorporate a flexible procedural design environment into our existing data management system. Each object in this system may be described by data, such as its mask layout; by a local procedure, such as a parameterized cell; by generic synthesis tools, such as a channel router or a placement program; or by a combination of all of these techniques. By coupling the development of such a system with our advanced IC design activity, particularly in the signal processing area, and because our designers have already had considerable exposure to procedural design, we will be able to generate dense layouts. At the higher levels of design, our work in the area of design optimization and synthesis will be coupled to the procedural environment and should lead to further increases in density. We are exploring the relationship between graphics and procedure by considering both the graphical descriptions and associated procedural representations as different facets of the same object. One does not exist without the other.

Another way of reducing the potential for errors, as well as insuring the technology parameterization necessary at the lowest level of design, is to have the procedural design system generate symbolic layout rather than detailed geometry. A spacing program can then be used to guarantee a layout-rule-correct design and to convert the symbolic layout to mask artwork. A powerful and flexible symbolic layout system will provide the basis for our future synthesis work.

References

1. D. Johannsen, "Bristle Blocks: A Silicon Compiler," *Proc. 16th Design Automation Conf.*, 1979, pp. 310-313.
2. H. E. Schrobe, "The Data Path Generator," *Proc. Conf. Advanced Research in VLSI*, P. Penfield, ed., Massachusetts Institute of Technology, Cambridge, Mass., Jan. 1982.
3. R. Ayers, *VLSI: Silicon Compilation and the Art of Automatic Microchip Design*, Prentice Hall, Englewood Cliffs, N.J., 1983.
4. A. R. Newton, D. O. Pederson, A. L. Sangiovanni-Vincentelli, C. H. Sequin, "Design Aids for VLSI: The Berkeley Perspective," *IEEE Trans. Circ. and Syst.*, Vol. CAS-28, No. 7, July 1981.
5. K. Keller, "An Electronic Circuit CAD Framework," Electronics Research Laboratory, Memo No. UCB/ERL M84/54, University of California, Berkeley, June 1984.
6. C. H. Sequin and A. R. Newton, "Description of STIF 1.0," *Design Methodologies for VLSI*, P. G. Jespers, C. H. Sequin and F. VanDeWeile, eds., Noordhoff Publishing, Jan. 1982, pp. 147-171.
7. S. A. Ellis, K. Keller, A. R. Newton, D. O Pederson, A. Sangiovanni-Vincentelli and C. H. Sequin, "A Symbolic Layout Design System," *Proc. 1982 IEEE Int'l Symp. Circ. and Syst.*, Rome, Italy, May 1982, pp. 670-674.
8. *EDIF Specification: Electronic Design Interchange Format* Version 1 0 0, Feb. 1985.
9. K. H. Keller, "An Electronic Circuit CAD Framework," Electronics Research Laboratory, Memo No. UCB/ERL M84/54, University of California, Berkeley, July 1984.
10. G. H. Mah and A. R. Newton, "PANDA: A PLA Generator for Multiply-Folded PLAs," *Proc. IEEE Int'l Conf. CAD*, Santa Clara, Calif., Nov. 1984, pp. 122-124.
11. A. Sangiovanni-Vincentelli, M. Santamauro and J. Reed, "A New Gridless Channel Router: Yet Another Channel Router the Second (YACR-II)," *Proc. IEEE Int'l Conf. CAD*, Santa Clara, Calif., Nov. 1984.
12. M. Arnold and J. Ousterhout, "LYRA: A New Approach to Geometric Layout Rule Checking," *Proc. 19th Design Automation Conf.*, Las Vegas, Nev., June 1982.
13. C. Marino, "Smalltalk on a RISC-CMOS Implementation," MS Report, Department of Electrical Engineering and Computer Sciences, University of California, Berkeley, Feb. 1985.
14. B. Cmelik, "EQNTOTT User's Guide," Department of Electrical Engineering and Computer Sciences, University of California, Berkeley, 1981.
15. R. Brayton, G. Hachtel, C. McMullen and A. Sangiovanni-Vincentelli, *Logic Minimization Algorithms for VLSI Synthesis*, Kluwer Academic Publishers, 1984.
16. R. Rudell, "ESPRESSO-IIC User's Manual," Department of Electrical Engineering and Computer Sciences, University of California, Berkeley, 1984.
17. D. Ryan, "Interactive Routing Toolbox for IC Layout," MS Report, Department of Electrical Engineering and Computer Sciences, University of California, Berkeley, Feb. 1985.

tiques on circuits designed in a dynamic NMOS process with two phase clocks. Its rule base allows critiques from straightforward static checks (e.g., beta ratios) to complex clocking-capacitive coupling checks (e.g., dynamic node glitches and feedback race conditions). A conscious design decision in Rubicc was to *not* perform either logic or circuit simulation. Rather, Rubicc's critique method is based on heuristics of VLSI design as obtained from the knowledge base of experienced designers. By reasoning in this manner, the system can find errors that are highly elusive or impossible to find through worst-case simulation.

Open questions in the use of such rule-based systems for design verification are the determination of a sufficient set of rules to guarantee a correct design. To achieve this goal, we must first characterize MOS design styles and the requirements for a correct design. To begin with, we plan to study static design styles and the requirements for correct static behavior. We hope then to extend the analysis to ac parametric characteristics.

Automating the design process

Ever since a handful of transistors could be integrated on a single chip, engineers and scientists have been developing ways for automating the layout process. Nowadays, the gate array and standard cell design approaches are used routinely and new floor plans, which offer higher performance for specific applications, are emerging with associated computer aids. Fixed macrocell,[19] parameterized macrocell,[20] and microcomputer-based floor plans[21] are examples of such systems.

PLA synthesis systems. In a complex digital circuit, many of the modules can be designed using regular techniques and a straightforward design style (e.g., RAM, ROM, and data path). However, the control logic is often the 10 percent of the chip that takes 90 percent of the time to design. Alternately, if a fast but simplistic approach is used for design, a very inefficient implementation will often result. For these reasons, there is an increasing interest in synthesis systems that can generate a compact logic design from a high-level specification.

The first attempts at automatic generation of a random logic design and its actual layout from high level description have met with limited success. In almost all cases, the performance of the final design did not even come close to that achieved by manual techniques. In particular, the use of a fixed floor plan and the often simplistic logic synthesis techniques used for the control part of the chip are the culprits for such poor performance.

Five years ago, we chose the PLA as a target layout style to see how well we could do in the area of synthesis of control logic for ICs. We have made contributions at all levels of the PLA design process, from system-level description, through logic minimization and folding, to the actual mask layout of PLA. We believe that this experiment demonstrates the potential of automatic synthesis for specific design styles and that even better gains are possible with more flexible design styles.

After completing this work, we moved on to the problem of generating optimal finite-state machines, or FSMs. To solve this problem, we needed an efficient approach to state assignment. An approach for performing an optimal assignment of binary codes to a set of symbolic inputs has been developed recently.[22] The idea is to view this assignment problem as a multiple-valued minimization problem. The multiple-valued minimization problem can be solved using a standard binary-valued minimization algorithm, such as that used in Espresso-IIC, or, preferably, it can be solved directly with a multiple-valued minimization algorithm, such as the Espresso-mv program which is under development.

The program Kiss was developed to determine a state assignment based on encoding the symbolic present-state input variable of an FSM. This was shown to be a successful approach, but we are still investigating an extension to this technique which will allow for the complementary output-encoding problem, and eventually solve both problems simultaneously.

Multilevel digital logic. We are now studying the automated synthesis of multilevel digital logic, especially the random logic seen in the control sections of VLSI digital integrated circuits. This problem includes partitioning the logic into manageable pieces and decomposing it into a design which best meets a stated trade-off between area and performance in a particular design style.

Automatic synthesis of such control logic has been attempted before and is currently an active area of research. Certain control logic components have a two-level representation containing as many as 2^n components even after minimization, where n is the number of primary inputs of the logic block. In addition, even if a two-level representation does contain a reasonable number of terms, there are cases in which a multilevel representation can be implemented in considerably less area.

A number of algorithms for factorizing Boolean functions have been proposed in the past, but they require an exhaustive search, which is out of the question for the complexity of the control logic of interest in today's designers. Some other algorithms suffered from the lack of understanding of the technological constraints associated with particular implementation of the logic. New algorithms are appearing that are effective in partitioning complex logic functions and that can take into consideration the constraint of a particular circuit design style.[23] The Logic Synthesis System of IBM[24] uses a prototype expert system to accomplish the mapping of combinational logic in random form (from an initial high-level description) into a gate array implementation.

We have developed a prototype system for the synthesis of multilevel combinational logic in the form of gate-matrix-like domino CMOS logic.[25] To characterize this problem, we have first characterized the CMOS domino design style using both simulation and analytical techniques to predict the performance of arbitrary, multilevel domino networks. These models of the technology are then used to drive a rule-based, logic-level optimization program, Mimic, which considers not only performance issues, such as circuit speed and area, but also other design constraints including charge redistribution and feedthrough effects. The results of the Mimic optimization are then fed to a topological optimization program, Twist, for the symbolic design of the final circuit. Twist employs algorithms similar to those used in our PLA folding program, Pleasure, with additional considerations necessary to obtain an efficient result using the modified gate-matrix design style. The output of Twist is then used by a tile-based electrical design and layout program, Tailor, which generates the final mask layout for the random logic function.

Two-pronged approach. We plan to continue our two-pronged approach: we will develop an algorithmic approach based on results in multilevel decomposition and on multivalued logic minimization, and we will develop a more general rule-based approach using an expert system whose rules will be carefully designed to take into consideration all the electrical design aspects of the control logic and of the technology used. We are particularly interested in evaluating and comparing the two approaches. By developing a set of metrics to judge the quality of the results, we hope to characterize the design styles and technologies best suited to each approach. It is our contention that rule-based systems make the most sense in situations where special cases are common. If there are few special cases, and a good heuristic algorithm is known, it should be used. We believe that in some design styles to achieve a high efficiency in the final design, quite a few special cases must be considered. Ultimately, we hope to combine the best of the two approaches into a flexible, multilevel logic design system where the rule base is used to customize the system for a specific technology or design style.

Often, rule-based systems are not analyzed in terms of the power of the rules they embody. In particular, for logic design, it is important to determine whether the rules developed to map a multilevel representation of a logic function into another are complete enough that all possible representations of the given logic function could be generated by the application in a certain sequence of the rules. In this way, we can prove that the rules have not prevented us from exploring some representation which may turn out to be the optimal one. We hope to develop rules which are guaranteed to explore the entire domain of the optimization being considered.

Floor-plan layout, placement, and routing

Macrocell design style. An emerging design style for large ICs is the so-called macrocell design style where a designer uses large functional modules, which can be generated by hand or automatically by synthesis programs, to implement the desired function. Some of the synthesis programs, such as our PLA synthesis system, can generate the functional blocks with constraints involving shape, aspect ratio, and pin positions. An open question at this time is how to identify the best assignment of these parameters so that area and performance of the chip are optimized. In particular, it is difficult to estimate the routing area as well as the actual dimensions of the blocks themselves. Because of the many variables involved in solving this optimization problem, very few heuristics have been developed, let alone optimal solutions. In addition, to perform the trade-offs on multilevel optimization, it is important to integrate floor-planning tools as well as global placement and routing tools. Chips designed by our designers at Berkeley use two metal layers, and there is a definite trend towards the use of more layers of interconnect. At this moment, very few multiple layer routers have been developed for IC design.

Our goal is to develop new algorithms and integrated tools that can support a wide variety of design styles and that can place and route macrocell-based chips with two or more levels of interconnect.

The problem of optimal floor-planning and placement of macrocells have several features in common. In each case, the cost function for an optimal solution is difficult to determine and difficult to compute; many criteria and trade-offs must be taken into account. In addition, even if the cost function is simplified, the combinatorial optimization problems obtained belong to a class of problems which are difficult to solve, that is, the class of NP-complete problems. Heuristics, generally the greedy type, are used to solve these problems. However, heuristic algorithms tend to perform poorly when many local optima of the cost function exist and these local optima may be quite far from the global optimum.

Simulated annealing, As proposed by Kirkpatrick et al.,[26] simulated annealing is a new algorithm for the solution of combinatorial optimization problems, whose basic feature is that it allows "hill climbing" moves to escape from local optima. These moves are accepted according to certain criteria which take the cost into consideration and not blindly as randomizing algorithms. The controlling mechanism is based on the observation that combinatorial optimization problems with a large configuration space exhibit properties similar to physical processes with many degrees of freedom. This method, when applied to combinatorial optimization, generates moves randomly and checks whether the cost of the new configuration satisfies an acceptance criterion based on temperature. The initial temperature, the number of moves generated at each temperature and the rate of decrease of temperature are all important parameters that affect the speed of the algorithm and the quality of the final configuration.

We have carried out experiments with simulated annealing in the area of placement and global routing of integrated circuits. In particular, we have developed Timberwolf,[27] an integrated set of placement and routing optimization programs based on the simulated annealing approach. Three basic optimization programs have been included in Timberwolf: a standard cell placement program, a standard cell global router program, and a generalized gate array placement program. In the standard cell case, Timberwolf reduced the area of the largest example tried (from 800 to 2700 cells) by 37 percent to 65 percent over an existing place-and-route program in use at AMI. The results obtained were about 15 percent better than the estimate of the area taken by a manual layout.

The results stated above have been obtained at the expense of a great deal of computer time (a 2700-standard-cell placement problem can take as much as 84 CPU hours on a Vax 11/780). A mathematical analysis of the algorithm is very important to understand the essential features which make the algorithm work well and to suggest techniques for controlling its operation. We are investigating this algorithm and its mathematical properties in depth.

Mathematical model. Markov chains can be used as a mathematical model of simulated annealing. It has been shown, using time-homogeneous Markov chains, that under certain assumptions on the number of moves generated by the algorithm at each temperature, simulated annealing produces asymptotically the optimum solution of combinatorial optimization problems with probability one.[28,29] This proof has underlined the essential properties of the algorithm, so a class of probabilistic hill-climbing algorithms has been derived that have the same asymptotic properties of simulated annealing.[29]

However, these results assumed that an infinite number of iterations had to be taken so that a stationary probability distribution could be reached at each temperature. This result is clearly not adequate to guide the choice of annealing schedules. The analysis we are carrying out in collaboration with Debasis Mitra of Bell Laboratories, is based on time-inhomogeneous Markov chains. We have been able to derive a number of new results using this approach.[30]

Further work is focused on the analysis of the algorithm under the assumption that the number of iterations used by the algorithm is much smaller than the number of the possible configurations of the optimization problem. We are also studying adaptive control schema for the annealing schedule. Because of its flexibility, a floor-planning tool can be developed based on simulated annealing. That allows the designer to specify for each block the various aspect ratios he can choose from, the pins that can be moved and the ones that cannot, and constraints on the positions pins can take (e.g., a group of pins can be shifted from one side of the block to another but the relative positions of the pins inside the group cannot change).

Channel routing. Channel routing is an essential component of any layout system. The goal of a channel router is to complete the interconnections in minimum area. The number of "through vias," the length of nets, and therefore the relative delays among the nets are also important to evaluate. These criteria are often considered secondary objectives but we believe that they should be given more emphasis because of their effects on the reliability and speed of state-of-the-art circuits.

We have developed a channel router (YACR2) that can route channels with cyclic constraints and that uses a virtual routing grid.[31] On a number of industrial examples, YACR2 has performed substantially better than existing channel routers.

In the routing part of our work we will develop algorithms for channel definition, global routing, and detailed routing—including channel, maze, and river routing—for multiple levels of interconnects. We will base our future work on detailed routing on the results we obtained in building YACR2. In particular, we expect to extend the results of YACR2 in such a way that the router will accept as a parameter the number of layers of interconnect, their relative preference for routing, and their layout rules.

Testing

Testing a VLSI circuit is becoming increasingly difficult. Manual test pattern generation is almost impossible for circuits of large complexity, and the running time for fault simulation can be prohibitive. Automatic test pattern generation techniques suffer from an exponential increase in computational complexity with the number of elements to test. Our early activity in the testing area has included testability analysis and some test pattern generation work.[32]

Test generation. Recently we have extended our testing work to include self-test and test pattern generation strategies for macrocell-based design styles. In particular, test pattern generation algorithms for macrocells such as PLAs, RAMs, ROMs, and general regular arrays are included in our module generators. For example, for PLA (folded and unfolded), a program called PLATPG has been developed which exploits both biased random testing and the PLA minimization techniques of Espresso-IIC to implement an efficient, deterministic PLA test-generation algorithm. Preliminary results of PLATPG show high performance with respect to runtime, memory requirements, and test set minimality. Test generation is also being added to the CMOS domino logic synthesis system described earlier.

Self-testing architectures. Complementary to this approach is the development of self-testing architectures for each of the macrocells generated automatically by our module generators. We anticipate that the module generators will automatically generate the self-testing structures needed and the necessary testing sequences to enable these structures.

Use of special-purpose hardware for CAD

As we approach the fundamental performance limits of uniprocessors, it is clear that only new, multiprocessor computer architectures can offer the large performance improvements needed to solve the complex problems of our time. Throughout the world, a great deal of research is in progress on the development of new, often unconventional computer architectures for both symbolic and numerical processing.[1]

However, such machines will not be able to exploit the parallelism available in problems unless new algorithms are developed that are well suited to a multiprocessor environment. We are continuing our work in the area of multiprocessor-based algorithms for CAD and we plan to extend the scope of our work to include a variety of new problems. In the past, it has often been assumed that advanced compiler technology would be sufficient to translate a conventional computer algorithm for optimal use on a special-purpose machine. In the case of circuit simulation, even the most advanced compiler technology, used in conjunction with a number of computer consultants, has shown poor speedup on pipelined machines (about 12 to 15 percent hardware utilization on the Cray 1).[2]

There are several compute-intensive problems which face CAD of VLSI circuits today. In each case, we are investigating new algorithms suited to the new generation of multiprocessing computers that will soon be with us. Our initial work with circuit simulation has shown that factors of more than 1000 times speedup are possible when the combination of computer and algorithms are well matched. Not only do we hope to produce new algorithms but through this work we also plan to investigate the characteristics of multiprocessors that are important for our CAD problems. These results should also help in the design of new machines.

Circuit simulation. Our first project in this area dealt with circuit simulation. To increase the performance of the circuit simulation programs described earlier, we have investigated the use of special-purpose machine instructions for simulation,[3] advanced algorithms for vector machines such as the Cray 1 and FPS 164,[2] as well as special-purpose multiprocessors for sparse matrix solution.[4] We are also investigating the use of special-purpose, single-board coprocessors for use in a personal computing environment.

Both the ITA and waveform relaxation methods are particularly amenable for use on data-driven multiprocessors because the computational method already decomposes the problem. We have developed a distributed form of the ITA algorithm, called DITA, and have implemented a prototype DITA simulator, the MSplice program. Running MSplice on a 10-processor multiprocessor, we obtained more than 70 percent utilization of the available hardware for the analysis of large, digital circuits. MSplice is a first step, and much work remains to determine the convergence properties and performance limits of such approaches. In particular, by generalizing the event-driven selective trace notion used for dynamic subcircuit ordering, it is possible to consider both spatial fanouts (adjacent subcircuits) *and* temporal fanouts (the next timepoint at which this subcircuit is to be solved) in a uniform manner. As a result, when a subcircuit is processed and found to change state, all its active fanouts are scheduled—including its next timepoint, which may also be located on a different processor. By using such a shared spectrum model of the circuit and its waveforms, new partitioning techniques become possible that not only permit decoupled subcircuits to be allocated to separate processors but also permit decoupled (or almost decoupled) segments of each waveform to also be allocated to separate processors. Natural divisions, such as different clock phases or different instructions in a microprocessor simulation, as well as more complex, path-based decompositions, will be explored here. Such an approach is also quite natural for the waveform relaxation method, and it is possible that the two approaches may merge into a unified model on the multiprocessor.

Circuit decomposition. The first step in either case is to decompose the circuit to be simulated into loosely coupled subcircuits. This step is not only crucial for the parallel implementation of the algorithm but also for its sequential implementation.

This two-pronged approach accommodates both designers concerned with silicon area usage as well as designers with stringent testing requirements. For the test pattern propagation, we are using our experience with fast testability analysis algorithms introduced in Victor[32,33] with the help of the logic manipulation techniques used in Espresso-IIC.

In the past four years, effort has been placed on testability analysis of analog circuits. The problem of testing analog circuits is generally much more difficult than the one of testing digital circuits because of the many possible faults that may occur. We have discovered a number of techniques for the fault diagnosis of analog circuits especially in the nonlinear time-domain case,[34-36] an open question for a number of years. We are now in the process of transferring our theoretical results to practice to provide an effective tool for the testability evaluation of a wide class of analog circuits. This effort parallels the one driven to the development of an automatic synthesis tool for analog circuits.

Parametric optimization, statistical design, and analog synthesis

When high-performance cells or analog circuits are designed, the circuit designer generally uses a two-phase iterative process: first, an initial circuit configuration is selected and then the values of circuit parameters (e.g., resistor and capacitor values, and device geometries such as bipolar transistor area and MOSFETs lengths and widths) are determined that satisfy a set of specifications to optimize a set of possibly competing design objectives. This entire process is repeated until a satisfactory design has been achieved. The most creative part of the design process is, in general, the selection of the circuit topology. For large, nonlinear circuits, the selection of values of the design parameters is often time-consuming and is usually stopped short of even a local optimum of the design objectives. This is due to the fact that the specifications and objectives may depend on ac, dc, and transient responses, which in turn depend on many design parameters. Thus, it is usually difficult for designers to predict the effect of parameter changes on circuit performance without numerous circuit simulations.

Delight_Spice optimization. Over the past five years we have developed Delight_Spice,[11] the result of the union of the Delight interactive optimization-based, computer-aided design system developed in collaboration with Elijah Polak, and the

---

If a tightly coupled section of a circuit were to be mistakenly broken up into two subcircuits, it will take many iterations to converge when the relocation algorithm is applied. And if the partitioning is too conservative, the subcircuits will be very large and the advantages of decomposition will be lost.

The problem of partitioning a large system into smaller pieces occurs frequently in engineering, and there are good heuristics for solving a variety of partitioning problems. Most of these existing techniques examine the structure of the problem. For example, there are a variety of techniques for decomposing systems so that the number of interactions across subsystem boundaries is minimized. However, the objective of these techniques does not quite match the problem of finding the decomposition that leads to the fastest relaxation convergence. This is because it is not the number of connections but the strength of the interactions among the blocks that is important. We are developing a new algorithm that is particularly suited to partitioning large MOS digital circuits for parallel implementation of relaxation methods. The algorithm is based on estimating the strength of interactions and partitioning so that the interactions across subcircuit boundaries are weak.

Layout and fault simulation. Based on our experience with the circuit simulation problem, we are exploring the relationships between algorithms and multiprocessor architectures for other CAD algorithms. We are researching the block placement problem for both boards and ICs, and the fault simulation problem for large, digital systems, in particular. Our experience with placement and global routing based on the simulated annealing approach for standard cells has been excellent.[5]

The basic problem with these techniques is the long running time needed to approach the ideal conditions under which convergence to the global optimum with probability one is guaranteed. To cope with the complexity, we believe that parallel processing is certainly needed. The challenge in devising parallel versions of the simulated annealing algorithm is to preserve the statistical properties of the sequential version. To do this, a clever partitioning of the problem must be devised. Since the computational cost of a move is low, we believe that machines with a very large number of processors of limited complexity such as the connection machine developed at MIT could be used effectively in this area.

Fault simulation is another example of a computationally intensive activity that continues to be necessary for the verification of complex, digital systems. In concurrent fault simulation, an event-driven selective-trace approach is used to guide the analysis, as is used in the ITA circuit simulation method mentioned above. However, rather than evaluating the current and voltages associated with a circuit, each event in concurrent fault simulation concerns the updating of a variety of faulty machines. We believe that this approach, with some modification, will make an excellent starting point for the development of new fault simulation techniques for use on multiprocessors.

References

1. C. L. Seitz, "Concurrent VLSI Architecture," *IEEE Trans. Comp.*, Vol. C-33, No. 12, Dec. 1984, pp. 1247-1265.
2. A. Vladimiresu, "LSI Simulation on Vector Computers," Electronics Research Laboratory, Memo No. UCB/ERL M82/75, University of California, Berkeley, Oct. 1982.
3. E. Cohen, "Performance Limits of Integrated Circuit Simulation on a Dedicated Minicomputer System," Electronics Research Laboratory, Memo No. UCB/ERL M81/29, University of California, Berkeley, May 1981.
4. H. F. Ko and A. Sangiovanni-Vincentelli, "BLOSSOM: An Algorithm and Architecture for the Solution of Large-Scale Linear Systems," *Proc. Int'l Conf. Comp. Design*, New York, Oct. 1983.
5. C. Sechen and A. L. Sangiovanni-Vincentelli, "The Timberwolf Placement and Routing Package," *Proc. 1984 Custom Integrated Circuit Conf.*, May 1984.

Spice circuit analysis program. Optimization performed by Delight_Spice frees the designer from the task of selecting values of the design parameters, making it possible for him to concentrate on more creative aspects of the design process. Delight_Spice uses new, powerful optimization algorithms and is heavily based on interaction. Using Delight_Spice to optimize industrial analog circuits, improvements of over 50 percent are often observed in the objective functions defined for circuits which were considered state-of-the-art manual designs. Drawbacks with the present version of the systems are:

- the long running time needed to complete a reasonable number of iterations especially when cost or constraints are expressed in the time domain; and
- the need for the user to understand the algorithms and the problem formulation supported by Delight_Spice.

Present work is centered around improving the efficiency of the program as well as the user interface.

The goal of the new user interface to Delight_Spice is to take advantage of the uniform user-program environment available through the Hawk graphics editor and the Squid data management system. A menu-driven input to Delight_Spice will reduce the apparent complexity of the many possible actions from which a designer can choose at different states in the optimization process.

Our design methodology based on parametric optimization is at an advanced state. However, this methodology has been developed thus far assuming that desired and realized values of design parameters are one and the same. This can be viewed as deterministic modeling of the realized values in which they are identically equal to their desired values. In the real world of integrated circuit processing, however, dimensions are scaled down and other economies are pursued until violation of this assumption becomes a major limit to cost-performance progress. In the area of statistical design, our goal is to extend this methodology so that this assumption is unnecessary. In particular, we intend to extend it to the case in which actual realized parameter values are modeled probabilistically.

Extension to statistical design. There have been a number of mathematically distinct formulations of optimization problems intended to deal with the problem. All of them are aimed at enlarging the set of cost-versus-performance points attainable through design, as opposed to simply maximizing some measure of performance. They include only some of the cost effects and ignore variations in expected revenues from the product. The first major objective of our research is to attempt a new formulation in which the manufacturer conducts parametric design with the goal of maximizing the economic gain attributable to the product.

Using probabilistic models, we expect to be able to express expected defect yield in terms of the same parameters as for the parametric yield, since the defect yield can be related to the total device area. It is known that scaling up all device dimensions increases parametric yield but decreases defect yield. Yet no existing formulation addresses this conflict.

Our emphasis is placed on economic modeling. First, we are studying the feasibility of developing a first-order model relating price, demand, and the values of at least some salient subset of the specifications of integrated circuit products. In this case, the term model means only the form of a set of mathematical relationships which would potentially apply to a wide range of types of products, particularly analog circuits. Numerical values within the model would depend on the product being designed and would be determined by the users of our methodology.

The development of a production cost model is under way. This model consists of the sum of all additive contributions to cost which depend either deterministically or stochastically on a set of continuous design parameters, be they electrical fabrication process parameters, plus a constant cost covering all contributions not dependent on those parameters. The sophistication of the model exceeds that of any model published heretofore. On the other hand, the goal is not to develop a detailed accounting of cost for all the cost-reduction purposes a manufacturer might have, but just to capture the essence of the manufacturer's cost considerations as they pertain to the formulation of the appropriate optimization problem. This reformulation is expected to have a major theoretical impact on our algorithm development work.

Analog silicon compiler. In addition to this work on parametric optimization and statistical design, we are now working on a synthesis system for analog circuits. We are developing an analog silicon compiler for a particular class of circuits: switched capacitor filters. The input to the program will be a high-level functional description of the specifications of the filter. The program will go through a step where the topological configuration of the filter will be created, a parametric optimization step where the component values will be adjusted to satisfy the specifications and improve performance, and a layout step where the components will be placed and routed.

As in the logic synthesis part of our work, we will pursue two avenues to build such a system: one based on an algorithmic view of the problem, the other based on an expert system view. Once more we are very much interested in comparing the two approaches both from an experimental point of view and from a theoretical point of view.

For the parametric optimization step, we expect to use simplified models and fast approximate algorithms to provide a "quick and dirty" solution. This solution will be optionally used by the designer as the starting point for a Delight_Spice interactive optimization run. Under study for the placement of analog blocks is an algorithm based on simulated annealing. The flexibility and extensibility of simulated annealing is ideal for analog layout because of the many criteria, such as thermal considerations, symmetry, and cross-talk, that have to be followed to come up with an acceptable placement. The place and route work will form the basis for a more general analog synthesis system where many different blocks will be automatically generated. In parallel to this work, we are developing a macro-modeling methodology that will allow the simulation of large-scale analog circuits where the approximations taken will be characterized by bounds on the error of the simulation.

Summary

The primary goal of our research is to forge an advanced and flexible IC design environment, coupling design methods with state-of-the-art, IC-oriented design and verification tools for the design of advanced integrated circuits and systems. Another major aspect of our research is the relationship between computer architecture and the CAD algorithms best suited to such architectures. For our work to continue to be effective, we believe that the close, working relationship we have with our Berkeley collegues in the IC/CAD, design, technology, and architecture areas is essential. In addition, the close, working relationships we have developed with many industrial groups has provided important feedback and direction for our research.

In this overview, we have described briefly some of the research projects with which we are involved. This work would not be possible without the support of our many government and industrial sponsors. In particular, the Defense Advanced Research Projects Agency and the Semiconductor Research Corporation have provided the base of support needed for such a broad program. In addition, NSF, ARO, and JSEP have supported a number of our major projects. Without the support of the sponsors of our Berkeley CAD/CAM Center, the Berkeley Microelectronics Association, as well as our other industrial sponsors, we would not have the computing resources nor sufficient funds to support our excellent group of research students and staff.

References

1. A. R. Newton, D. O Pederson, A. L. Sangiovanni-Vincentelli, C. H. Sequin, "Design Aids for VLSI: The Berkeley Perspective," *IEEE Trans. Circ. and Syst.*, Vol. CAS-28, No. 7, July 1981.

2. K. Keller, "An Electronic Circuit CAD Framework," Electronics Research Laboratory, Memo No. UCB/ERL M84/54, University of California, Berkeley, June 1984.

3. M. W. Bales, "Layout Rule Spacing of Symbolic IC Artwork," Electronics Research Laboratory, Memo No. UCB/ERL M82/72, Department of Electrical Engineering and Computer Sciences, University of California, Berkeley, May 1982.

4. E. Lelarasmee, A. Ruehli and A. Sangiovanni-Vincentelli, "The Waveform Relaxation Method for Time Domain Analysis of Large-Scale Integrated Circuits," *IEEE Trans. CAD for IC*, July 1982, pp. 131-145.

5. J. Kleckner, R. Saleh and A. R. Newton, "Electrical Consistency in Schematic Simulation," *Proc. IEEE Int'l Conf. Circ. and Comp.*, Oct. 1984, pp. 30-34.

6. A. R. Newton, "Techniques for the Simulation of Large-Scale Integrated Circuits," *IEEE Trans. Circ. and Syst.*, Vol. CAS-26, No. 9, Sept. 1979, pp. 741-749.

7. J. E. Kleckner, "Advanced Mixed-Mode Simulation Techniques," Electronics Research Laboratory, Memo No. UCB/ERL M84/48, University of California, Berkeley, June 1984.

8. E. Cohen, "Performance Limits of Integrated Circuit Simulation on a Dedicated Minicomputer System," Electronics Research Laboratory, Memo No. UCB/ERL M81/29, University of California, Berkeley, May 1981.

9. A. Vladimirescu, "LSI Simulation on Vector Computers," Electronics Research Laboratory, Memo No. UCB/ERL M82/75, University of California, Berkeley, Oct. 22, 1982.

10. J. T. Deutsch and A. R. Newton, "A Multiprocessor Implementation of Relaxation-Based Electrical Circuit Stimulation," *Proc. ACM/IEEE 21st Design Automation Conf.*, Albuquerque, N.M., 1984.

11. W. Nye, A. Sangiovanni-Vincentelli, J. Spoto and A. L. Sangiovanni-Vincentelli, "DELIGHT_SPICE: An Optimization Based System for the Design of Integrated Circuits," *Proc. 1983 Custom Integrated Circ. Conf.*, Rochester, N.Y., May 1983.

12. C. Marino, "Smalltalk on a RISC-CMOS Implementation," MS Report, Department of Electrical Engineering and Computer Sciences, University of California, Berkeley, Feb. 1985.

13. H. F. Ko and A.Sangiovanni-Vincentelli, "BLOSSOM: An Algorithm and Architecture for the Solution of Large-Scale Linear Ssytems," *Proc. Int'l Conf. Comp. Design,* New York, Oct. 1983.

24. J. Darringer, W. Joyner, C. Berman and L. Revillyan, "Logic Synthesis through Local Transformations," *IBM Journal of Research and Development*, Vol. 25, No. 4, July 1981.

25. M. Hofmann, "A Comparison of Static and Dynamic CMOS Design Styles," internal memo, Department of Electrical Engineering and Computer Sciences, University of California, Berkeley, July 1984.

26. S. Kirkpatrick, C. D. Gelatt and M. P. Vecchi, "Optimization by Simulated Annealing," *Science*, Vol. 220, No. 4598, May 1983, pp. 671-680.

27. C. Sechen and A. L. Sangiovanni-Vincentelli, "The Timberwolf Placement and Routing Package," *Proc. 1984 Custom Integrated Circuit Conf.*, May 1984.

28. M. Lundy and A. Mees, "Convergence of the Annealing Algorithm," unpublished manuscript, Cambridge University.

29. F. Romeo and A. Sangiovanni-Vincentelli, "Probabilistic Hill-Climbing Algorithms: Properties and Applications," Electronics Research Laboratory, Memo No. UCB/ERL M84/34, University of California, Berkeley, March 1984.

30. D. Mitra, F. Romeo and A. Sangiovanni-Vincentelli, "Convergence and Finite-Time Behavior of Simulated Annealing," Electronics Research Laboratory, Memo No. UCB/ERL M85/3, University of California, Berkeley, Jan. 1985.

31. J. Reed, "YACR: Yet Another Channel Router," Electronics Research Laboratory, Memo No. UCB/ ERL M85/12, University of California, Berkeley, Feb. 1985

32. I. Ratiu, A. Sangiovanni-Vincentelli and D. O. Pederson, "VICTOR: A Fast VLSI Testability Analysis Program," *Proc. 1982 Int'l Test Conf.*, Philadelphia, Nov. 1982, pp. 397-404.

33. R. S. Wei and A. Sangiovanni-Vincentelli, "PLATPG: A PLA Test Pattern Generation Program," submitted for publication.

34. V. Visvanathan and A. L. Sangiovanni-Vincentelli, "Diagnosibility of Nonlinear Circuits and Systems, Part I: The DC Case," *Special Joint Issue of IEEE Trans. Comp. and IEEE Trans. CAS on Design for Testability*, Nov. 1981, pp. 889-898.

35. R. Saeks, A. L. Sangiovanni-Vincentelli and V. Visvanathan, "Diagnosibility of Nonlinear Circuits and Systems, Part II: Dynamical Systems," *Special Joint Issue of IEEE Trans. Comp. and IEEE Trans. CAS on Design for Testability*, Nov. 1981, pp. 898-906.

36. V. Visvanathan and A. Sangiovanni-Vincentelli, "A Computational Approach for the Diagnosibility of Dynamic Circuits," *IEEE Trans. CAD of ICAS*, Vol. CAD-3, No. 3, July 1984, pp. 165-172.

14. L. W. Nagel, "SPICE2: A Computer Program to Simulate Semiconductor Circuits," Electronics Research Laboratory, Report No. ERL-M520, University of California, Berkeley, May 1975.

15. R. S. Gyurcsik et al., "Language Comparison for Circuit Simulation on Desktop Computers," *Proc. 1984 Int'l Symp. Circ. and Syst.*, Montreal, 1984.

16. K. Mayaram and D. O. Pedersen, "Circuit Simulation in LISP," *Proc. 1984 Int'l Conf. CAD*, Santa Clara, Calif., Nov. 1984, pp. 24-26.

17. A. R. Newton and A. L. Sangiovanni-Vincentelli, "Relaxation-Based Circuit Simulation," *IEEE Trans. ED*, Vol. ED-30, No. 9, Sept. 1983, pp. 1184-1207.

18. C. Lob, "RUBICC: A Rule-Based Expert System for VLSI IC Critique," Electronics Research Laboratory, Memo No. UCB/ERL M84/80, University of California, Berkeley, Sept. 1984.

19. A. E. Dunlop, "Automatic Layout of Gate Arrays," *Proc. 1983 IEEE Int'l. Symp. Circ. and Syst.*, Newport Beach, Calif., May 1983.

20. S. Trimberger, "VITComposer—A Powerful Graphical Chip Assembly Tool," *Proc. IEEE Int'l Conf. CAD*, Santa Clara, Calif., Nov. 1984.

21. M. Buric, C. Christensen and T. Mateson, "The Plex Project: VLSI Layouts of Microcomputers Generated by a Computer Program," *Proc. IEEE Trans. CAD IC and Syst.*, 1983, pp. 49-51.

22. G. De Micheli and A. Sangiovanni-Vincentelli, "Multiple Constrained Folding of Programmable Logic Arrays: Theory and Applications," *IEEE Trans. CAD IC and Syst.*, Vol. CAD-2, No. 3, July 1983, pp. 151-167.

23. R. K. Brayton and C. T. McMullen, "The Decomposition and Factorization of Boolean Transformations," *Proc. 1982 ISCAS Symp.*, Rome, Italy, May 1982, pp. 49-54.

A Design Methodology and Computer Aids for Digital VLSI Systems

STEPHEN W. DIRECTOR, FELLOW, IEEE, ALICE C. PARKER, MEMBER, IEEE, DANIEL P. SIEWIOREK, FELLOW, IEEE, AND DONALD E. THOMAS, JR., MEMBER, IEEE

Invited Paper

***Abstract*—The current status of a research program at Carnegie-Mellon University aimed at the formulation of a hierarchical design methodology for digital VLSI circuits is described. In addition, this paper describes a set of computer aids which supports this methodology. One of the goals of this work is to provide a design environment which allows for a significant reduction in time between the initial concept of a complex digital system and the generation of masks. Another goal is to allow the designer to efficiently explore a number of design alternatives.**

Manuscript received September 20, 1980; revised February 15, 1981. This work was supported in part by the National Science Foundation under Grant ENG 78-25755 and by the U.S. Army Research Office under Grant DAAG/29/79/C/0213.

S. W. Director, D. P. Siewiorek, and D. E. Thomas, Jr. are with the Department of Electrical Engineering, Carnegie-Mellon University, Pittsburgh, PA 15213.

A. C. Parker was with the Department of Electrical Engineering, Carnegie-Mellon University, Pittsburgh, PA. She is now with the Department of Electrical Engineering, University of Southern California, Los Angeles, CA.

I. Introduction

IT HAS become generally recognized that the computer is an essential tool for designing very large scale integrated circuits. However, even with the aid of the computer, complexity of VLSI technology is on the verge of outstripping the engineer's ability to design. The root of the problem is that changes in technology have transformed integrated circuit design into digital *system* design. Hence, integrated circuit designers are now forced to consider the design of a complete digital system, not just a small component of a system which performs a single specialized function. Towards this end, research at Carnegie-Mellon University has been aimed at developing an integrated circuit design methodology which is capable of dealing with the design of a total digital VLSI based

Reprinted from *IEEE Trans. Circuits Syst.*, vol. CAS-28, no. 7, pp. 634–645, July 1981.

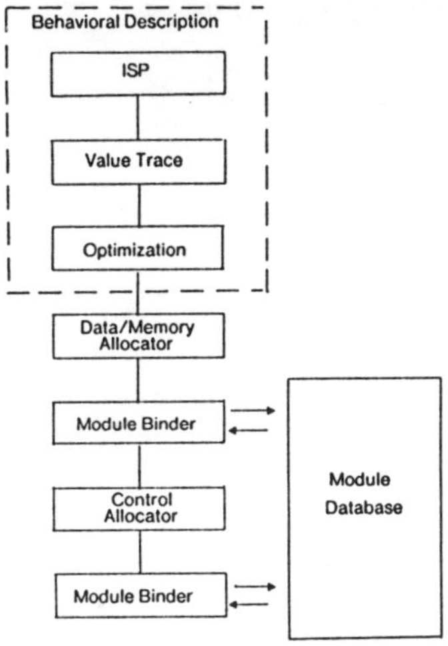

Fig. 1. The levels of design.

system, as well as developing a set of design aids which supports this methodology.

One of the principle goals behind the design aids being developed is that they be able to significantly reduce the time between the initial concept of a complex system and the generation of IC masks. Cutting design time by one to two orders of magnitude is more important than realizing designs which are superior to manually produced designs according to some metric, such as chip area. A second equally important goal is to allow the designer to efficiently explore design alternatives. Since design aids can consider a large number of design tradeoffs per unit time, it is even conceivable that for sufficiently complex designs the computer aids could help produce designs superior to those produced manually.

The CMU design methodology, described more fully in Section II is *hierarchical.* Complexity is handled by describing a design at higher levels of abstraction which eliminates unnecessary detail, particularly during the early phases of a design. In addition to taking a hierarchical approach, this methodology employs the concept of a *design space*. Given a set of metrics which characterize the quality of a design, the design space is defined as a Euclidean space whose coordinate axes correspond to these metrics. Alternative designs are represented by points in the design space. Through the study of a design space, the designer can efficiently direct his use of the design aids to produce better designs.

II. A Hierarchical Approach to Design

The basic design methodology which is proposed for VLSI circuit implementations of digital systems is shown in Fig. 1. This methodology is a blending of design techniques which have proven useful for digital system design with those that have been successful in integrated circuit design.

A. High-Level Behavioral Description

As indicated in Fig. 1, the design process begins with a behavioral specification of the digital system to be designed. Such a specification provides a model which accurately characterizes the input–output behavior of the system without reflecting any internal structure. This level of specification might be expressed in terms of a flowchart, or as we choose to do below, in terms of a high-level hardware description language.

Transformations of the formal description can be considered as a means of producing a more efficient hardware implementation. The philosophy behind this step is similar to that used in compiler design. Since concise expression of an algorithm in a high-level programming language does not usually lead to optimal machine code, compilers include a large collection of heuristics which can be used to improve the efficiency of the generated code. It has been found that many of these heuristics are best applied to the source program before translation, since the original structure of the program and the interactions of its various parts are still clear at that level [1], [2]. Similar heuristics can be used to "optimize" the formal behavioral description of the digital system being designed.

B. Synthesis

The next step in the design process is to synthesize from the behavioral description a structure in terms of physical modules which will implement the required behavior. In a digital system this step can be partitioned into two separate synthesis problems: the synthesis of the system data paths and the synthesis of the controller which will evoke the operations in the data paths. In general these two parts of the system are tightly coupled and their design should proceed concurrently. However, in order to make the design process tractable and provide us with a chance to gain a deeper understanding of the synthesis procedure, we perform the data path synthesis first followed by the controller synthesis.

B.1 Data Path Synthesis:

Synthesis of the data path is separated into two steps:

1) Generation of a register-transfer level structure of the data path, called the data path graph.

2) The specification of the interconnection of physical modules which implement the data path functions.

The first step is referred to as *data path allocation* while the second step is referred to as *module binding*. The result of the second step is a *bound path graph*.

The data path allocator employs various algorithms and heuristics for generating the register-transfer structure. Depending on which algorithms and heuristics are chosen, different *design styles* will result. Possible design styles include:

1) A *distributed style* which is characterized by duplicate hardware for arithmetic and logic operations, a low frequency of register sharing, and many independent paths between registers and operators with multiplexers to control resource sharing.

2) A *bus style* which is characterized by a high degree of register and logic operation sharing, typically with functional units tied to one common data path.

3) A *pipeline style* which is characterized by a high

frequency of register and data path usage allocated to sustain constant flow of data.

The characteristics of designs built in these styles are different as are the algorithms and heuristics needed to allocate the data path. Other design styles can also be identified [3].

The output of the data path allocator is a register-transfer level description of the necessary data path. The module binding step specifies an interconnection of physical blocks contained in a *module database* to implement the data path logic. The module binder is the first step where detailed module set information is available; thus it is important that the module binder be technology relative. That is, it should be responsive to new modules and module sets and be capable of selecting appropriate physical modules from the new (or old) entries in the database. In addition, the module binder should be capable of:

1) specifying several modules to implement a node;

2) combining several nodes for implementation by a single module;

3) synthesizing all logic produced by an allocator even though a module set may not have a specific module to match;

4) specifying the control requirements for the bound modules. This information is passed to the control allocator.

B.2 Control Synthesis:

As in the case with data path synthesis, control synthesis is separated into two steps:

1) *Control allocation* which takes the description of the digital hardware data path and description of the desired behavior and produces a description of a control engine which evokes the data path devices in an order consistent with the behavioral description.

2) *Module binding* in which one obtains an implementation of the specified control engine in terms of the physical modules contained in the module database mentioned above.

Just as the knowledge and heuristics embodied in the data path allocator determined the resulting hardware style, there are several styles which a control allocator could reflect. The styles are all variations of the synchronous sequential-state machine controller and include:

1) *The microcode style*. This style encodes the next-state information and the data path select and evoke signals in a read-only memory which is addressed by the present state address.

2) *The PLA style*. This style generates the next-state, data path select and evoke signals from a present state register and a programmed logic array.

In general the inputs to the control allocator include:

1) A description of each of the devices which will be used to implement the data paths of the digital system. This description contains a name for the particular device, the operations which the device may perform, a concise description of how the particular device may be controlled for each operation, and how these devices are interconnected.

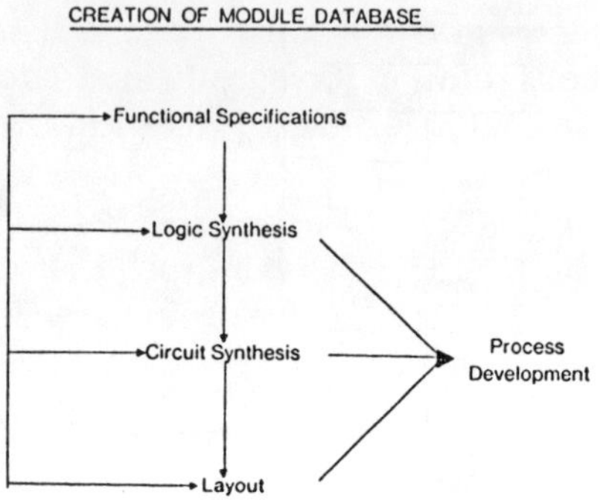

Fig. 2. Steps needed to create module database.

2) A description of the control sequence. This control sequence is contained in the microsequence table, which is part of the path graph. This table contains register-transfer instructions and control-flow directives. Each of the register-transfer instructions specifies exactly which devices in the data path graph should be used for an operation, and what operation each should perform. The micro-sequence program also includes the control directives in the form of conditional and unconditional branch statements. Each of these is used to specify control for the controller itself.

Under the microcode style, the allocator must then determine the microinstruction format, control store wordwidth and length, clustering of microoperations into microinstructions, and controller hardware parameters.

The second step of control allocation is module binding which is the same as module binding for the data path.

C. Module Database

The module database contains information regarding each of the modular units that may be used in order to realize either the data path or the control part of the digital system. This database could either be fixed, i.e., defined once and for all such that any system realization must use the modules which exist in the current database, or else it can be extendable, in which case the designer is allowed to create new modules if he so desires.

Each module needs to be described on a variety of levels, the highest level being the *functional level* in which the register-transfer operation of the module is specified. An example of a functional block is an adder. The next lower level, or *logic level*, is the one at which the module is described in terms of logical units such as flip-flops and gates. Below the logic level is the *circuit level* in which the module is described in terms of electrical components such as transistors and capacitors. Finally, the lowest level is the *physical layout level* in which the module is described in terms of the mask dimensions.

Creation of a module database involves a number of synthesis and optimization steps as indicated by the block diagram shown in Fig. 2. We use the term synthesis here to mean the generation of the next lower level of detail while optimization implies the generation of alternative designs that might better conform to the designer's objectives.

As indicated above, new entries may have to be created in the module database. These new modules may be suggested by human intuition or by usage statistics gathered during synthesis activities. Frequently these circuits have to meet multiple conflicting constraints such as area, speed, power, and yield.

Once an interconnected network of modules has been created, it must be translated into a physical set of IC masks. Activities at this level include placement of modules, routing of interconnects between modules, compacting the geometric shapes to occupy minimal area, and design rule checking.

BEHAVIORAL DESCRIPTION

```
Mini :=
      Begin
      **Memory.State**
      Mp\Primary.Memory[0:255]<0:11>,

      **Processor.State**

      PC\Program.Counter<0:11>,
      Acc\Accumulator<0:11>,
      IR\Instruction.Register<0:11>,
         Op\Operation<0:2>:= IR<0:2>,
         Ibit\Indirect.Bit<> := IR<3>,
         Adr\Address<0:7> := IR<4:11>,

      **Effective.Address.Calculation**

      Z\Effective.Address<0:7>:=
         Begin
         Decode Ibit =>
            Begin
            0 := Z ← Adr,
            1 := Begin
                 If Adr Eql 0 => Z ← Mp[0] ← Mp[0] + 1;
                 If Adr Neq 0 => Z ← Mp[Adr]
                 End
            End
      End,


**Instruction.Cycle**

IExec\Instruction.Execution:=
   Begin
   Decode Op =>
      Begin
      0\AND:= Acc ← Acc And Mp[Z()],
            1\TAD:= Acc ← Acc + Mp[Z()],     ! 2's Complement Add
            2\ISZ:=                          ! Increment and Skip if Zero
                  Begin
                  Mp[Z] ← Mp[Z()] + 1 Next
                  If Mp[Z] Eql 0 => PC ← PC + 1
                  End,
            3\DCA:=
                  Mp[Z()]@Acc ← Acc@#0000, ! Deposit and Clear Acc
            4\JSR:=                              ! Jump to Subroutine
                  Begin
                  Mp[0] ← Mp[0] + 1 Next
                  Mp[Mp[0]] ← PC Next
                  PC ← Z()
                  End,
            5\JMP:= PC ← Z(),                              ! Jump
            6\RET:=                        ! Return from subroutine
                   Begin
                   PC ← Mp[Mp[0]] Next
                   Mp[0] ← Mp[0] - 1
                   End,
            7\CTL:=
                   Begin
                   If IR<3> => PC ← PC + 1 Next
                   If IR<4> => Acc ← Not Acc Next
                   If IR<5> => Acc ← Acc + 1 Next
                   Decode IR<6:7> =>
                          Begin
                          '10   := Acc ← Acc SR0 1,
                          '01   := Acc ← Acc SL0 1,
                          Otherwise := No.Op()
                          End Next
                   If IR<8> => If Acc Lss 0 => PC ← Pc + 1;
                   If IR<9> => If Acc Eql 0 => PC ← PC + 1;
                   If IR<10> => If Acc Gtr 0 => PC ← PC + 1 Next
                   If IR<11> => Stop()
                   End
      End
   End,


                ICycle\Interpretation.Cycle:=
                      Begin
                      Repeat
                            Begin
                            IR ← Mp[PC] Next
                            PC ← PC + 1 Next
                            IExec()
                            End
                      End
                End
```

Fig. 3. The ISP description of a PDP-8 like computer.

III. Computer Aids Which Support Hierarchical Design

In this section we describe design aids which either have been, or are currently being, developed to support the design approach described in Section II.

Presently, each program operates in a stand-alone mode. The programs typically provide for user interaction in the decision process as well as providing an automatic mode which can be viewed as a semi-intelligent "surrogate" designer. By interacting with these design programs, we are better able to understand a level of design. The automatic mode allows for the generation of multiple designs that populate a design space (i.e., different design points represent tradeoffs between cost and performance). By studying the resulting design space, we can develop heuristics for improving on the automatic mode of operation.

Currently, each program interfaces to the others through an ASCII database. Although an ASCII representation does not provide optimal information density, it is useful for portability purposes. Moreover, this technique allows for the development of a program that uses a data structure before the development of the program that produces the data structure. Specifically, the data structure may be generated with a standard text editor.

A. Description at the Algorithm Level

The design methodology described in Section II begins with a behavioral specification of the system. We have chosen the ISP (for *I*nstruction *S*et *P*rocessor)[1] language to provide this behavioral specification.

A.1 The ISP Hardware Description Language and Simulator:

The ISP language was initially introduced to describe digital computers [4], but has since been augmented so that it can now be used to describe the functional behavior of any digital system. While a detailed discussion of ISP is beyond the scope of this paper (see [5]) we illustrate it by example.

Fig. 3 is the ISP description of a PDP-8 like minicomputer. This description has three major format divisions: Memory declarations, Formats and Operations, and Instruction Definitions and Interpreter. *Memory declarations* include a description of the main memory (Memory State), the processor registers (Processor State), the interface with the external world (External State) and the implementation variables. Memory usually consists of a two-dimensional array of bit storage elements. The first dimension, delineated by square brackets, gives the number of memory words and the second dimension, offset by angle brackets, the number of bits per word. In the example, the primary memory (Mp) consists of 4096 words (number 0 through 4095), each of 12 bits (numbered, left to right, 0 through 11). The processor state consists of a program counter (PC), an accumulator (ACC), and an instruction register (IR). All are 12 bits in length. The IR is an implementation variable that is included for notational convenience. The

[1]ISP is a generic name for a family of languages. The first implementation, ISPL has been superceded by an improved implementation, ISPS. Throughout the remainder of the paper we will use the generic term ISP and do not make a distinction between the specific dialects.

external state (e.g., lights, switches, interrupts, etc.) is not depicted for simplicity.

The second major division, *Formats and Operators*, contains information about the instruction format, address calculation, and service facilities. In the example, the instruction format is defined in terms of the IR. There is a three bit Operation Code (Op), a bit that indicates indirect addressing (Ibit), and an eight-bit address (Adr). The section "Effective Address Calculation" gives the algorithm for calculating the final eight-bit address (Z) that addresses memory. If there is no indirection (Ibit=0), Adr is taken as the value of Z. If there is indirection (Ibit=1) and Adr is not zero, the effective address is contained in the memory location pointed to by Adr. Indirection through memory location zero uses the contents of location zero as Z followed by autoincrementation. The autoincrement Addressing mode is used for stepping through an array of data. Service facilities (e.g., handling interrupt, exception, and console conditions) are omitted to simplify the example.

The last major division, the *Instruction Set Definition and Interpreter*, is usually the largest. In the example, the instruction set is defined in the Instruction Execution (IExec) section. The three-bit operation code (Op) is used to select the appropriate instruction algorithm. For example, if Op is 1, the contents of the accumulator is added to the contents of the memory location pointed to after the effective address (Z) calculation. Finally, the instruction interpreter section (ICycle) summarizes the operation of the computer. Assume a binary representation of a program and its data has been entered into Mp (e.g., corresponding to initial conditions in a circuit simulator). The ICycle indicates that the memory word pointed to by the program counter is fetched to the instruction register, the program counter is incremented, and the instruction set definition function (IExec) called. The cycle repeats indefinitely, IExec selects the appropriate instruction function based on the value of the operation code (a subfield of the instruction register). Instruction Functions, in turn, may invoke the effective address calculation function.

The behavior of the desired digital system can be simulated by using the ISP level of description. The simulator which exists for this purpose allows a designer to measure the effect of algorithm level features such as the trade offs of including certain instructions and addressing modes in the design of a processor [7]. The simulator also includes a fault insertion mode to evalute the specification's behavior under fault conditions. Results of the simulation may be used to determine if modification in the specification is required.

A.2 The Value Trace:

While the ISP language is useful for describing the desired behavior of a system, such a description implies a certain data flow, through the use of assignment statements, and a certain control flow, through the use of loops, conditionals and procedures. These flows can impact the resultant cost/performance characteristics of the design. (In this context, performance can be equated with speed.)

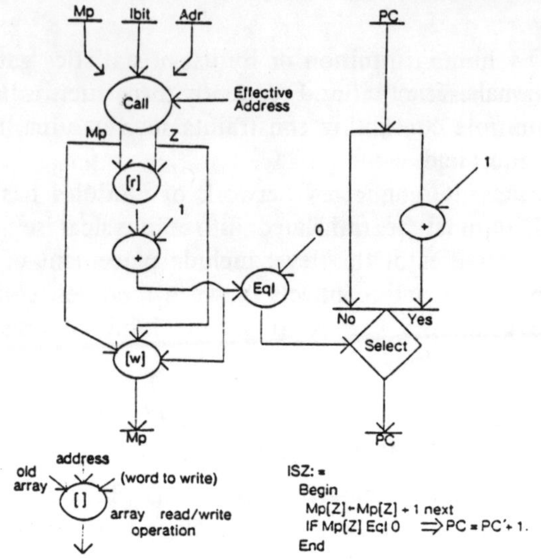

Fig. 4. The VT body for the ISZ instruction of the PDP-8 computer.

In order to remove the bias which is inherent in an ISP description, the ISP description is transformed into an alternative representation called the Value Trace or VT [8], [9]. A Value Trace is a connected graph made up of a series of blocks, called VT-bodies [9]. Each VT-body represents a control environment in the ISP description which has a single entry point. These environments include procedure bodies, labeled blocks, and loop bodies. VT-bodies are composed of a list of inputs (i.e., all those ISP variables changed in the VT-body), and a data flow graph, which shows the actions performed by the body. Each node in the data-flow graph represents one action, such as an addition. There is an arc from a node *A* to a node *B* if and only if a *value* produced by *A* is used as an input to *B*.

It is important to note that all notions of a "variable" are removed from the value trace when it is generated, since the arcs between nodes represent "values." The point here is that a variable implies the storage of a value in a physical memory device, and a state transition to load the memory. However, a value does not imply storage, hence the designer (or computer aid) is free to decide which values should be assigned to which memory devices. More specifically, a VT optimizer can be developed which is free to temporarily ignore variables declared in the ISP description, and to ignore the designers ordering of operations (although values must still be generated before they are used).

The flexibility provided by the VT allows a designer the freedom to produce alternate representations of a system. By exploring the behavioral level of design, a designer can optimize according to criteria such as cost, speed, or data flow partitioning.

An example of the value trace for the ISZ instruction in our PDP-8 minicomputer is shown in Fig. 4. The effective address calculation procedure is called. Within the context of ISZ, the procedure call is just another action which uses Mp, Ibit, and Adr and produces (potentially) a new value for Mp (e.g., due to side effects of the procedure execution) and Z. The data flow of ISZ can be analyzed without specific reference to the actions within the Effective Address procedure.

Fig. 4 also illustrates the ability of the VT to handle conditionals. The "If" and "Decode" constructs in ISP are replaced by a "select" in the VT. As its name implies, the Select picks one of the Parallel data flow paths as its value.

Since the VT deals with values, not variables, the value "Mp[Z( )]+1" (which has no specific name in the ISP description due to its appearance within an expression) is used both as an input to the array write ([w]) and equal (Eql) operations. Once created, the value of "Mp[Z( )]+1" exists for all time to be accessed by whatever operator is appropriate.

The VTGEN program, which translates a compiled ISP description into a VT format, has been written. Once generated, the VT may be transformed by the VTOPTIMIZE program discussed in the next section.

A.3 The VT Optimizer:

Once a Value Trace has been generated from an ISP description, we can consider various transformations on it for the purpose of realizing a more efficient hardware implementation. The VTOPTIMIZE program uses heuristics similar in intent to those employed by optimizing compilers. Specific heuristics include:

1) *Dead Code Elimination*, whereby statements which do not effect any other part of the program are dropped.

2) *Constant Propagation*, whereby references to a variable may be replaced by references to the constant until the variable is redefined.

3) *Redundant Subexpression Elimination*, whereby the results of the first instance of a computation are stored in a temporary variable and saved for later use, avoiding the extra steps for the recomputation.

4) *Code Motion*, whereby an operation which is invariant in a loop may be moved outside the loop so that it will be executed less frequently.

In addition to the above transformations which are aimed at reducing redundant and unnecessary operations, VTOPTIMIZE can be used to alter the parallel/serial nature of the design, even to the extent of transforming the design into a pipeline implementation. Such transformations allow the designer to consider a variety of alternate implementations.

Transformations which are used by VTOPTIMIZE to increase speed include:

1) transformation to a maximally parallel VT;

2) replacement of identical computations by a single computation and extra storage;

3) replacement of procedures by "in-line" code to remove procedure-calling overhead.

Alternatively, VTOPTIMIZE can decrease the cost of a design by performing transformations on the Value Trace that serialize the data flow, allowing subsequent allocators to share hardware. Specifically, VTOPTIMIZE can perform the following manipulations:

1) Serialization of operations.

2) Transformation of n-way branches into log(n) two-way branches.

3) Replacement of common sequences with subroutines.

Note that the manipulations described above for speed and cost improvement, as well as for elimination of inefficiencies, involve changes in the control structure of the "program." Thus transformations at this level of design can have sweeping effects on the final implementation of the digital system. A program has been written that generates a Value Trace from an ISP description. The various efficiency, speed, and cost transforms have been implemented. Applications of these transforms is currently done in an interactive mode. The automatic, goal-oriented application of transformations is a current research topic. A formal semantic definition of ISP and the Value Trace is almost complete. This work will allow us to verify that the global optimizations preserve desired behavior.

B. Synthesis

The behavioral description as represented by the Value Trace is an abstract set of data operations and values, and a sequence in which the values are produced. Synthesis is the process of choosing functional units which, when structurally interconnected, will implement the required behavior. Specifically, synthesis involves making a number of decisions which include:

1) The assignment of operations to operators. This process involves two related decisions: the *number* of operators required for each operation type and the choice of *which* logical operator is assigned to perform each operation specified in the behavior.

2) The assignment of values to registers and memory. This allocator function should allow for the possibility of two values using the same register in different time periods.

3) Specification of interconnections to route data between registers and operators. This step encompasses the construction of multiplexers at register and operator inputs and gating at bused outputs to control the movement of data.

4) The addition of registers (such as memory-address and memory-data registers) implicit in the behavioral description, and deletion of registers for storage of variables/values which are temporary.

5) The synthesis of a control unit and associated control signals to evoke the operations specified by the behavioral description.

As discussed earlier, we partition synthesis into two separate problems: synthesis of the system data paths, and synthesis of the controller which will evoke the operations in the data paths. Each of these synthesis steps is further subdivided into an allocation step and a module binding step. We now describe the programs which have been developed to implement these tasks.

B.1 The Data Path Allocator (DPA):

The data path allocator, or DPA, takes the output from the Value Trace Optimizer, VTOPTIMIZE, and generates a register-transfer level structure of the data paths, called a data path graph. There have been numerous papers that describe approaches to synthesis [2], [10]–[13]. The approach which we have taken (which is based on the register-transfer module allocator of Barbacci [14] and the allocation algorithm developed by Rege [15]) is different than some of the others in that it does not assume a fixed technology or module set. The allocator might be compared to a two-pass compilation in which the first pass is a syntax or feasibility check. From a Value Trace the allocator constructs data structures analogous in function to symbol tables, and enforces constraints necessary to ensure that the data-storage locations, logical mappings, and input/output interface characteristics specified in the description can be implemented in hardware. If no errors are encountered, an allocation is made for the basic data-storage structures called for in the description, and any additional data paths, storage, and operators necessary to implement the variable-accessing schemes described by the ISP. The second pass is the semantic phase, with activity of code generation replaced by the allocation of data paths, operators, and storage.

While the allocator described above accepts the Value Trace as input, at present the allocator which has been developed [16] uses an ISP description as input. The allocator employs the following heuristics during the first pass:

1) If a variable has been declared in the description or inserted as a temporary register earlier in the allocation process, a register of that bit width is created.

2) If an array has been declared in the ISP description, a memory of the correct size is allocated, along with a memory address register and memory data register of the required bit widths.

3) If a virtual array or virtual register has been declared and mapped onto an array in the ISP description, it could have different dimensions from that array. If it does, the extra hardware necessary to access and store virtual words into the actual memory is designed.

During the second pass the following heuristics are employed:

1) For each occurrence of an operation in the ISP description, an operator is allocated.

2) Operators are linked to the storage for source variables which feed them.

Allocation is then completed by adding of multiplexors where required, i.e., if an operator source or register has more than one input, the inputs are gated by means of a multiplexer.

A highly interactive allocator which accepts the Value Trace as input is under construction. This interaction will allow, for example, the designer to constrain a particular variable to a particular register and operations on that variable to a particular logic unit. In addition, any part of a designed circuit may be redesigned, without perturbing the remaining hardware, thus permitting iterations on a particular design. A bus style data path allocator is also under development.

B.2. The Control Allocator:

Synthesis of control hardware can begin when the data paths of the system under design have been fully defined, device primitive operation times are known and sequences of control signals necessary for each primitive operation are derived. An automated control allocator takes a description of some digital hardware data path (the data path graph) and a procedural description of the desired behavior (the microsequence table) and produces a description of a control engine which evokes the data path devices in the order specified by the behavior description.

A control allocation [37] program has been developed which maps each member of the microsequence table into one or more basic data path operations (device primitives). Combinations of the primitive device operations are then used to define microoperations. Once this has been done for each control sequence step, the result is a microoperation (or control) graph at a level of detail sufficient for the completion of the control allocation process. The problem is then one of determining the form of the controller.

The allocator selects a particular controller configuration by using parameters measured from the control graph. Once selected, the controller hardware is added to the path graph description. The control allocator then evaluates the potential for combining microoperations into microinstructions using data and hardware dependencies as a guide. After the microinstructions are formed, the size requirements for the microstorage are known and this information is placed in the data path graph description. The final step of the control allocation process is the conversion of the microinstructions into bit patterns which will be stored in the microprogram ROM. This step is fairly simple and is similar to the assembling of a machine language program.

The method which the control allocator uses to combine microoperations into microinstructions is based upon *attraction weights* [17]. These weights are based on the probability that two microoperations could be performed in parallel. The attraction weights are calculated for every combination of microoperations throughout the control graph and the pair with the highest weight is placed in the same microinstruction first. It is then necessary to recalculate the potential combinations and attraction weights since the action of combining two may restrict the range of other microoperations in the control graph. When the new attractions weights are known the pair with the highest value are combined and the process is repeated until there are no more potential combinations or the size of the microword is too wide. If the width is the limiting factor, then the routine creates a new microinstruction format, modifies the format-selection field and continues the process. Using this procedure, the microinstructions will contain collections of microoperations which could have occurred in parallel the most often throughout the complete control graph.

In any given microinstruction, only a subset of the data path's control signals are active. Thus control words can be compacted. The canonical controller uses a bit steering (format selection) field in a control word to determine the functions that other bits in the control word must perform. That is, a bit steering field "steers" other control word bits to their proper destinations in the data paths. Decoders outside the microstore "steer" the evoke control signals to the correct place. (Some signals do not need to be steered and are simply overlaid, since they do not cause a change of state in an inactive part of the digital system.)

If all the control signals of a particular design are partitioned into disjoint subsets such that no member of one subset is ever active in the same microinstruction as any member of any other subset, then a bit-steering field can select the active subset and the control word needs to be only as wide as the largest subset plus the width of the bit-steering field.

In this degenerate case we have the conventional vertical or horizontal microprogrammed control word. If all control signals can be active at the same time, the bit steering field is 0-bits wide and a conventional horizontal microprogrammed word results. If the activation of signals is mutually exclusive, the bit steering field is $\log n$ (n = total evoke control signals) wide and the partition has n blocks with one control signal in each.

B.3 The Module Binder:

The outputs of the data path allocator and control allocator are a logic-structural description of the necessary data paths and a sequence table of operations needed to evoke the appropriate register transfers. The module binder is a program which searches a database for physically realizable modules and binds these to the logic-structural description (path graph). The output of this program is a specification of the interconnection of the physical modules which will implement the data path functions.

The module binder steps through the path graph one node at a time. At each node, the module database is searched for candidate modules to implement the function of the node. A list of all modules which can be used to perform this function is produced. Each alternative in this list is then evaluated in terms of various criteria. Presently used criteria include: dollar costs, chip area, power dissipation, and propagation delay. This list could be expanded to include other considerations such as geometry.

Obviously, this selection process takes a narrow view of the optimization process in that internode dependencies are not considered. Future research will include the study of more global approaches for an optimization strategy.

If no modules are available to perform the requested function, the module binder can propose alternate, but equivalent, functions. These alternate functions which are derived from a set of axioms [18], [19] are specified in terms of a BNF-like notation [20].

The application of these axioms provides a dimension of extra capability to the binding process although they also increase the complexity of the search process. In fact it is interesting to note that these axioms could be used to design a system using only NAND gates. One useful aspect of these axioms is to log their use and provide feedback to the standard module (module set) designer on which functions would be appropriate to add to the cell library. This feedback on module set utilization would lead to major improvements in the overall design of a system.

Once candidate modules have been chosen, the next step of the module binding phase is the manipulation phase of the path graph to reflect the particular choice of several modules and their interconnection. The control sequence table is modified to include any extra control steps or signals needed to evoke the newly bound modules.

C. Physical Modules

The module database contains the physical blocks which are used by the modular binder to realize the data path graph and control graph. Each block must be describable in terms of entities such as registers, PLA's, or gates. These units in turn must be realizable in terms of mask geometries. Computer aids which are needed to develop a module database include logic and circuit synthesis and optimization programs, automated and/or interactive layout aids, functional level as well as logic and circuit level simulators, design rule checkers, schematic extractors, process simulators, etc. While there has been much effort devoted towards developing such computer aids, the work at CMU has been aimed at circuit level synthesis and optimization, mixed level simulation, and geometric mask manipulation.

C.1 Circuit Synthesis and Optimization:

We can view circuit level synthesis as the realization of some specified logic function by a set of electrical components such as transistors, resistors, and capacitors. The resulting circuit, in addition to realizing a certain specified logic function, may be required to meet a variety of other constraints such as a limit on the power dissipation, a limit on propagation delay, and/or a limit on chip area. During the past dozen years or so, various mathematical programming methods have been employed to determine an appropriate set of values for the designable parameters of a circuit (e.g., mask geometries) to meet a given set of design specifications. For the most part these methods have met with limited success because they do not address one of the central problems of design, namely, trying to meet multiple specifications which often compete with one another. While the existence of competing design objectives has been discussed in literature, it has received little attention by circuit designers. In our work we examine the use of multiple criterion optimization (MCO) for electronic circuit design.

If some (usually all) of the criteria are competing, there will be no point that simultaneously minimizes all of them i.e., there is no "optimal solution." Instead of optimality, the concept of noninferiority [21], [22] is used to characterize a solution to the MCO problem. In general there are an infinite number of noninferior points for a given MCO problem. The collection of noninferior points is the *noninferior set*. The image of the noninferior set is called the *noninferior solution set* or, the *noninferior* or *tradeoff surface*. Fig. 5 illustrates these concepts. Observe that a noninferior point is the same as the intuitive notion of an optimum tradeoff solution (design). That is, a design is noninferior if an improvement in any objective requires a degradation in at least one of the other objectives. Clearly, if a designer was able, he would not want to choose an inferior design. Thus the problem of trying to generate

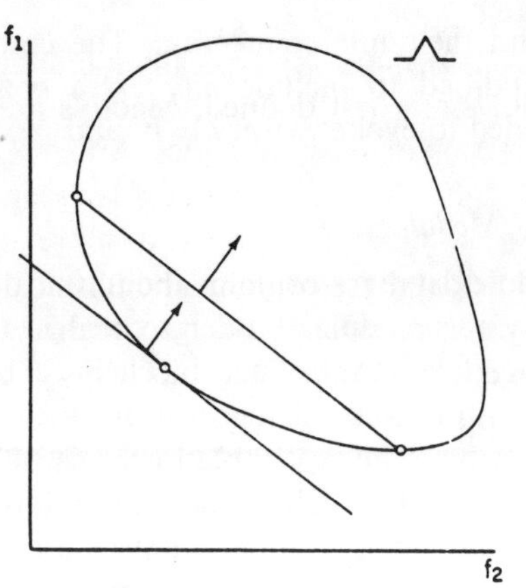

Fig. 5. A noninferior surface for a two-criterion optimizing problem.

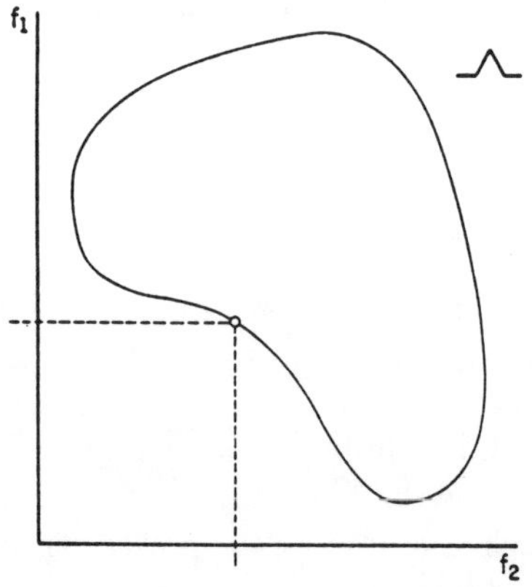

Fig. 6. A nonconvex, noninferior surface.

noninferior solutions to an MCO problem is the problem the designer faces when trying to reach a final design.

One of the most widely used techniques for solving an MCO problem is the *weighted sum* method in which a scalar objective function problem is formed by adding a weighted combination of the individual objective functions together. This scalar optimization problem is then solved. This difficulty with the weighted sum technique arises when the lower boundary of the noninferior surface is not convex, as illustrated in Fig. 6, because not every noninferior solution is also a solution of this scalar optimization problem.

In order to determine all noninferior solutions, regardless of the convexity of the noninferior surface, we have developed a *weighted max-norm* technique. Furthermore, an interactive scheme for weight selection has been proposed and good results have been obtained [22].[2]

In addition to designing modules at the circuit level so as to best meet a given set of specifications, of prime concern is that the yield be high. The yield, which is usually much less than 100 percent for VLSI circuits, is defined as the ratio of the number of manufactured circuits which meet the desired performance objectives to the total number of manufactured circuits. In order to maximize the yield, the circuit designer needs to take into consideration the fluctuations that are inherent in the manufacturing process. Our work in this area has been to take into account these fluctuations and modify design strategies so as to be able to simultaneously maximize yield while meeting the design objectives. It is interesting to note that in many situations there is a tradeoff between the performance characteristics and the statistical characteristics of a circuit. Such a situation may arise because oftentimes the extremes of circuit behavior, e.g., low-power dissipation or fast switching times, occur at the extreme limits of some parameter values which makes the design very sensitive to parameter variations—just what statistical design tries to avoid. Therefore, a natural extension of the MCO of electronic circuits would be to include an objective function which reflects statistical design criteria in the formulation of the MCO design problem. Towards this end a *nominal value* of the designable parameters is specified and the *feasible* or *acceptable region* in the parameter space is defined.

Preliminary work has shown that this formulation of the design problem is reasonable [23], [24]. However, further work in this area is still needed.

In order to relate the variations in the electrical behavior of a given circuit to fluctuations that are inherent in the manufacturing process, one has to be able to model the fabrication process itself. Towards this end a process simulation strategy is being developed which can be tailored to a specific manufacturing process which can be used to relate fluctuations in electrical parameters to the fluctuations in manufacturing parameters. More specifically, we are incorporating into the statistical design procedure the process simulator FABRICS (*FABR*ication Process of Bipolar *I*ntegrated *C*ircuits *S*imulator) [25], [26].

FABRICS generates samples of the electrical parameters which are typical for bipolar IC elements (i.e., samples of random variable X, e.g., resistances of the resistors, current gain coefficients and Early voltages of n-p-n and p-n-p transistors, etc.) and in-line parameters (e.g., sheet resistances and junction depths of diffusion layers). The input parameters to FABRICS are of two types: constants (e.g., times and temperatures of the thermal operations, element dimensions in the photolithographic masks) and random variables representing the main disturbances in the process and simulated by random number generators (RNG). This simulator accounts for fluctuations in the thickness and resistivity of the epitaxial layer, variation of the boron and phosphorus diffusivities due to crystal defects and thermal stresses, enhancement of boron diffusivity caused by the phosphorus diffusion, segregation of impurity atoms on the boundary of Si and SiO_2, inaccuracy in photolithographic process and imperfections of Si surface which are modeled by surface recombination properties. The process models allow for the calculation of impurity profile parameters in terms of times and temperatures of all thermal operations.

[2]At present we have not yet employed the MCO heuristics to design at levels above the circuit level since these heuristics depend on bounding the space with designs that optimize each constraint. However, this is an area of further research.

C.2 Multilevel Simulation:

We are currently engaged in the development of a mixed-level simulator known as SAMSON (*S*ystem for *A*ctivity-directed *M*ixed *S*imulation of *N*etworks). This simulator is based on the principle of the exclusive simulation of activity (or ESA) [27]. This concept is used in most logic simulation algorithms where activity is measured with a Boolean metric: the presence or absence of logic events. Activity at the circuit-level is identified with the rates of change of network variables, and is gauged by the sizes of the integration time steps selected to maintain truncation errors within prescribed limits. The new simulator views an electrical network as a collection of interconnected subnetworks each of which can be described on the logic or circuit level. Those described on the circuit level are allowed to have their own integration formula order and step size. At any instant of time during the course of the simulation circuit level subnetworks are classified as either dormant (low activity) or alert (highly active) based on the relative magnitudes of their step sizes. Dormant subnetworks participate in the simulation as passive entities whose outputs are extrapolated and propagated to permit the solution of the alert subnetworks, and whose inputs are monitored for significant changes that might call for their activation. Regardless of the modeling level, ESA attempts to exploit the temporal sparseness of large networks in order to reduce simulation time. ESA based mixed simulation, therefore, allows a designer to apply the most appropriate models to different parts of a network, and to achieve sufficiently accurate simulation results at a reasonable cost.

The input phase of the mixed simulator, SAMSON, is now working. It has facilities for defining circuit-level and logic-level models in terms of a Pascal-like language, and for using such models in the construction of an interconnected set of subnetworks which is fed, afterwords, to the simulation phase. Testing of the simulation phase is currently underway, and preliminary results are encouraging.

C.3 Geometric Mask Manipulation:

The geometry level can be viewed as an interconnection of objects that has to be physically laid out on two or three dimensions while satisfying a set of spacing constraints known as design rules. The optimization function, of course, is to minimize the total layout area.

An automatic mask compaction program, called SLIM [28] has been developed that takes a design at the circuit level, uses the relative positioning information of the circuit elements to create a "legal" initial placement, and iteratively removes inter-element space. Space is removed by identifying compaction ridges. A compaction ridge extends completely through a mask. Space between compaction ridges can be removed if no design rules are violated by the removal.

Other work supports manually laid out designs. Under development is a hierarchical design rule checker (DRC) [29]. Mask descriptions are typically hierarchical in nature. Basic modules are first defined. Macro's are subsequently defined in terms of basic modules. Much like the parameterized macros in a programming language, these mask macros can be used to specify a more complex mask without the designer worrying about the detailed implementation of the macro. Design rule checking is also hierarchical. First the macro is checked for violations. Subsequently, only the spacing between macros needs to be checked, not the spacing between every mask feature.

The theory of computational complexity is also playing a significant role at the geometric level. Transformations of the design rule checking problem into other problems with known efficient solutions have been sought. Bentley [30] has implemented a DRC which looks for intersections of two-dimensional polygons. Intersection detection can be efficiently implemented as a polygon-dimension sort followed by a test. The result is an algorithm requiring $N \log_2 N$ run time where N is the number of polygons. Further efficiency is being sought by optimizing the algorithm according to the expected statistical behavior of IC masks [31].

Finally, significant progress has been made in mapping of special purpose algorithms directly onto regular two dimensional arrays [38]. Data, such as matrices to be multiplied is fed to the array several entries at a time. As the data makes its way through the array, it meets and is combined with other data. The partial results leave the array, several at a time. Data flows through the array, journeying to another cell each clock tick. A methodology has been developed for taking algorithms, expressible as recurrence equations (e.g., equations of the form $f_i = F(f_{i-1}, f_{i-2}, \cdots, f_0)$ directly into systolic array designs.

IV. Experimental Results

To this juncture several of the design aids described above have been tested to ensure proper operation and to determine the quality of the algorithms. This section briefly outlines some of these results.

The ISP compiler and simulator were used in the selection of an instruction set for the Military Computer Family [32]. Factors of two in architectural efficiency were measured.

The distributed Data Path Allocator was used to produce a path graph for the PDP-8. TTL 7400 series modules were bound to the data path using an early version of the module binder. The resultant design was 30-percent larger in gate count than the commercial PDP-8/E [33] with over an order of magnitude savings in designer's time. Given that this was the first sizable design processed by the design system, and that the design system had no optimizing capabilities at the time, the results were quite encouraging.

Output of the distributed Data Path Allocator was fed to a translator (i.e., a one-to-one binding table, not the more general module binder) for binding into Sandia CMOS Standard Cells. The MP2D cell placement and channel router [34] was used to produce an implementable chip. A design for a Voter (used to tie triplicated LSI-11's together into a fault tolerant system [35]) was automatically synthesized and compared to a hand design fabricated by Sandia Laboratories. The automatically synthesized design was within 20 percent in cell area of the manual design [36], again with a significant savings in design time.

When given the data path graph and microsequence table (normalized to remove interrupt and service functions) for the PDP-11/40, the microprogrammed allocator synthesized one design that exactly matched the parallelism of the commercial design while requiring 14 percent more bits in the control word [17].

The automatic module binder was tested against a set of four human designers on two separate designs, each using a different module set: the TTL 7400 series of modules, and the Sandia Standard Cells. The module binder's surrogate designer varied from 6 percent better than the four designer average to 23 percent worse than average on cost and 10 percent better to 38 percent worse on delay [19]. Over 64 designs were generated by using the module binder in a automatic mode in one eight-hour work day. The module binder has also been used to demonstrate the redesigning of old designs with new module sets.

The weighted max-norm technique for finding the trade-off surface for multiple criterion optimization problems has been used to design some logic gates such as NAND gate [22]. Based on these results it seems that while the application of MCO ideas is still in its early stages, the results are promising.

In summary, substantial progress has been made in producing design aids at all levels in the heirarchy from behavioral to geometry. The challenge remains to integrate the various aids into a completely automated design system that has access to fabrication facilities.

References

[1] C. M. Geschke, *Global Program Optimization*. Ph.D. dissertation, Dep. of Computer Science, Carnegie-Mellon Univ., Pittsburgh, PA, Oct. 1972.

[2] G. Huet and B. Lang, "Proving and applying program Transformations Expressed with second-order patterns," *Acta Informatica*, vol. 11(1), pp. 31–55, Jan. 1978.

[3] D. E. Thomas and D. P. Siewiorek, "Measuring designer performance to verify design automation systems," *IEEE Trans. Comput.*, vol. C-30, pp. 48–60, Jan. 1981.

[4] C. G. Bell and A. Newell, *Computer Structures: Readings and Examples*. New York: McGraw-Hill, 1971.

[5] M. R. Barbacci, "Instruction set processor specifications (ISPS): The notation and its applications," *IEEE Trans. Comput.*, vol. C-30, pp. 24–40, Jan. 1981.

[6] M. R. Barbacci, G. E. Barnes, R. G. Cattell, and D. P. Siewiorek, "The ISPS computer description language," Tech. Rep., Computer Science Dep., Carnegie-Mellon Univ., Pittsburgh, PA, 1977.

[7] W. E. Burr, et al., "Overview of the military computer family architecture selection," in *Proc. AFIPS Nat. Comp. Conf.*, vol. 46, 1977.

[8] E. A. Snow, D. P. Siewiorek, and D. E. Thomas, "A technology relative computer-aided design system: Abstract representations, transformations, and design tradeoffs," in *Proc. 15th Annual Design Automation Conf.*, pp. 220–226, June 1978.

[9] M. C. McFarland, "Global transformations on abstract hardware descriptions: A formal approach," Internal Carnegie-Mellon Univ. Rep., Oct. 1979.

[10] Y. Chu, "CDC description and simulation of logic design," Tech. Rep. TR-418, Dep. of Computer Science, University of Maryland, Oct. 1975.

[11] G. Zimmerman, "The MIMOLA design system: A computer-aided digital processor design method," in *Proc. IEEE 16th Annual Design Automation Conf.*, 1979.

[12] J. A. Darringer and W. H. Joyner, Jr., "A new look at logic synthesis," in *IEEE Proc. 17th Annual Design Automation Conf.*, 1980.

[13] G. S. Miranker, *The use of conflict in the translation and optimization of hardware descriptive languages*, Ph.D. dissertation, Dep. Electrical Engineering and Computer Science, Massachusetts Institute of Technology, May 1979.

[14] M. R. Barbacci, "Automated exploration of the design space for register-transfer (RT) systems," Ph.D. dissertation, Dep. of Computer Science, Carnegie-Mellon University, Pittsburgh, PA, Nov. 1973.

[15] S. Rege, "Designing variable data format digital modules with cost speed tradeoffs," Ph.D. Dep. of Electrical Engineering, Carnegie-Mellon University, Pittsburgh, PA, Aug. 1974.

[16] L. J. Hafer and A. C. Parker, "Register-transfer level automatic digital design: The allocation process," in *Proc. IEEE 15th Annual Design Automation Conf.*, vol. 15, 1978.

[17] A. Nagle, "Heuristic optimization of microcontrollers," in *Proc. IEEE 18th Ann. Design Automation Conf.*, vol. 18, 1981.

[18] D. E. Thomas and G. Leive, "A technology relative design system," in *Proc. IEEE Int. Conf. Circuits and Computers*, New York, 1980.

[19] G. Leive, "The Design, implementation, and analysis of an automated logic synthesis and module selection system," Ph.D. dissertation, Dep. of Electrical Engineering, Carnegie-Mellon University, Pittsburgh, PA, Jan. 1981.

[20] P. Naur, *Concise Survey of Computer Methods*. New York: Petrocelli Books, 1974.

[21] D. L. Fraser and S. W. Director, "Multiple objective function optimization of digital MOSFET circuits," in *Proc. 1978 Int. Symp. on Circuits and Systems*, pp. 1114–1115, Apr. 1978.

[22] M. R. Lightner and S. W. Director, "Multiple criterion optimization for the design of electronic circuits," *IEEE Trans. Circuits Syst.*, vol. CAS-28, pp. 169–179, Mar. 1981.

[23] R. K. Brayton, S. W. Director, and G. D. Hachtel, "Yield maximization and worst case design with arbitrary statistical distributions," *IEEE Trans. Circuits Syst.* vol. CAS-27, pp. 756–764, Sept. 1980.

[24] M. R. Lightner and S. W. Director, "Multiple criterion optimization with yield maximization," *IEEE Trans. Circuits Syst.*, vol. CAS-28, to be published.

[25] W. Maly and A. J. Strojwas, "Simulation of bipolar elements for statistical circuit design," *Proc. 1979 Int. Symp. Circuits Syst.*, pp. 788–791, May 1979.

[26] W. Maly, A. J. Strojwas, and S. W. Director, "Fabrication based statistical circuit design of monolithic IC's," in *Proc. Int. Symp. on Circuits and Systems*, Apr. 1981.

[27] K. Sakallah and S. W. Director, "An activity directed circuit simulation algorithm," in *Proc. IEEE Conf. Circuits and Computers*, pp. 1032–1035, Oct. 1980.

[28] A. E. Dunlop, "SLIM—The translation of symbolic layouts into mask data," in *Proc. IEEE 17th Annual Design Automation Conference*, 1980.

[29] R. Hon, "The hierarchical analysis of VLSI designs," Ph.D. dissertation, Proposal, Computer Science Dep. Carnegie-Mellon Univ., Pittsburgh, PA, Dec. 1980.

[30] J. L. Bentley and T. Ottman, "Algorithms for reporting and counting geometric intersections," *IEEE Trans. Comput.*, vol. C-28, pp. 643–647, Sept. 1979.

[31] J. L. Bentley, D. Haken, and R. Hon, "Statistics on VLSI designs," Tech. rep. Computer Science Dep., Carnegie-Mellon Univ., Pittsburgh, PA.

[32] M. R. Barbacci and D. P. Siewiorek, "Evaluation of the CFA test programs via formal computer descriptions," *Computer Magazine*, vol. 10, Oct. 1977.

[33] A. C. Parker, D. E. Thomas, D. P. Siewiorek, M. Barbacci, L. Hafer, G. Leive, and J. Kim, "The CMU design automation system: An example of automated data path design," in *Proc. 16th Annual Design Automation Conf.*, pp. 73–80, June 1979.

[34] A. Feller, "Automatic layout of low-cost, quick-turnaround, random-logic custom LSI devices," in *Proc. IEEE 13th Annual Design Automation Conf.*, June 1976.

[35] D. P. Siewiorek and V. Kini, "Reliability in multiprocessor systems: A case study of C.mmp, Cm*, and C.vmp," Tech. Rep., Carnegie-Mellon Univ., Pittsburgh, PA, 1978.

[36] J. H. Kim and D. P. Siewiorek, "Issues on IC Implementation of High Level, Abstract Designs," in *Proc. IEEE 17th Ann. Design Automation Conf.*, June 1980.

[37] R. J. Cloutier, "Control allocation: The automated design of digital controllers," Design Research Center Tech. Rep. DRC-01-05-80, Carnegie-Mellon Univ., Pittsburgh, PA, Apr. 1980.

[38] H. T. Kung and C. E. Leiserson, "Systolic arrays for VLSI," in *Introduction to VLSI Systems*, (C. A. Mead and L. A. Conway, eds.), Reading, MA: Addison-Wesley, 1980.

Integrated CAD, CAM, and CAT of VLSI Circuits and Systems: The CMU Perspective

Stephen W. Director, Wojciech Maly, Rob A. Rutenbar, John P. Shen, Daniel P. Siewiorek, Andrzej J. Strojwas, and Donald E. Thomas, Carnegie-Mellon University

The primary goal of the Semiconductor Research Corporation-Carnegie Mellon University Research Center for Computer-Aided Design is to carry out a comprehensive research program that will result in the development of an integrated methodology and associated computer aids for the design, manufacture, and test of VLSI circuits and systems. Currently the research center has 10 faculty members associated with it. In addition to the authors, they include Randall E. Bryant, Zolton Cendes, Sarosh Talukdar, and Robert G. Wedig. There are also about 40 graduate students pursuing research in the center as well as a number of visiting researchers.

Approximately half of the $2.5 million operating budget for the center comes from the Semiconductor Research Corporation. The remaining funds, and technical guidance, come from the National Science Foundation, the Army Research Office, AT&T Bell Labs, Digital Equipment Corporation, General Electric, General Motors, GTE, Harris Semiconductor, Hewlett-Packard, IBM, Intel, ITT, Northern Telecom, Phillips, Raytheon, and United Technologies.

Three disciplines are involved in the realization of a VLSI circuit: design, manufacture, and test as Figure 1 shows. Each of these disciplines involves a number of activities and requires the use of various computer aids. Because computer-aided design, computer-aided manufacture, and computer-aided test traditionally have been viewed as separate disciplines, the aids that have been developed for them are by and large unrelated and noninteractive, as shown in Figure 2.

In our view, it is essential that an integrated approach be used for the design and implementation of computer aids for these three disciplines. Figure 3 represents our view of the ideal integrated CAD, CAM, and CAT world in terms of a comprehensive set of generic tasks and computer aids. The relative placement of the blocks shown in the figure roughtly indicate the relationships and interactions between the various tasks. We may view each of the three disciplines as a subset of these tasks. Clearly the three subsets are not disjoint.

The SRC-CMU Research Center for CAD has a broad-based program aimed at developing an integrated approach to CAD, CAM and CAT. Special emphasis has been placed on the investigation of AI techniques. A number of tools have already resulted from these activities. Table 1 shows these tools in relationship to the task they perform and the level at which they perform it. We anticipate the release of a number of additional tools in the near future as well as an environment in which these tools will communicate with each other.

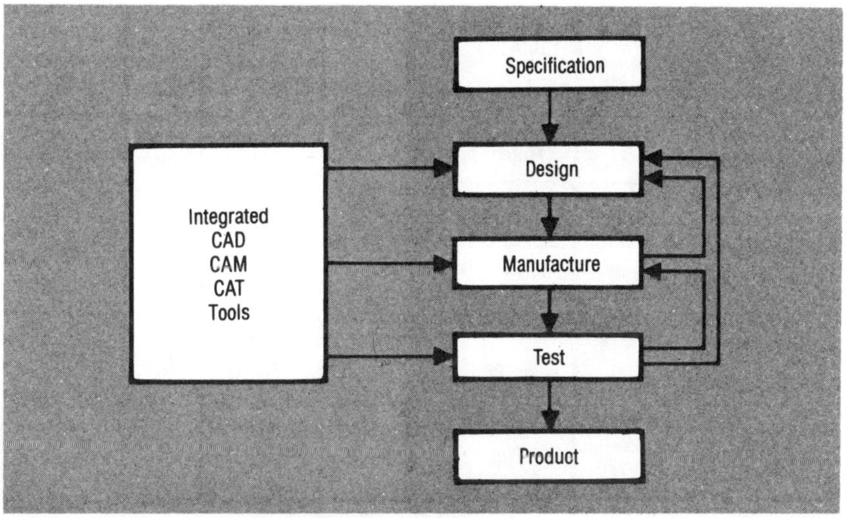

Figure 1. VLSI design, manufacture, and test.

Reprinted from *IEEE Design and Test of Computers*, pp. 87–100, June 1985.

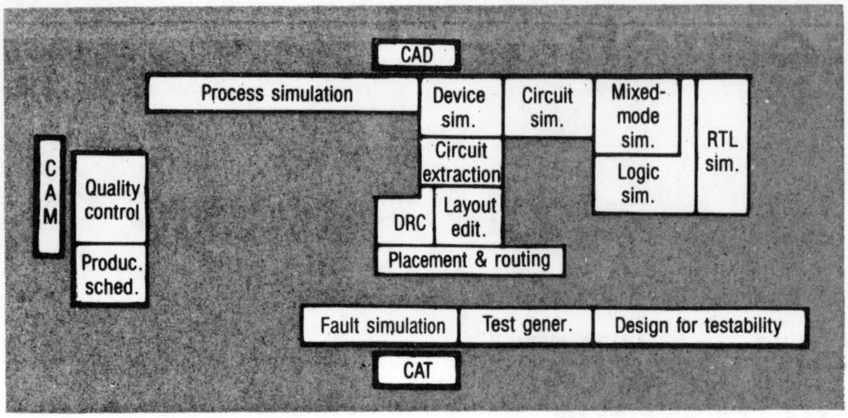

Figure 2. Traditional view of CAD, CAM, and CAT. Individual boxes indicate the types of tools, or activities, usually employed in the areas of design, manufacture, and test.

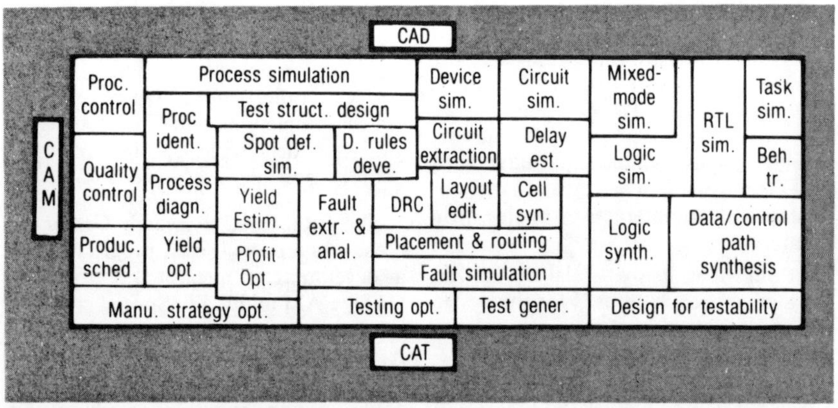

Figure 3. An integrated view of CAD, CAM, and CAT showing the interrelationships of tools and activities.

Level	Synth.	Opt.	Sim.	Test	Man.
Task	Demeter				
Architecture	Micon				
Behavioral	DAA Facet Emucs		Coral		
Functional	Moby		MOSSym	IFA	
Logic			Wasim Primo Samson	IFA	
Transistor	Talib				
Mask	Mason Weaver		Fred	IFA	Vlasic
Process	Ped	Promise	Fabrics Ped		Prod

Table 1. Computer aids developed at CMU.

Computer-aided design

The goal of our CAD research is to develop a design automation system that involves a design methodology and a supporting set of computer aids that can take a behavioral description of the desired VLSI circuit, a description of the target technology, and a set of constraints, and then produce a set of masks that could be used to fabricate the circuit. The design produced should be testable and comparable in cost, speed, and reliability with manually produced designs. Availability of such a design automation system would significantly reduce the time required to design a VLSI system as well as produce designs which are correct the first time and manufacturable with an acceptable yield.

While the stated goal is admirable, we recognize that such an ideal design automation system is not likely to be realized in the near future. However, it is realistic to assume that an interactive design automation system in which the human designer plays an active role can be developed within the next five years.

The illustration on the next page identifies our research activites, which can be categorized around several main thrusts.

- An investigation of knowledge-based expert systems as an aid in synthesis and analysis. (See box on page 90 for a description of expert systems.)
- Development of a consistent multilevel representation for our programs to communicate and for multilevel analysis aids such as multilevel simulation.
- Development of top-down synthesis methods at the system and physical design levels that can take advantage of bottom-up information to arrive at an implementable solution.
- Development of simulation tools to span all levels of the hierarchy.
- Development of special-purpose hardware for CAD tasks.

These research thrusts are being applied across the hierarchy of VLSI design. We view the design hierarchy to be composed of eight levels:

Task level. At this level a designer is concerned with the conception of a new system in terms of broad description of the product and its position in the marketplace.

Architecture-definition level. At this level the system is described in terms of the main building blocks such as processors, controllers, and buses.

Behavioral level. At this level the system is described in terms of its functionality in an implementation-independent manner.

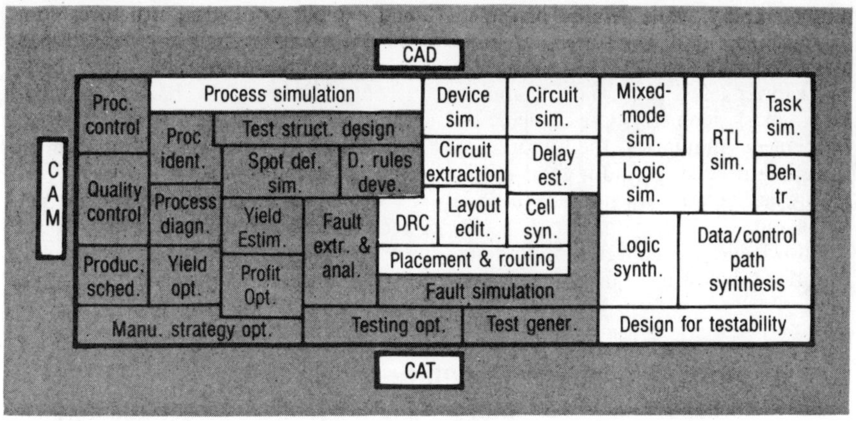

Tasks associated with computer-aided design.

Functional-structure level. At this level the system is described in terms of data paths, involving abstract components such as registers, multiplexers, and such as ALUs and control sequences.

Logic-structure level. At this level the system is described in terms of a detailed interconnection of logic components.

Transistor or switch level. At this level the system is described in terms of individual devices.

Mask-layout level. At this level the system is described in terms of geometry and other physical characteristics of the design.

Process level. At this level the fabrication process needed to manufacture the system, or chip, is described.

A given system is represented differently at each of these levels. In order to tie these representations together, we have developed the Design Interchange Format.[1] DIF is a port-oriented, hierarchical representation for describing the logical structure of a design. Behavioral and physical attributes of the design are characterized through attribute fields in the structural representation. For instance, the implementation-independent behavioral description used by the synthesis tools described next is linked directly into DIF. A connection is formed between the system architecture and the logical and physical implementation. This is a major step beyond other interchange formats such as EDIF.[2]

Synthesis

A synthesis tool takes a description of a desired system at one level in the design hierarchy and generates a description at the next lower level in the hierarchy. A number of synthesis tools have been developed or are under development now.

Behavioral-to-structural synthesis. Recently, we completed work on two tools that synthesize functional-structure level representations from a behavioral-level description: DAA[3,4] and Facet.[5] The behavioral description was in the form of a programming language (ISPS[6]) that did not explicitly describe the hardware nor the specific control sequence of the design. The synthesis tools were free to come up with alternate implementations at the functional structure level.

A connection is formed between the system architecture and the implementation.

The Design Automation Assistant (DAA) employs a knowledge-based expert system to perform synthesis. The rule base, which consists of 314 rules, was developed by asking industrial designers to critique the designs of a 6502 microprocessor produced by DAA. After the rules were developed, the behavioral description of an IBM-370 was given to DAA and the resulting design was critiqued as being comparable to an above-average design by an engineer at IBM. Since then, work on DAA has continued at AT&T Bell Laboratories.[4] This work represents one of the first uses of an expert system as a VLSI design, rather than analysis, tool.

Facet employs a graph-theoretic-based algorithmic approach to the same synthesis task as DAA. More specifically, the assignment of behavioral-level operations (the operations specified on the various lines of the behavior description) to structural-level operators (such as ALUs and adders) was achieved through the solution of a clique-partitioning problem.

In some cases, the algorithmic approach produced designs with fewer components than the expert-system approach. However, component count is not the only means for determining the quality of a design. In fact, the designs produced by the expert-system approach were found to be acceptable by industrial standards. Since the quality of a data path is a complex quantity to measure, we expect that future synthesis tools will employ both algorithmic and expert-system approaches, combining the best aspects of both.

As a follow-up to the DAA project, we have been working on Sugar,[7] an expert system that organizes design decisions differently than DAA does. At any given point in the design process, DAA makes the best possible choice it can, even if it does not have a complete set of information on which to base a decision. On the other hand, Sugar recognizes when complete information is not available and delays the decision-making as long as possible. Although Sugar is not complete and cannot yet be compared to other synthesis tools, preliminary results (published in the 1985 Design Automation Conference proceedings) shown how the delayed binding of decisions can improve the control sequence of a design, increasing its performance.

With the above experience, we have begun developing a behavioral-level workstation. All the currently marketed CAD workstations require the designer to draw a structural diagram of the system to be designed. In the behavioral-level workstation, the designer enters a behavioral description of a desired system and is given the capability to transform it. He can thus investigate various control and data-path implementations of a design.

Architecture-definition level synthesis. Micon[8] is an expert system to design single-board computers. The objective was to provide a workbench that allows users to design a customized single-board computer from a high-level specification in a short time. The designs produced by Micon required approximately 25 percent fewer chips than commercial designs—due primarily to the fact that Micon could choose from a larger set of VLSI support chips than existed when the commercial designs were built.

Physical design

Research in physical design includes both interactive and automatic design aids for layout synthesis and analysis. Interactive tools allow the designer to rapidly explore good design alternatives. Automatic layout tools synthesize various pieces of final masks; these tools can be invoked explicitly by designers or by larger automated synthesis systems.

Floor planning. During the floor-planning task, the chip-level topology of modules and signal flow is determined. Mason[9] is a tool we have developed to support interactive floor-planning. Mason employs new partitioning heuristics to permit individual modules to have several allowable shapes, and to account for the global orientation of these modules. This tool quickly produces a set of floor plans satisfying designer-specified chip boundary constraints. Conventional partitioning schemes do not adequately account for variable modules shapes or orientations. By considering trade-offs among cut-set size, partition size, module shape, and global orientation, Mason can synthesize better, more globally optimum floor plans.

Data-path layout synthesis. Synthesis tools like DAA reduce a behavioral description to a structural description of a data path. A necessary next step is to transform this structural description into a physical layout. We are currently developing a tool to move between the structural and layout levels, thus completing the lower levels of the overall chip synthesis task. Data-path layout synthesis proceeds in three steps: clustering of related operations into separate data paths; placement of functions and internal buses along each data path; and layout of final silicon based on a library of procedurally defined data-path cells. Unlike simple approaches employing a single, fixed-topology data path, this method can produce more flexible data-path layouts.

Knowledge-based cell synthesis and routing. Two physical design tools, Talib[10] and Weaver,[11] illustrate the applicability of expert-system techniques to layout tasks. Despite using a rule-based approach, each implements a substantially different problem-solving architecture.

Talib synthesizes cells consisting of up to 100 transistors. From a netlist of transistors and gates, and boundary constraints on cell size and signal locations, Talib generates a complete NMOS layout in CIF. Talib attacks the layout task by using artificial intelligence planning techniques in a hierarchy of abstraction levels. For example, circuit clustering and area partitioning are activities associated with higher levels in the planning hierarchy, while detailed placement of transistors and wires occur at lower levels. High-level plans are refined toward lower levels, eventually leading to actual mask geometry. Use of plans and refinement minimizes the amount of searching needed to find a good cell topology in the large design space of possible layouts.

Knowledge is organized around design situations, allowing Talib to recognize and exploit topological structures common in many layouts. Its rule base contains more than 2000 rules; Talib typically outperforms novice IC designers, and routinely produces practical IC layouts with comparable or better area than expert human designers do.

Weaver is an expert system that performs detailed routing of channels and switchboxes. This system employs a blackboard architecture that integrates

Exploiting artificial intelligence in CAD

Perhaps the most widely employed AI approach centers on knowledge-based expert systems. Such systems codify knowledge about a particular domain into a set of IF-THEN actions called production rules. Rule-based systems have been particularly successful for CAD applications where decisions must be made in the face of complex and conflicting constraints, where planning must occur at several levels of detail, where several areas of expertise must be integrated, or where an optimal solution is not necessary, but where a solution that avoids making obvious mistakes is all that is required.

During the knowledge acquisition phase that begins an expert-system project, it is not sufficient to produce a set of unrelated rules, each of which captures some small portion of knowledge about some specific CAD task. Rather, emphasis must also be placed on how this domain knowledge will be organized to gain leverage in design tasks. While we have produced several useful expert-system tools, we have also studied the appropriateness of several problem-solving architectures for these rule-based tools.

Many domains. It is not uncommon to find two rule-based tools with substantially different architectures attacking what our intuition might indicate are very similar CAD tasks. Nor is it uncommon to find two tools with very similar problems solving architectures attacking very different CAD tasks. For example, Talib uses planning in a hierarchy of abstraction spaces to synthesize NMOS cell layouts. Sugar, operating at higher levels of the design hierarchy, also uses planning techniques to synthesize data-path architectures from abstract behavior descriptions. However, Sugar delays commitment to a particular data-path component so it can gather more evidence on the best decision.

This approach contrasts to that used in DAA, which also synthesizes data-path architectures, but opportunistically makes decisions based on bottom-up technology-sensitive information stored in a database. Weaver, a detailed channel and switchbox router operating at the mask level like Talib, integrates several different routing experts to account for conflicting routing constraints.

Ulysses[1] is not envisioned as a single CAD tool but as a complete rule-based VLSI design environment; its domain of expertise is how to choose and apply particular CAD tools in the correct sequence to get from a high-level hardware description to a final layout. Ulysses, like Weaver, integrates several communicating experts operating at different design levels.

Micon designs complete single-board microcomputer systems by elaborating abstract templates for individual microprocessors, memory and I/O subsystems. ArchGen is a tool under development that generalizes this approach to the task level, permitting designers to interactively synthesize complete single-chip architectures. Both tools exploit the fact that designers tend to incrementally improve upon known structures.

Algorithmic tools also used. Although many of these tools are entirely rule-based, not all our AI-based tools are written exclusively as rule-based expert systems. Some expert-system tools also use several communicating routing experts. Algorithmic routers usually concentrate on optimizing only a single metric such as wire length or congestion. Weaver employs several rule-based experts in areas such as interconnection length, congestion, and via usage, and allows them to communicate among themselves across a conceptual blackboard visible to all. This integration of several routing approaches enables the system to produce results that simultaneously satisfy many critical routing requirements. One indication of the success of this approach is that Weaver has routed switchboxes that were unroutable by previous algorithmic means.

Simulation

Simulators developed at CMU span most of the levels in the design hierarchy. One trend in the design of CMU's simulators is the use of abstractions. Abstractions are classical mechanisms for controlling detail. Abstraction can define a new level which approximates the details of a lower level but with a simulation cost of a higher level. An example is Wasim,[12] a waveform simulator that allows device and circuit properties at the cell level.

Another trend is the use of hierarchy. Hierarchy can be used within a single design level by taking advantage of the natural hierarchy of the system under design. The hierarchical statistical simulator achieves several orders of magnitude performance increase over a traditional flat simulator. Hierarchy can also be used to mix two or more design levels. Samson[13] bridges the circuit and logic levels with more than two orders of magnitude performance increase over a traditional circuit simulator.

A third trend is the increased use of statistical techniques. Fabrics[14] uses statistics to model process variations. The hierarchical statistical simulator summarizes detailed behavior with a probability distribution function. A fourth trend is the use of symbolic techniques. MOSSym[15] produces Boolean equations from switch-level circuit descriptions. Equivalence testing can prove design correctness with several orders of magnitude performance improvement over traditional exhaustive simulation. Following are descriptions of the CMU simulators listed in the top-down order.

Hierarchical statistical simulator. At the architecture-definition level, the hierarchical statistical simulator focuses on performance prediction of hierarchically structured multiprocessors. An address trace or an analytical model is used to generate a memory reference pattern for a uniprocessor. The reference pattern is captured in a histogram which is used to produce an inverse probability density function, or IPDF. The IPDF is accessed via a uniform random number generator to produce a memory reference pattern with statistical properties close to that of the original address trace. The process can be recursively applied. For example, the processor IPDF can be used to generate an IPDF for a processor-cache pair, which can then be used to generate an IPDF for a single-bus multiprocessor composed of a large number of processor-cache pairs.

Architectural simulation. The architecture-definition level also has two simulators used to evaluate system reliability. Stars is a recursive, hierarchically structured, event-driven simulator. Component failures and repairs are scheduled in the event list according to their probabili-

algorithmic techniques. For example, Sugar, DAA, and Ulysses incorporate algorithmic subsystems written in C. Talib includes an algorithmic compaction step, but is in this case written as a rule-based system to fit more easily with the overall structure of Talib. Other tools that are almost entirely algorithmic include small rule-based subsystems. For example, the fabrication line diagnostic system Prod incorporates a very small rule-based system to aid its statistical pattern-recognition tasks.

In contrast, Cleopatra[2] is a natural-language interface to CAD circuit simulation tools that does not use a rule-based approach at all. It exploits AI parsing techniques based on frames for individual language constructs and CAD tool properties. The second phase of the Demeter project will use a frame-based knowledge representation scheme to create a network of frames storing attributes of design data and tools. The goal is to provide a uniform, extendable representation for CAD data, tools, and methodologies.

AI in CAD. Several new projects continue our thrust into the application of AI in CAD. One new project explores the synthesis of register-transfer-level designs from behavioral descriptions by integrating several knowledge sources with expertise on different implementations styles. This approach is promising because it may allow us to more readily add new design styles or modify existing styles. Another new project is studying how the template-based techniques similar to those employed by Micon and ArchGen can be extended into the new domain of analog component synthesis. This is a domain that heavily exploits known structures (such as A/D topologies) but must deal with an entirely new sets of strongly interacting performance, layout, and process constraints.

Most of our rule-based CAD tools have been written in the OPS5 programming language.[3] OPS5 is a production-rule language that can arbitrarily represent complex rules and execute them very efficiently. However, as our knowledge bases continue to increase in size, there is considerable interest in accelerating the execution of these rule-based tools. We have explored several architectural solutions to this problem, including studies that demonstrate how existing machine architectures affect the implementation of OPS5's basic rule-interpretation algorithms, studies of the speedups obtainable by designing an optimized production-rule-processor architecture in a fast technology like gallium arsenide, and studies of how multiprocessor systems can be used to accelerate OPS5-based tools.[4]

References

1. M. L. Bushnel and S. W. Director, "Ulysses: An Expert-System-Based VLSI Design Environment," *Proc. ISCAS,* June 1985.
2. T. Samad, *A Natural-Language Interface for CAD Applications,* PhD dissertation, Carnegie-Mellon University, Dept. of ECE, 1985.
3. C. L. Forgy, "OPS5 User's Manual," Research Report CMU-CS-81-135, Carnegie-Mellon University, Dept. of CS, July 1981.
4. C. Forgy et al., "Initial Assessment of Architectures for Production Systems," *Proc. Nat. Conf. AI,* August 1984.

ty distributions. When a component changes state (that is, from working to failed or vice versa) the simulator tests all subsystems employing the component to see if the subsystem state has changed (that is, from working to idle or vice versa). Stars provides reliability, availability, mean time to failure, mean up-time, and mean down-time evaluations. Mark1 simulates a Markov model of a system if given its transition matrix. Mark1 provides availability, mean time to failure, and state occupancy information.

The mixed-level functional representation is a modeling methodology used at the instruction-set level, a level unique to digital computers positioned between the architecture-definition level and the register-transfer level. The target system is partitioned into entities that are an encapsulation of functions representing behavior. Example entities are processor, physical memory, memory management, error detection, operating system, and user. By focusing attention on those entities required to answer a specific question (such as a uniprocessor address trace of a user program), details of the other entities are suppressed, leading to more than an order of magnitude speedup over conventional simulators.

Switch-level simulation. MOSSym simulates the behavior of a MOS circuit represented as a switch-level network. Inputs to the simulator can be set to 0, 1, or a Boolean variable. The simulator then computes the behavior of the circuit as a function of the past and present input variables. By using heuristically efficient Boolean function manipulation algorithms, the verification of a circuit by symbolic simulation can proceed much more quickly than by exhaustive logic simulation.

Wasim attempts to exploit the modularity of cell-based VLSI designs. Input/output waveform models are generated for modules in a cell library. Each cell is simulated at the device level. Polynomial curve-fitting is used to summarize the cell behavior in a macromodel. An input logic/circuit description is represented by an interconnection of macromodels. An event-driven simulator then predicts the output waveform as a function of simulation inputs.

Circuit-level simulation. Samson2[13] is a mixed circuit/logic simulator that uses event-driven simulation techniques to exploit the inherent temporal sparseness in VLSI circuits. At the circuit level, Samson2 can be more than an order of magnitude faster than Spice2 while providing comparable accuracy. This performance is achieved by partitioning a circuit into subcircuits that can have separate integration step sizes reflecting their activity level. The accuracy of the solution is maintained by controlling the errors due to decoupling among the subnetworks as well as by controlling the local truncation within each subnetwork.

The overhead in the event-scheduling process is small even if the circuit lacks temporal sparseness. Because of the event-driven basis of the algorithm, it is possible to simulate a VLSI circuit in which some subnetworks are characterized in terms of logical behavior and other subnetworks in terms of the circuit-level components, such as MOSFETs and capacitors. Samson2 has been tested for many MOS circuits and demonstrated high efficiency and excellent convergence properties.

Process-level simulation. To model the stochastic nature of the manufacturing process, we have developed an IC process/device simulator called Fabrics II.[14] This simulator consists of two major parts, a processor simulator and a device simulator. The models implemented in Fabrics II allow the accurate simulation of typical semiconductor devices manufactured by a variety of fabrication processes (NMOS, CMOS, and bipolar). Fabrics II employs efficient numerical and analytical models that are solutions of the partial differential equations that describe each fabrication step, under a set of restricted or simplifying conditions that yield reasonable results.

Conventional partitioning schemes do not adequately account for variable module shapes or orientation.

The device simulator in Fabrics II uses the physical parameters generated by the process simulator, in combination with the layout of the devices, to produce device model parameters. The model parameters generated by Fabrics II are compatible with the device models implemented in Spice. Individual device dimensions are derived from a CIF file using the circuit extractor Ace[16] and the interconnect extractor Fred.[17] This layout information is divided into two parts: device dimensions, which the device simulator uses to calculate device model parameters, and interconnections, which produce equivalent parasitic resistances and capacitances. The CPU time consumed by Fabrics II for the process/device simulation is usually a small fraction of the time consumed in the circuit-simulation stage. As an example, the complete process and device simulation of an MOS transistor requires less than 0.4 seconds of CPU time on a VAX 11/780.

Special hardware for CAD

CAD machines are an increasingly practical way to attack the increasing complexity of VLSI systems. Hardware accelerators for simulation tasks are commercially available, and this migration of software CAD tools into hardware appears to be an extremely promising approach in other CAD applications such as physical design.[18,19] CMU's CAD machine research takes a variety of forms. Where existing parallel hardware or multiprocessor systems are available, we study how to restructure CAD algorithms to execute on these machines. When parallelism alone does not provide the necessary performance, we study new architectures and dedicated hardware to accelerate these CAD tasks. Work in this area combines features of basic CAD research, computer architecture, and systems integration.

Two new projects explore the use of CAD machines to accelerate computationally demanding CAD tasks. The first project is studying multiprocessor approaches to simulated-annealing algorithms. The second is studying special hardware for IC mask checking.

Unlike traditional iterative-improvement algorithms, annealing algorithms can avoid becoming trapped in a locally optimal but globally inferior solution. Annealing algorithms have been extremely successful in place and route applications, but consume enormous amounts of computer time.[20,21] We are experimenting with restructuring methods to map annealing-based IC placement tasks onto shared-memory multiprocessor machines. Parallel algorithms are being implemented in a programming language with explicit multitasking constructions. This approach has two key advantages:

First, it separates the computational aspects of the algorithm from concurrency control aspects, allowing us to study them individually. Simulators implemented in this fashion can be used to verify convergence properties, estimate speedups accruing to multiple processors, and measure contention for important shared resources like memory.

Second, these implementations more closely resemble the code that might run on a real parallel machine. We will ultimately port the more successful parallel annealing strategies to a multiprocessor to measure practical performance improvements.

An alternative implementation approach for CAD machines is to build special hardware. Special hardware is often most cost-effective when used to implement mature, stable algorithms. We are investigating the potential benefit of moving basic IC mask-checking operations into hardware. Our interest centers on scanline algorithms, which form the basis for most geometry operations in DRC and circuit-extraction tools. Many excellent algorithms are known, but their application can still be time-consuming,[22] especially when used on flat versions of entire chips, or when checking unique cells in chips with little regularity, or when trying to interactively check large single cells. This project is exploring scanline-processor architectures, and attempting to predict the effects of making such hardware primitives available to higher-level software tools.

References

1. D. P. LaPotin et al., "DIF: A Framework for VLSI Multilevel Representation," *Integration,* 1984.
2. J. D. Crawford, "EDIF: A Mechanism for the Exchange of Design Information," *IEEE Design & Test,* February 1985.
3. T. J. Kowalski and D. E. Thomas, "The VLSI Design Automation Assistant: An IBM System/370 Design," *IEEE Design & Test,* February 1984.
4. T. J. Kowalski, *The VLSI Design Automation Assistant: A Knowledge-Based Expert System,* PhD dissertation, Carnegie-Mellon University, Dept. of ECE, April 1984.
5. C. J. Tseng and D. P. Siewiorek, "Facet: A Procedure for the Automated Synthesis of Digital Systems," *Proc. 20th DA Conf.,* June 1983.
6. M. R. Barbacci, "Instruction Set Processor Specifications (ISPS)," *IEEE Trans. Comp.,* January 1981.
7. J. V. Rajan and D. E. Thomas, "Synthesis by Delayed Binding of Decisions," *Proc. 22nd DA Conf.,* June 1985.
8. W. P. Birmingham and D. P. Siewiorek, "Micon: A Knowledge-Based Single-Board Computer Designer," *Proc. 21st DA Conf.,* June 1984.
9. D. P. LaPotin, *Mason: A Global Floor-Planning and Routing Estimation Tool,* PhD dissertation, Carnegie-Mellon University, Dept. of ECE, 1985.
10. J. Kim, J. McDermott, and D. P. Siewiorek, "Exploiting Domain Knowledge in IC Cell Layout," *IEEE Design & Test,* August 1984.
11. R. Joobani and D. P. Siewiorek, "Weaver: A Knowlege-Based Routing Expert," *Proc. 22nd DA Conf.,* June 1985.
12. S. R. Nassif, *Wasim: A Waveform-Based Simulator for VLSI Circuits,* PhD dissertation, Carnegie-Mellon University, Dept. of ECE, 1985.
13. K. A. Sakallah and S. W. Director, "An Event-Driven Approach for Mixed-Gate and Circuit-Level Simulation," *Proc. ISCAS,* May 1982.
14. S. R. Nassif, A. J. Strojwas, and S. W. Director, "Fabrics II: A Statistically Based IC Fabrication Process Simulator," *IEEE Trans. CAD,* January 1984.
15. R. E. Bryant, "Symbolic Verification of MOS Circuits," *Proc. Chapel Hill Conf. VLSI,* May 1985.
16. A. Gupta, "Ace: A Circuit Extractor," VLSI Document V105, CSD, Carnegie-Mellon University.
17. M. Trick, A. J. Strojwas, and S. W. Director, "Fast RC Simulation for VLSI Interconnect," *Proc. ICCAD-83,* November 1983.
18. R. A. Rutenbar, T. N. Mudge, and D. E. Atkins, "A Class of Cellular Architectures to Support Physical Design Automation," *IEEE Trans. CAD,* October 1984.
19. R. Rutenbar, "Future Directions for DA Machine Research," *Proc. 22nd DA Conf.,* June 1985.
20. S. Kirkpatrick, C. D. Gelatt, Jr., and M. P. Vecchi, "Optimization by Simulated Annealing," *Science,* May 1983.
21. C. Sechen and A. L. Sangiovanni-Vincentelli, "The Timberwolf Placement and Routing Package," *Proc. Cust. IC Conf.,* May 1984.
22. P. T. Chapman and K. Clark, Jr., "The Scanline Approach to Design Rules Checking: Computational Experiences," *Proc. 21st DA Conf.,* June 1984.

Computer-aided manufacture

We envision that a CAM system for VLSI chips should be able to handle a variety of tasks (see illustration), such as optimal design of an IC process and the diagnosis and control of a fabrication line to detect faults and minimize yield to reduce manufacturing costs. This work should provide an attractive extension to the CAM systems currently being introduced into VLSI fabrication processes. Some of the specific results of our work thus far have been the development of tools for statistical process/device simulation, process specification/synthesis (process compilation), process characterization (statistical parameter extraction), process optimization (yield maximization), process diagnosis, and adaptive process control/quality control.

Statistical process/device simulation. Statistical process/device simulation is of key importance in the proposed CAM system. To achieve predictive capabilities essential for statistical design or diagnosis purposes, the models have to be physical and computationally efficient. The set of models currently implemented in Fabrics II satisfies these requirements. Process models produce two-dimensional impurity distributions and layer thicknesses (recently, an efficient model for two-dimensional oxidation was added to the simulation system). The input to such process models consists of process parameters, such as times and temperatures of diffusion steps; doses and energies of ions in implantation; output parameters of previous modeling steps; and process disturbances.

The disturbances are modeled as physical parameters, such as diffusivities of impurity atoms and misalignments between photolithographic masks. These disturbances are independent random variables with probability distributions determined by the particular fabrication process. Furthermore, to account for local (intra-die) and global (inter-die) variations in device parameters, we employ a multilevel structure[1] of process disturbances, where the disturbances are generated by hierarchically defined random number generators at levels that correspond to natural divisions: the lot, wafer, chip, and device levels.

The device simulator in Fabrics II uses the physical parameters generated by the process simulator, in combination with the layout of the devices, to produce device model parameters. In the current implementation, the device simulator

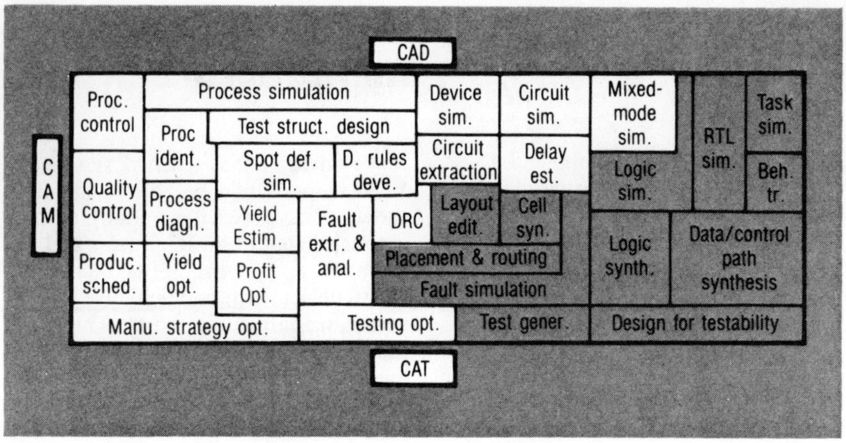

Tasks associated with computer-aided manufacture.

parses a model template and determines a list of parameters that have to be calculated. The model parameters are then computed using functions stored within the device model library, and are used to fill a model template that is passed on to the circuit simulator. The current version of the device model library contains models of the following devices: MOS transistors (including small-geometry effects), bipolar transistors (n-p-n and p-n-p, or lateral and substrate), MOS capacitors, diffused resistors, and interconnections (parasitic resistances and capacitances for arbitrary geometry). Such models can be efficiently tuned to a particular IC fabrication process to obtain excellent agreement between simulated and measured data.

Process specification/synthesis. To facilitate process synthesis, an interactive process compiler has been developed for Fabrics II. This process compiler uses the graphical process editor Ped and allows the user to interactively define a process by:

(1) specification of the sequence of process steps and control parameters;

(2) graphical definition of the lithographic masks for all types of devices manufactured in the process;

(3) use of a process interpreter to perform simulations of the previously entered process steps; and

(4) display of the simulation results such as impurity profiles and device section views.

The user may then repetitively adjust the process-step specification until the desired results are obtained. Extensive error-checking is performed during this stage to detect possible inconsistencies in the process-step definitions. When the user is satisfied with the simulated process, he may invoke a process compiler to create a new process supervisor that is automatically linked to Fabrics. Thus synthesis of a new process and specification of process control parameters can be performed quite efficiently.

Synthesis of a new process and specification of control parameters can be performed quite efficiently.

Process characterization. To obtain accurate simulation, Fabrics II, as do most simulators, must be tuned to a particular IC fabrication process. Fabrics II requires the determination of a set of process disturbances that model fluctuations inherent in IC manufacturing. Since the process disturbances cannot be obtained from direct measurements, the probability-density functions describing process disturbances have to be identified through in-line and test-structure measurements. This identification problem can be formulated as a nonlinear optimization problem. Since this problem is of high dimensionality, efficient decomposition techniques must be applied. A system for the automatic extraction of the probability distribution of process disturbances, called Prometheus,[2] has been successfully applied to tune Fabrics II to particular fabrication processes. In addition, in order to facilitate the IC characterization process, we have developed a set of novel test elements, each of which is sensitive only to a few process disturbances.[3]

Yield maximization. A design verification/optimization package can be realized by combining Fabrics II and a circuit simulator (such as Spice or Samson). With such a package, a number of design optimization tasks can be performed, including:

- Nominal design in terms of process parameters (for example, minimization of power-delay product of a dynamic RAM in terms of gate-oxidation parameters[4]).
- Worst-case design in terms of statistically independent process disturbances.[5] This allows for realistic worst-case predictions and precise estimation of the probability of occurrence.
- Statistical design of ICs in the space of layout and process parameters (such as the optimal cell design[6]).

Recently, a yield optimization tool called Promise[7] was added to the system. This package uses design-centering methods to maximize the production yield in terms of process controls.

To obtain a realistic estimate of the final production yield, the probability distribution of spot defects (shorts or breaks due to local lithographic errors) must be taken into account.[8] A layout simulator, called Vlasic,[9] can be used to predict the decrease in yield due to spot defects. For circuit design purposes, all the random factors affecting IC layout have to be translated into layout design rules. To this end, a statistical design rule developer is now being implemented.

Two types of rules are developed: geometric and electrical.

Geometric rules specify the minimum distance between two edges to avoid (with a certain probability) shorts or breaks in IC layout due to lateral effects (line-width variations, misalignments, lateral diffusion, and depletion-layer expansion) simulated by Fabrics II and spot defects generated according to the identified probability distributions.

Electrical rules specify minimum device dimensions or bounds for the inverter ratios to guarantee acceptable performance. The design rule developer uses the statistical hypothesis testing approach to derive rules which maximize total yield.

Redundancy impact on yield and reliability. The use of redundancy to enhance yield of large semiconductor chips has become common practice. One of the oldest forms of redundancy is triplication and majority voting. With triplication, no external repair procedure, such a laser cutting of fuses, is required. Faults, if they exist, are masked. Furthermore, redundancy that was not used in the manufacturing process can be used to tolerate additional failures during operational life.

The classical model of triplicated systems, which assumes the system fails when two or more copies fail, is pessimistic. Indeed, there are many cases where all three copies can have failures but where their majority vote is still correct.

The theoretical model has been extended to take into account these compensating failures.[10] This analytical model has been coupled with a fault simulator to predict yield and reliability of actual designs. A prototype system, which reads design data directly from a CAD database, has given designers early feedback on yield.

Process diagnosis. A diagnostic system called Prod[11] was recently developed to identify the process disturbances that caused a drop in yield for a particular production run. The approach used in Prod is based on the analysis of the probability distribution of the performance measurements (such as delay times and voltage levels) from manufactured chips or scribe lane test structures, and employs statistical pattern-recognition techniques.[12]

The operation of Prod can be divided into two phases: learning and fault identification. In the learning phase, Fabrics II is used as a fault simulator to determine the probability distribution of the performances that correspond to artificially introduced single faults or fault combinations. The learning phase continues until a number of most significant fault combinations have been simulated and stored in a library. To identify faults in an operating fabrication line, the same performances are measured and the resulting probability distribution is compared to those stored in the library.

Prod is a complete integrated system and includes a database using the Unix file system. To verify its utility in identifying faults that cause drops in yield in an IC fabrication line, the system was tested on a number of examples ranging from test transistors to the critical path of a dynamic RAM.

Quality/adaptive process control. The aim of the manufacturing control system under development is to enable a fabrication line to meet the required demand at minimal cost. The approach presented below is a generalization of the cost-minimization problem described in Maly and Pizlo.[13]

The costs in IC manufacturing are the material cost, the processing costs at each step, the testing and assembly costs, the maintenance cost, and the penalty incurred if the demand is not met by a prescribed time. A normal process can be defined as one in which the yield value is satisfactory. The process can be characterized in terms of the in-line measurements (such as layer thickness and sheet resistances) taken after each major process step.

If the probability distribution of these in-line parameters in a given lot (or wafer) is significantly different from the distribution for the normal process (shifted mean values or increased variances), a decision has to be made whether this lot (or wafer) is to be rejected from further processing to avoid continued spending on a lot that would result in an unacceptably low yield.

These quality-control decisions depend on the relation between the observable parameters and specified circuit performances—and hence process yield. The cost of the processing steps taken so far, the cost involved in further processing steps, and the cost of implementing the corrective measures in the subsequent steps must be taken into account.

We have established a formulation for the cost function that is applicable in any IC manufacturing process and can be used for quality-control decisions. A quality-control problem has been formulated in terms of statistical hypothesis-testing techniques; we are now developing the software tools to implement the testing procedures. The quality-control decisions can be made efficiently if analytical models exist that relate observable in-line parameters with circuit performances.

A software system has been developed in which polynomial regression models are automatically extracted from the simulation results. If the quality of the obtained models is not sufficient, off-line simulations are performed to verify the hypotheses about lot or wafer rejection. If the deviations in the measured in-line parameters are not beyond control, the quality-control system may invoke an adaptive control system to find adjustments in process controls in the subsequent steps that would increase the yield in this lot. This problem can be formulated as the control of a stationary stochastic process; we are developing a set of algorithms that can be applied for coarse and fine adjustments in the process controls.

Such an adaptive control approach seems especially attractive for custom VLSI chip manufacturing where the time constraints on parts delivery are extremely critical. Moreover, with the introduction of computer-aided manufacturing systems, adjusting control parameters of individual processing steps becomes a feasible and risk-free procedure. Initial experiments show that the computational efficiency of Fabrics is sufficient to perform the off-line simulations necessary in such a system.

References

1. W. Maly and A. J. Strojwas, "Statistical Simulation of the IC Manufacturing Process," *IEEE Trans. CAD,* July 1982.
2. C. J. Spanos and S. W. Director, "Prometheus: A Program for VLSI Process Parameters Extraction," *Proc. ICCAD-83,* September 1983.
3. I. Chen and A. J. Strojwas, "A Methodology for Optimal Test Structure Design," *Proc. Cust. IC Conf.,* May 1985.
4. S. R. Nassif, A. J. Strojwas, and S. W. Director, "Fabrics II: A Statistically Based IC Fabrication Process Simulator," *IEEE Trans. CAD,* January 1984.
5. A. J. Strojwas, S. R. Nassif, and S. W. Director, "A Methodology for Worst-Case Design of Integrated Circuits," *Proc. ICCAD-83,* September 1983.
6. A. J. Strojwas, S. R. Nassif, and S. W. Director, "Optimal Design of VLSI Minicells Using a Statistical Process Simulator," *Proc. ISCAS,* May 1983.
7. D. J. Giannopoulos and S. W. Director, "IC Fabrication Process Optimization," *Proc. ICCAD-84,* November 1984.
8. W. Maly, "Modeling of Lithography-Related Yield Losses for CAD of VLSI Circuits, *IEEE Trans. CAD,* July 1985.
9. H. Walker and S. W. Director, "Yield Simulation for Integrated Circuits," *Proc. ICCAD-83,* September 1983.
10. G. York, D. P. Siewiorek, and Y. X. Zhu, "Compensating Faults in Triple Modular Redundancy," *Proc. FTCS-15,* June 1985.
11. P. Odryna and A. J. Strojwas, "Prod: A VLSI Fault Diagnosis System," *Proc. ISCAS,* June 1985.
12. A. J. Strojwas and S. W. Director, "A Pattern-Recognition-Based Method for IC Failure Analysis," *IEEE Trans. CAD,* January 1985.
13. W. Maly and Z. Pizlo, "Tolerance Assignment for IC Selection Tests," *IEEE Trans. CAD,* April 1985.

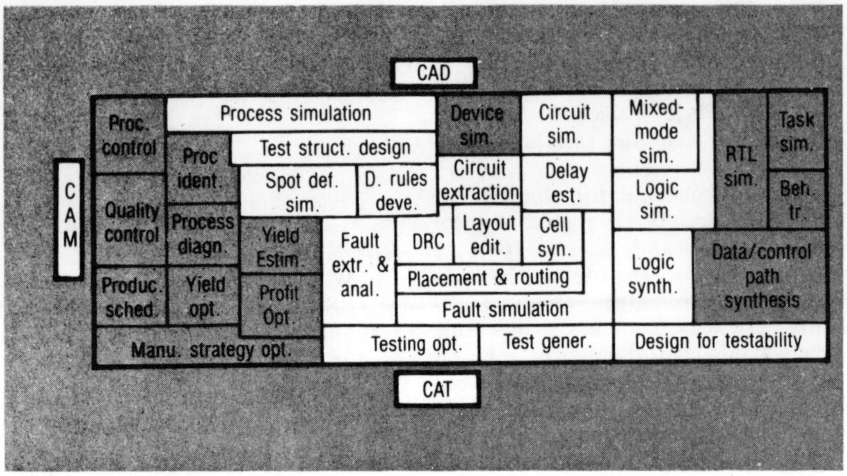

Tasks associated with computer-aided test.

Computer-aided test

Most widely used testing methodologies and tools (see illustration for tasks associated with CAT) in the industry today were originally developed more than a decade ago for testing printed circuit boards containing SSI/MSI TTL packages. The traditional approach to testing possesses a number of attributes that are becoming very undesirable.

Typically, testing is not considered until the entire design process has been completed. Interaction between designers and test engineers is often nonexistent, and then adversarial. In the past, off-line fabrication testing and online error detection/correction were considered two separate disciplines. In traditional testing, the unit or device under test is considered to be a black box by the tester who makes little use of the functionality of the unit under test.

We believe that a new and more integrated approach to testing is necessary. We consider the following three dimensions of integration to be essential:

- **Integration of design and test.** Testing issues must be considered early in the design phase. In addition to optimizing cost and performance, testability must be considered as one of the design criteria. Designers must be aware of the testability issues and test engineers should be able to influence the design.
- **Integration of off-line and on-line testing.** The off-line testing and on-line testing issues must be jointly addressed. Typically, the hardware added to aid off-line testing is not used in the operational environment. We believe it is possible to design built-in-test circuitry that supports both off-line and on-line testing while minimizing the overall overhead required.
- **Integration of unit under test and unit performing test.** With the incorporation of built-in-test circuitry and self-testing features, the traditionally distinct boundary between the unit under test and the tester is becoming ambiguous. It makes sense to take advantage of the intrinsic functionality of a complex chip to test itself. Furthermore, one way of reducing the horrendous cost of a sophisticated tester is to build some of its functions into the chip being tested.

The project seeks to exploit certain regular and iterative structures possessed by many circuits.

Design for testability. It is our belief that design-for-testability techniques should be developed for all levels of design abstraction.[1] At the functional level, easily testable structures are being investigated. At the logic and circuit level, a methodology has been developed to automate the incorporation of scan structures in semicustom CMOS VLSI circuits. We are also looking into techniques for modifying circuit layouts to reduce the likelihood of certain faults that are very difficult to test.

Easily testable structures. Many useful circuits possess certain regular and iterative structures. This project seeks to exploit this structure for testing.[2,3] Several classes of array-structured circuits that can be designed to be very easily testable have been studied.[4]

For example, two-dimensional array multipliers have been designed and shown to require only a small number of test patterns, regardless of the array size. An array multiplier is called C-testable if all its adder cells can be exhaustively tested while requiring only a constant number of test patterns. A modified design of the conventional carry-save array multiplier is shown to be C-testable and requires only 16 test patterns. Similar results are obtained for several other array multipliers, including a modified design of the Baugh-Wooley two's-complement multiplier, which requires 55 test patterns, and a modified design of the TRW sign-magnitude/two's-complement multiplier, which requires only 33 test patterns.

Automated scan design. A methodology and associated CAD tools to facilitate the automated design of easily testable semicustom VLSI circits have been developed, in conjunction with GE Microelectronics Center, for the CMOS gate-array technology. Extension to standard cell technology appears feasible. This methodology systematically groups the sequential elements to form multiple scan paths, and uses multiplexers to allow the sharing of I/O pins by mission and test inputs and outputs.

The multiple-scan-path design enhances controllability and observability. The multiplexed scan paths reduce the long testing time required by most scan designs without increasing the number of I/O pins. Standard designs from gate-array designers are automatically translated into easily testable designs.

The entire procedure is transparent to the designer; however, a set of possible testable design choices can be presented to allow him to make the best testability versus overhead trade-off. Unlike other in-house scan design tools, it is anticipated this package will be commercially available.

On-line self test. There is much interest in developing on-line concurrent testing or monitoring schemes to detect intermittent and transient faults. Traditional on-line testing methods tend to be very costly or limited in their effectiveness. We believe that effective and inexpensive means for on-line testing of VLSI processor can be developed by taking into consideration the functional and structural features of a processor.

A technique called signatured instruction streams involves encoding the instruction streams of an application program at compile time. At run time, the same encoding is regenerated using added hardware circuitry and compared with the pre-computed encoding, or signatures, to perform error-checking. On-line testing of systolic arrays is also being investigated.

Processor self-monitoring. A self-monitoring technique using signatured instruction streams has been developed.[5] Using this technique, a VLSI processor can detect, with very short detection latency, all transient errors that affect the program control flow. The implementation requires about 10 percent overhead for built-in-test circuitry and incurs about five percent performance penalty. A demonstration system using an MC68000 processor has been designed and built.

A general-purpose hardware fault inserter at CMU can be programmed to inject faults of varying duration into various bit lines of the system bus to simulate the occurrence of transient errors. Using this fault inserter and the demonstration system, extensive fault insertion experiments are being performed to determine the transient error coverage of the signatured instruction streams self-monitoring scheme. Many transient faults of various durations and types will be inserted. Initial results indicate that the system can detect more than 99 percent of the faults that affect program control flow, and about 85 percent of all the randomly inserted faults.

Efficient and effective MOS VLSI circuit testing requires new approaches and tools.

System self-monitoring. An extension of the self-monitoring technique, called roving monitoring, uses a watchdog-type monitoring processor to concurrently monitor multiple application processors.[6,7] By time-sharing the roving monitor processor, the overall hardware overhead is reduced at the expense of increasing the error-detection latency. The valid control flows in each application program are abstracted into a signature graph. The roving monitoring processor uses the signature graphs to check the validity of the sequence of run-time signatures generated by each application processor.

A special-purpose monitoring processor is required. A roving monitoring processor, called RMP1, with a novel architecture combining an interesting ar-

Inductive fault analysis

An example of crossing multiple levels of the design hierarchy and involving elements of CAD, CAM, and CAT is our work on inductive fault analysis. IFA is a systematic fault-extracting procedure that spans the process, layout, transistor, and logic levels. A higher-level fault model is developed based on conclusions drawn from examining particular defects at the lower level; hence the word "inductive." Using IFA, all the transistor and logic-level faults likely to occur in a given MOS integrated (sub)circuit can be extracted from physical defects at the layout level using process-level information.

The four major steps of the IFA procedure are: (1) generation and placement, on the circuit layout, of physical defects using statistical data obtained from the fabrication process; (2) extraction of transistor-level faults caused by these defects; (3) abstraction, if possible, of these faults to the logic level; and (4) classification of fault types and ranking of faults based on the likelihood of occurrence. Given the layout of an integrated (sub)circuit, a customized and very accurate fault model and an associated ranked fault list can be automatically generated that take into account the technology, layout, and process characteristics.

Initial application of the IFA procedure to an example circuit led to a number of very interesting observations.* Only 64 percent of the extracted faults can be modeled by the traditional single- and multiple-line stuck-at fault models. Not all faults are equally likely, and unusual faults, such as the creation of new transistors, are possible. Only three percent of the faults involve transistors stuck at open, whereas 30 percent of the faults are bridging faults. The figure below illustrates the various levels involved in the IFA procedure.

*W. Maly, J. Ferguson, and J. P. Shen, "Systematic Characterization of Physical Defects for Fault Analysis of MOS IC Cells," *Proc. Int'l Test Conf.*, October 1984.

Spot defect at the layout level

(a)

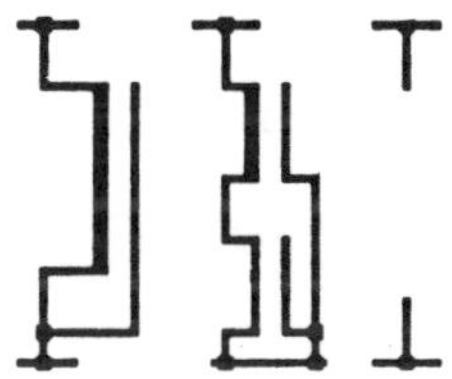

Possible faulty behaviors at the transistor level

(b)

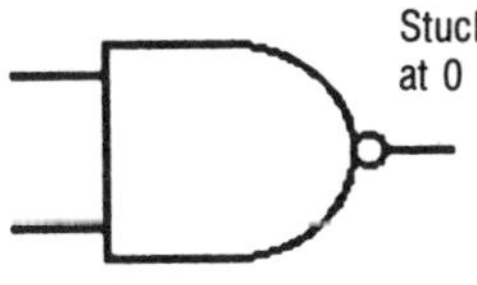

Possible faults at the logic level

(c)

A physical defect at (a) the layout level involving a missing spot in the implant region can possibly produce (b) one of a number of faulty behaviors at the transistor level which may appear as (c) a line stuck-at fault at the logic level.

chitectural feature called compiled graph programming and a three-stage pipelined hardware structure can monitor up to 16 8-MHz MC68000 processors simultaneously. The RMP1 uses high-speed TTL building blocks. Experiments using the RMP1 are now under way to evaluate the effectiveness of various roving strategies.

Self-testing systolic arrays. Special-purpose cellular architectures for signal processing have produced processing rates far in excess of general-purpose architectures. Systolic arrays (a pipelined cellular architecture) developed at CMU have been applied to a large spectrum of problems.[8]

A one-dimensional systolic array, called Warp, is under construction to investigate testing and replacement of faulty cells—a major problem for large arrays. A technique has been developed that allows spare cells to migrate through the array. The spare can serve in a duplicate-and-match mode for testing. If a cell is faulty, bypass registers allow for its replacement. An augmented Warp cell allows testing, spare motion, and cell replacement without degradation of the array's throughput. Extensions of the technique have been made for two-dimensional systolic arrays.

Using signatured instruction streams, a VLSI processor can detect all transient errors that affect the program control flow.

A new methodology for fabrication test. CMU has started a research effort to develop a new methodology and associated software tools for the testing of MOS VLSI circuits. Most of the existing fault modeling and test-generation techniques are based on the traditional scenario of putting many TTL SSI/MSI packages on a printed circuit board. It is believed that the efficient and effective testing of MOS VLSI circuits requires the development of new approaches and tools. We believe the following three problems associated with traditional testing must be addressed.

- **Technology-independent fault model.** Regardless of the implementation technology, the single-line stuck-at fault

Design for reliability

System reliability has been a major concern since the beginning of the electronic digital computer age. The earliest computers were constructed of components such as relays and vacuum tubes that failed to operate correctly as often as once every million cycles. This error rate was far too great to ensure correct completion of even modest calculations requring tens of millions of operating cycles. Our dependence on computing systems has grown so great that it is becoming impossible to return to less sophisticated mechanisms. Today, interest in reliability pervades the computer industry, from large mainframe manufacturers to semiconductor fabricators, who produce not only reliability-specific chips (such as for error-correcting codes) but also entire systems.

Even with this increased interest in reliability, designers have had to use intuition and ad hoc techniques to evaluate design trade-offs since there have been few, if any, reliability-oriented computer-aided design tools. If such a tool existed, it usually was stand-alone (that is, not integrated into the CAD database) and required a reliability expert to operate. Over the past 10 years, researchers at CMU have developed a design methodology and supporting CAD tools for reliable design.

Pervades process. Design for reliability starts at the highest level (task description) and pervades the entire design process. A unique aspect of design for reliability is the number of factors that affect the design, including: hardware error detectors, software error handlers, mechanical design, and environmental parameters. Reliability analysis tools must have access to a wide variety of databases; design for reliability is an integrating factor in the design process. In addition to Stars and Mark1, described in the accompanying reports, a number of reliability analysis tools have been developed.

The Adviser program computes the symbolic system reliability/availability expression given the interconnection graph (PMS diagram) of the system, the reliability/availability of each class of identical system components, and a simple statement of the system functionality requirements.*

Lambda is a program that predicts hard-failure reliability of electronic systems. Based on *Mil Handbook 217,* Lambda calculates failure rates from a CAD database of parts and environmental parameters. Redundant structures are also evaluated. Output includes failure rate, reliability, mean time to failure, and mission time. Lambda has been extensively used in industry. Finally, Relicalc calculates availability, mean up-time, and mean down-time for recursive, hierarchical defined systems. Systems are described in terms of series, parallel, or *m*-of-*n* (that is, *m* units required to be functional out of *n* total units) structures.

Integration emphasized. These tools have emphasized integration into CAD databases while minimizing the level of reliability expertise required to generate meaningful analysis. By making reliability analysis available at the push of a key, the logic designer will be able to make early and repeated design trade-offs that make design for reliability a minor extension of the design process.

To demonstrate that a knowledge base exists for reliable design, Micon is being upgraded to incorporate a subset of the *Catalogue of Reliability Techniques* that has been assembled. Thus design for reliability will be included in Micon's domain of single-board computer design knowledge.

*V. Kini and D. P. Siewiorek, "Automatic Generation of Symbolic Reliability Functions for Processor-Memory Switch Structures," *IEEE Trans. Comp.,* August 1982.

model is the most widely used fault model. We believe that, for VLSI testing, the fault model should be based on the physical defects and should take into account the technology, layout, and process characteristics.

• **Single-level approach.** Most existing testing methodologies and tools assume the logic-gate-level model. Athough MSI-level macros are often used in logic simulators, most systematic test-generation algorithms require the gate-level representation. We believe, just as in design, a new testing methodology should involve a multiple-level or hierarchical approach in order to ease the complexity burden.

• **Unranked fault list.** Traditionally, the relative importance of different faults is not considered. We believe that not all faults are equally likely or equally important. The fault list should be ranked and an ordered test set used to detect the most likely to occur faults first.

Based on this ranked fault list, very effective test sets can be generated.

Fault extraction. Our first project in this area addresses the problem of the inappropriateness of the traditional fault model. Our underlying assumption is that circuit-level faults are caused by physical defects inherent in the IC fabrication process. A technique called inductive fault analysis allows the transistor or logic-level faulty behaviors of a given MOS integrated circuit to be extracted from the layout level using process-level information concerning spot defects.[9] The extracted faults can then be grouped and ranked according to their likelihood of occurrence. Based on this ranked fault list, very effective test sets can then be generated (see box on inductive fault analysis on page 97).

Switch-level test generation. Efforts are also under way to develop a switch-level automatic test-generation program. The output from the inductive fault analysis procedure will include nonclassical faults. Hence, the ability to generate test patterns at a level lower than the logic-gate level is essential. Our initial effort involves modifying the classical D-algorithm to accommodate bidirectionality in the switch-level representation of a MOS circuit. This is a difficult problem and it is still unclear how successful this approach will be.

References

1. S. P. Tomas and J. P. Shen, "A Survey of Functional-Level Testing and Testability Measures," Research Report, Carnegie-Mellon University, Dept. of ECE, November 1983.
2. D. C. H. Lee and J. P. Shen, "Easily Testable (N,K) Shuffle-Exchange Networks," *Proc. Int'l Conf. Para. Proc.*, August 1983.
3. J. P. Shen and F. J. Ferguson, "Design of Easily Testable VLSI Array Multipliers," *IEEE Trans. Comp.*, June 1984.
4. F. J. Ferguson and J. P. Shen, "Easily Testable Array Multipliers," *Proc. 6th Comp. Arith. Conf.*, June 1983.
5. J. P. Shen and M. A. Schuette, "On-Line Self-Monitoring Using Signatured Instruction Streams," *Proc. Int'l Test Conf.*, October 1983.
6. J. B. Eifert and J. P. Shen, "Processor Monitoring Using Asynchronous Signatured Instruction Streams," *Proc. FTCS-14*, June 1984.
7. S. P. Tomas and J. P. Shen, "Detection of Control-Flow Errors in Multiple-Processor Systems Using Roving Monitoring," submitted to ICCD '85, June 1985.
8. H. T. Kung, "Why Systolic Architecture?," *Computer*, January 1982.
9. W. Maly, F. J. Ferguson and J. P. Shen, "Systematic Characterization of Physical Defects for Fault Analysis of MOS IC Cells," *Proc. Int'l Test Conf.*, October 1984.

Integrated environment

In order to achieve an interactive design automation system that is made up of a set of disparate tools, we must develop a computing environment that has

• the ability to allow various programs that have incompatible input and output languages to communicate with each other;

• the ability to manipulate a variety of hardware description languages and to be able to resolve incompatibilities that may result;

• the ability to organize the flow of information between programs to realize a specified design task;

• the ability to implement automatically a diverse set of design tasks and methodologies;

• the ability to allow the designer to arbitarily interrupt a design task (such as a specified program execution sequence) and restart or redirect the sequence of operations;

• the ability to explain its sequence of design activities to the designer and to explain the reasons for particular design decisions;

• the capability of easily introducing new computer aids or tools or substituting newer tools for older ones;

• the ability to keep track of all design attempts for a particular VLSI chip, and to make some comparison of these attempts based on quantitative measures.

We refer to a computing environment that has the above capabilities as a VLSI design environment. Such an environment, called Ulysses,[1] is under development and employs AI concepts to achieve the flexibility the above attributes require.

In Ulysses, each CAD tool—as well as the designer himself—is viewed as a knowledge source. All knowledge sources are considered to be self-activating, asynchronous, parallel processes Knowledge sources communicate with each other through a global database called the blackboard. The blackboard supports the many levels of representation necessary for a VLSI design.

Ulysses allows the arbitrary sequencing of CAD tools to realize a wide variety of design tasks and design methodologies. These tasks are described via a language called Scripts. A scheduler interprets these scripts and chooses which knowledge source or script should execute next. The scheduler may also automatically switch to another design point; however, the designer can overrule decisions made by the scheduler.

With the advent of a design environment such as Ulysses, we may view the role of the IC designer as that of an intelligent adviser communicating with an intelligent design environment. In such a situation, it would be very difficult for the designer to express his intentions traditionally. To allow the designer the greatest flexibility in using the design automation system, we feel a natural-language interface is required.

Natural-language interface. The task of building a natural-language interface for CAD is not, unfortunately, simply one of applying an existing AI technology to a new domain. Though the study of previous natural-language interfaces is a rich source of ideas, it also underscores the imperfections of these interfaces, as well as the need for a fresh approach.

The task of building a natural-language interface for CAD requires more than applying existing AI technology.

A key component of a natural-language interface is the parser, which analyzes a sentence or sentence fragment and produces some kind of structured representation for it. Previous natural-language interfaces in general have addressed this issue of parsing in rather piecemeal fashion. Their coverage of constructions in English (the natural language most interfaces have dealt with) has been motivated and dictated by the domain of application of these interfaces, and not by a systematic study of the structure of English. For the goals of some of these interfaces, these limitations have not been detractions, but have restricted the utility of the interfaces to other domains.

Since most natural-language interfaces have been toy ones, either restricted to simplified domains or by self-admission never intended for real use, their application to substantial problems has always been problematic.

In contrast to the domains for which most natural-language interfaces have been intended, the CAD domain is substantial and complex. We have developed a new approach to natural-language understanding. Based on this approach, we have implemented an initial version of a natural-language interface called Cleopatra[2] that deals with circuit-simulation postprocessing.

While Cleopatra is currently limited to the subdomain of circuit-simulation postprocessing, future extensions will gradually address other design tasks. Some of these tasks will not require too much additional programming effort. These include postprocessing for functional and behavioral simulators and for database query. Our ultimate goal, however, is a natural-language interface for our integrated CAD/CAM/CAT environment.

Demeter focuses on initial phases of design. More specifically, it is concerned with synthesis between the task level and the register-transfer level.[3] The initial phase of this project resulted in experimental software that explored the plausibility of integrating existing reliability analysis tools (see box on page 98 on design for reliability) into a unified environment[4] and synthesis from the architecture-definition level.[5] Phase 1, corresponding to the architecture-definition level, explores issues in information management (such as version control, translation, filtering, and merging of data from various tools) and databases. In addition to the traditional design database, there will be databases concerned with domain-specific design knowledge, historical design trends, and information on the various tools.

References

1. M. L. Bushnel and S. W. Director, "Ulysses: An Expert-System-Based VLSI Design Environment," *Proc. ISCAS,* June 1985.
2. T. Samad, *A Natural-Language Interface for CAD Applications,* PhD dissertation, Carnegie-Mellon University, Dept. of ECE, 1985.
3. D. P. Siewiorek, E. P. Giuse, and W. P. Birmingham, "Proposal for Research on Demeter: A Design Methodology and Environment," Research Report, Carnegie-Mellon University, Dept. of ECE, January 1983.
4. D. Giuse, D. P. Siewiorek, and W. P. Birmingham, "Demeter: A Design Methodology and Environment," *Proc. ICCD 83,* October 1983.
5. W. P. Birmingham and D. P. Siewiorek, "Micon: A Knowledge-Based Single-Board Computer Designer," *Proc. 21st DA Conf.,* June 1984.

Part III: The Industry Programs: USA

A VLSI DESIGN SYSTEM

Ajoy K. Bose
Basant R. Chawla
Hermann K. Gummel

Bell Laboratories
Murray Hill, NJ 07974

ABSTRACT

This paper discusses a system of Computer Aids used at Bell Laboratories for the design of VLSI circuits. Special emphasis is placed on tools for simulation, modeling, layout, and testing, and their integration to form a CAD system. The system has been in production use for the design of hundreds of chips, including some very complex chips such as a 32-bit microprocessor.

1. INTRODUCTION

The use of computer aids for the design of VLSI circuits at Bell Laboratories has evolved over many years. Most of the aids being used today have gone through an evolutionary process and some aids are now in their third or fourth generation. Collectively, they form a system that enables an efficient and reliable VLSI design process.

The rapid growth in circuit complexity and the emergence of new technologies over the last decade have placed enormous demands on CAD. Innovations in algorithms, design methodology, software architecture, design languages, and data bases, were required. As a result of these improvements, the design time has remained fairly constant or has grown at a small rate while the complexity of chips has grown by orders of magnitude.

The major components of the CAD system are: tools for simulation, layout, and testing; standard design languages that provide robust interfaces between various design phases; facilities to close local and global design loops; facilities that aid in the preparation of the enormous volume of input data for large-scale simulations; and tools for the generation and maintenance of design libraries. The primary goals for the CAD system are:

1. To exploit VLSI technology capable of hundreds of thousands of transistors on a chip.

2. Design integrity, as reflected in meeting design specifications with adequate margins, leading to good manufacturing yields.
3. Designer productivity - the ability to design chips with low design effort; avoidance of design iterations that are costly in terms of delays and development effort.

The system has to address several different technologies, circuit forms, and different design styles. Among the relevant technologies are NMOS (enhancement mode and depletion mode), CMOS, standard buried collector bipolar, oxide-isolated high speed bipolar, and CBIC (complementary bipolar). Circuit forms of interest today are custom and catalog digital circuits, memories, microprocessors, microcomputers, analog circuits (such as op amps), and hybrid circuits having significant analog and digital components (such as signal processing chips, codecs, etc.). Among the design styles or circuit realizations are standard cell (polycell) approaches, gate arrays, PLAs, symbolic and hand-packed layouts.

While some aids, such as a circuit simulator or an interactive graphics system, are of a very general nature and serve all technologies and design styles, others are application specific. The CAD tools are written portably and almost all of them are available on the same midi computer, allowing them to act as an integrated system. Some of the CPU-intensive aids, such as simulators, are also available on a super computer.

This paper addresses only the design aids for implementing integrated-circuit chips, given specifications for such chips. Not included are aids for high-level system design, such as architectural simulators, logic synthesizers, or printed circuit board aids. The remaining sections of this paper provide an overview of the capabilities in the areas of CAD interfaces, simulation and modeling, layout, testing and the generation of libraries. Since the number of individual aids discussed is very large, only a brief description of these aids is included. Wherever available, references are provided to publications where a more detailed description of the aids can be found.

2. CAD INTERFACES

Interfaces between different aids are achieved through standard languages, and through design libraries which are created from a common source. This approach offers simplicity and efficiency of operation, and lends itself to be usable in a mini and midi computer environment. This contrasts with approaches based on all-encompassing data bases which require vast resources, have built-in inefficiencies for reasons of security and data integrity, and are often inflexible and cumbersome to use.

Increasing complexity has often required development of new tools. Expedient development of such tools has been made possible by the standard language interfaces and by specifying the contents of the design library needed by the tool. With rapid advances in technology it has been necessary that the design libraries be adaptable to changes in the technology. To this end, design styles that are technology updatable are used.

The two major standard languages in the CAD environment are:

1. A connectivity description language LSL-LOCAL [1] (short form LSL for Logic Simulation Language), which is used by the logic, timing, and fault simulators, and by modular layout aids.

2. A mask geometry or artwork description language XYMASK [2].

Both of the above languages are hierarchical and offer an indefinite depth of nesting for defining a circuit in terms of its sub-circuit components. In the case of connectivity description, the basic components could be logic gates or modules such as polycells. In the case of artwork description, the basic components are shapes such as rectangles, polygons, or constant-width paths. To efficiently utilize the silicon area, the geometries are allowed to have non-rectilinear edges.

In addition to the textual description of circuits, a hierarchical

Reprinted from *IEEE Int. Symp. Circuit Syst.*, pp. 734–739, June 1983.

schematic entry program SCHEMA allows graphical entry of logic or circuit diagrams and produces the circuit connectivity description for the simulators and the modular layout aids. This eliminates the need for error-prone manual encoding of circuit descriptions.

Well-defined interfaces between design aids are important to eliminate the need for data transcriptions and the attendant potential for errors. Examples of data transmitted through such interfaces are: the layout capacitances generated by the modular layout aids which are directly used by the timing simulator, and the set of input stimuli for the various simulators and test aids.

3. SIMULATION AND MODELING AIDS

There exists a spectrum of simulation and modeling aids addressing the range from logic and timing simulation at the chip level to the modeling and simulation of device behavior and doping profiles. This section describes in detail these aids for logic and timing simulation, critical-path analysis, circuit simulation, device modeling, parameter extraction, device simulation, and process simulation.

3.1 Logic and Timing Simulation

While the memory and linear circuits are described primarily at the device level, the digital circuits can be described at the functional, gate, and device levels. The memory and analog circuits are typically simulated by circuit simulators (described in a later section) and digital circuits are typically simulated by logic and timing simulators. Digital circuits are described in terms of an interconnection of logic gates or modules in the LSL language.

The circuit designer needs to verify the chip design for logical correctness and timing. It is typical to separate these two verification steps. In the first step, the design is verified through a unit delay simulation where the logical behavior of the circuit is analyzed and the timing information is suppressed. Once the logic of the circuit has been verified, timing simulations are performed to ensure that the circuit can operate at the intended speed and that the circuit delays do not introduce errors in the logical operation. Prior to chip layout, the effects of routing capacitances are estimated. Following the layout, the actual routing capacitances are used. For sections of a circuit where timing is critical, the timing analysis is performed with a circuit simulator.

The MOTIS simulation system is used for logic verification, and for both pre-layout and post-layout timing verification. This system contains a unit delay simulator, a fine-grain timing simulator [3], a coarse-grain multiple delay simulator [4], and a mixed-mode simulator [5].

The MOTIS system allows simulation of both MOS and bipolar circuits. The MOS circuits can be explicitly specified in terms of load and driver transistors or transmission gates. These circuits can be simulated in unit delay, multiple delay or timing (or any combination of these using the mixed-mode simulator) with the delays for multiple delay simulation being computed automatically. The bipolar circuits are only specified in terms of logic gates. The intrinsic delays of these gates are stored in a library and the system contains a very rich set of commands to automatically adjust delays based on loading capacitances. RAMs, ROMs, PLAs, and Registers can be described in the simulator as functional elements. For post-layout simulation, the actual routing capacitances are fed back from the LTX layout system (discussed in a later section) to MOTIS.

The system also provides numerous capabilities for monitoring and controlling the simulation, setting different sections of the circuit in different simulation modes, to compare the results of several simulation runs, and to perform path delay analyses (discussed in the following section). The output of the MOTIS system can be viewed at a graphics screen or stored on disk for later viewing or used for different types of comparison. An example of the results of a mixed-mode simulation run containing some gates in the fine-grain timing mode and some gates in the coarse-grain multiple delay mode is shown in Figure 1. The MOTIS system has been used to simulate circuits with up to a few hundred thousand transistors.

3.2 Critical Path Analysis

Since comprehensive simulation can be a very time-consuming process, tools are provided to independently identify and analyze critical paths in a design. The MOTIS system provides a capability which identifies all clocked elements in the circuit, computes clock skews for the clocked elements, analyzes all paths between the clocked elements and identifies those paths that fail at the circuit's intended clock rate [6]. The delays used in the analysis are those computed for multiple delay simulation. Facilities are provided to show delays of all paths or details of selected paths. This feature can be used in the selection of higher speed modules to ensure that the path delays do not exceed the required value, and it allows the timing verification of a circuit to be carried out with minimal simulation.

3.3 Circuit Simulation

Circuit simulation is of vital importance for the design of integrated circuits. The circuit simulator used in the CAD system is ADVICE [7], which has evolved from SPICE-2 [8]. Besides passive circuit elements, ADVICE has models for diodes, bipolar transistors, MOS transistors, JFETs, Josephson junction devices, switches, and transmission lines. ADVICE is principally used in designing memory circuits, analog circuits, standard cells, and in analyzing paths in digital designs where timing is critical.

ADVICE contains a rich set of process-dependent models, particularly for MOS devices, in support of the changing needs of circuit designers as technological advances are made. Extensive facilities exist to support user-supplied process and device libraries. Besides the usual DC and transient analyses, ADVICE supports AC, noise, and Monte-Carlo analyses. The output of ADVICE can be viewed on a graphics screen, plotted on a variety of pen plotters, or stored for post-processing. On today's large computers, ADVICE can typically handle circuits having a few thousand transistors.

3.4 Device Models

The accuracy and speed of circuit simulation depend on the device models supplied to the simulator. There exist trade-offs between accuracy and speed. In general, the complexity of a device model increases with the physical effects it is called upon to model. Models of bipolar and MOS transistors used in ADVICE are described below.

The bipolar model in ADVICE is a form of the Gummel-Poon model [9] based on a charge control relation linking the junction voltages, collector current, and base charge. This compact model exhibits numerous first and second order physical effects, and it provides good accuracy both in frequency domain and large-signal transient analysis.

For MOS devices, ADVICE contains several models. These models have evolved with the technology. The early model is a long-channel model. More recent models include short-channel and narrow-channel effects. One of the most used MOS models is the CSIM (Compact Short-channel IGFET Model) model [10]. Enhancements have been made in this model to accurately represent the near-threshold behavior and to incorporate narrow-channel effects on threshold voltage. Another model, which has been used in designing memory circuits, is the IM model [11].

3.5 Parameter Extraction

Without tools for extraction of model parameters from measurements, the usefulness of compact models would be rather limited. Parameter extraction programs exist for extracting parameters for both MOS and bipolar models.

The parameters for the CSIM model for MOS transistors are extracted from electrical measurements by using the MOSPAC program. Since parameters can be extracted for many devices, capabilities exist to perform statistical analyses and make scatter plots and histograms. MOSPAC also analyzes characterization data from several lots to determine worst-case sets of parameters. Similar capabilities are available for bipolar technologies in programs BIPAC and PAREX.

3.6 Device Simulation

To develop compact models which are based on device physics, one needs to understand device behavior at very detailed levels. This requires tools which solve electron and hole transport equations, Poisson's equation, and continuity equations in a region of given doping profile and for given boundary conditions. An iterative scheme for solution of these equations considering a one dimensional transistor [12] has been the basis of device simulation in one and two dimensions. The tools used today are based on the simulators developed elsewhere. For performing one-dimensional device simulations, the SEDAN [13] program from Stanford University is used.

Programs GALATEA and MEDUSA [14] have been obtained from the University of Aachen, West Germany, and are in use for the multi-dimensional simulation of semiconductor devices. These programs have been adapted and extended for Bell Laboratories' use. These extensions include a two-dimensional small-signal ac simulator. A program for simulating MOS transistors in two and three dimensions is also available [15]. All of the above programs operate on a super computer and are used principally in understanding device physics and in developing compact models for circuit simulators.

3.7 Process Simulation

Device simulation requires doping profiles and boundary conditions. Experimental determination of doping profiles is a difficult task at best in one dimension and it is even more difficult in two dimensions. Process simulation is thus important in predicting one- and two-dimensional doping profiles which can be used in device simulators. A process simulator SUPREM [16], developed at Stanford University, has been used for one-dimensional simulation. A new program, BICEPS [17], has been developed for process simulation in one or two dimensions. BICEPS is capable of simulating a large variety of processing steps, including diffusion, ion implantation, predeposition, local and global oxidation, epitaxial growth, and depositing and etching steps. Like the device simulators, BICEPS operates on a super computer.

While much of the effort in process simulation has been concentrated on determining impurity profiles, effort is now being spent in understanding the development of resists under exposure of light and electron beams. A program, SAMPLE [18], from the University of California, Berkeley, is used for this purpose.

The combination of two-dimensional process simulation, realistic models for processing steps, and a powerful computing facility have proven to be a valuable asset in the development of new technologies.

4. LAYOUT AIDS

Most of the layout aids are specific to the design styles which they address. Some of these aids are now funneling into a third-generation semi-automatic layout system which preserves the hierarchy specified by the designer and supports various design styles at the module level. Another layout aid is, however, very general since it provides a general interactive graphics capability. Among the aids described below are a modular semi-automatic layout system, an interactive layout system, and module generation programs based on symbolic layout.

4.1 Modular Semi-Automatic Layout

Increasing complexity of integrated circuits and increasing volume of designs have forced the designers to use modular layout procedures which are amenable to automation. Two such styles that are used in custom and catalog logic parts are the polycells used in the MOS circuits and function blocks used in the bipolar circuits. The primary layout system in both cases is LTX [19,20]. This system has been used to design a large number of custom MOS and bipolar chips. An example of a complex custom CMOS chip [21] containing about 32,000 transistors is shown in Figure 2.

LTX provides both two-dimensional and one-dimensional placement capabilities. The success of two-dimensional placement depends on a good initial partial placement by the designer. One-dimensional placement algorithms perform in-row placement optimization. A dog-leg channel router which usually performs within ten percent of optimum is used to make interconnections [22]. Interactive editing of the routing can be performed on a graphics screen under program control. The layout generated by LTX is thus error-free.

Output of LTX is a partial mask description in the XYMASK language which is then used in the interactive graphics system described below to perform power bus routing, to add features required by the mask shop, and finally to create a magnetic tape for driving a pattern generator.

A new version of the LTX system provides extensive support for hierarchical layout [23]. Several new algorithms for placement and signal and power bus routing, combined with bottom-up and top-down design capabilities and on-line rule checking make this system a powerful tool for VLSI layout.

4.2 Interactive Layout

For specialized and non-structured layout tasks, an advanced, hierarchical graphical editor, GRED, is available. Among its key features is the capability for feature snaps that allow a designer to work at a relatively low magnification without the need for frequent windowing. A large set of commands allows experienced users to work with high productivity. The output of GRED is in the XYMASK language. In a typical usage pattern, a layout phase on GRED is followed by a design rule checking and characterization phase using HCAP (see below). Among the uses of GRED is layout of analog circuits and of NMOS polycells.

4.3 Symbolic Layout

Special forms of symbolic layout approaches which use a grid structure underlying the symbolic representations are used in CMOS digital circuits. A CMOS polycell generation aid CPGEN [24] allows creation of a CMOS polycell library in a technology updatable manner. As shown in Figure 3, the design style consists of a horizontal grid used for placement of contacts and interconnection metal. The grid structure assures that design rules will not be violated. The grid spacing depends on the design rules and changes with each new set of design rules. A complete library of cells in a symbolic form can thus be adapted quickly to a new set of design rules.

Another form of symbolic layout which has been found useful for microprocessors is the Gate-Matrix style [25]. Aids are available for the creation, editing, checking, and transformation of character symbolic layout blocks.

5. LIBRARY AIDS

A major problem in the layout and simulation of integrated circuits is the generation and support of module descriptions for the standard building blocks modules. Creation of simulation and layout data for a library of modules such as MOS polycells or bipolar function blocks can require enormous effort--if aids are not used in this process. Not only is a manual procedure more time-consuming, it is also highly error-prone. In addition to the difficulty of generating libraries, one is constantly faced with obsolescence resulting from the advances in technology.

A second-generation artwork characterization and verification system HCAP [26] aids in automatically generating libraries which are used for simulation by MOTIS and ADVICE, and for layout by LTX. It supports both MOS and bipolar technologies. The input to this system is the artwork description of a module in the XYMASK language with embedded textual data for linking names with shapes. Technology independence is achieved by using a command file which governs the Boolean graphical operations needed to perform device recognition and to determine circuit connectivity. The command file also contains the various parameters such as those for calculating parasitics.

Besides characterization of artwork and translations into formats accepted by other aids, HCAP performs design rule checking. The design rule checking in the HCAP environment benefits from the connectivity information deciphered by the system. Many false alarms are avoided and real errors which would not be uncovered by the usual clearances-only checkers are found. Also, certain application dependent checks are made possible such as checks when the tolerance in the x direction being different from that in the y direction.

While HCAP performs "intelligent" checks for design rules on layout blocks of limited size, a companion program LARC performs a subset of these checks on large chip level layouts. The program uses powerful algorithms for shape-merging and other Boolean operations prior to performing the checks. This program can efficiently handle complete chips having complexities of the order of a 32-bit microprocessor.

The artwork characterization and verification capabilities of the HCAP system have been used for circuits other than those requiring module libraries. Examples are portions of memory, digital signal processors, and microprocessor chips.

Among the various outputs produced by HCAP, are shaded plots of the mask data as shown in Figure 4. For standard cells and gate matrix blocks HCAP generates gate oriented circuit diagrams such as that shown in Figure 5, thus aiding the closing of the loop between the design intent and layout realization.

6. TEST AIDS

Part of the design and specification of an integrated circuit chip are the tests to be applied to it to verify the operation and performance after fabrication, and to separate good chips from faulty ones resulting from defects introduced in processing. This involves the use of a fault simulator which predicts the fault coverage of the circuit by a set of input test patterns. The designer adds vectors to the test set to bring the fault coverage to an acceptable level.

The MOTIS fault simulator [27] can model and simulate classical "stuck-at" faults for circuits described at the conventional logic gate level. In addition, the simulator provides the novel capability of modeling and simulating MOS circuits at the MOS device level. For these circuits, besides the classical stuck-at faults, nonclassical transistor stuck-closed and stuck-open faults can optionally be modelled and simulated. ROMs, RAMs and PLAs, if present in the circuits, can be directly simulated at the functional level as supported in design verification. Fault simulation can be carried out in the conventional unit delay mode or optionally using integer rise/fall delays as computed by the MOTIS multiple-delay simulator.

In general, the tests described above are not in a format which is acceptable to test hardware. Aids, such as a program called TPG, are used to convert these tests into a complete test program which is optimized for factory production. In addition, TPG generates test specifications which become part of the design transmittal package for the manufacturing line.

The results produced by a test set are typically voluminous, beyond the capabilities of an individual to analyze in a timely fashion. Usually, analysis is required to diagnose problems in design during development. A program ANAPAC is used to obtain wafer maps, histograms, or statistical summaries. Problems in design, specification, mask set, or processing may be uncovered.

7. CONCLUSION

A system of computer aids, for the design and testing of VLSI circuits, has been presented. Some of the aids are of a general nature and technology independent. Others exploit the advantages offered by modular design procedures and enable the quick turnaround which is needed to introduce advanced products in a timely manner. The tools are constantly evolving to keep pace with the rapidly advancing semiconductor technology and with advances in computing, communication, and test equipment.

REFERENCES

[1] H. Y. Chang, G. W. Smith, and R. B. Walford, "LAMP: System Description", Bell Sys. Tech. J., Vol. 53, No.8, October 1974, pp. 1431-1449.

[2] B. R. Fowler, "XYMASK", Bell Laboratories Record, Vol. 47, No.6, July 1969, pp. 204-209.

[3] B. R. Chawla, H. K. Gummel, and P. Kozak, "MOTIS - An MOS Timing Simulator", IEEE Trans. Cir. & Sys., Vol. CAS-22, No.12, December 1975 pp. 901-910. Also, P. Kozak, H. K. Gummel, and B. R. Chawla, "Operational Features of an MOS Timing Simulator", Proc. of 12th Design Automation Conf., Boston, Mass., June 23-25, 1975, pp. 95-101.

[4] H. N. Nham and A. K. Bose, "A Multiple Delay Simulator for MOS LSI Circuits", Proc. of 17th Design Automation Conf., Minneapolis, Minn., June 23-25, 1980, pp. 610-617.

[5] V. D. Agrawal, A. K. Bose, P. Kozak, H. N. Nham, and E. Pacas-Skewes, "A Mixed-Mode Simulator", Proc. of 17th Design Automation Conf., Minneapolis, Minn., June 23-25, 1980, pp. 618-625.

[6] V. D. Agrawal, "Synchronous Path Analysis in MOS Circuit Simulator", Proc. of the 19th Design Automation Conf., Las Vegas, Nevada, June 1982.

[7] L. W. Nagel, "ADVICE For Circuit Simulation", 1980 IEEE ISCAS, Houston, Tex., April 28-30, 1980.

[8] L. W. Nagel, "SPICE2: A Computer Program to Simulate Semiconductor Circuits", Electronics Research Laboratory Report No. ERL-M520, University of California, Berkeley, May 1975.

[9] H. K. Gummel and H. C. Poon, "An Integral Charge Control Model of Bipolar Transistors", Bell Sys. Tech. J., Vol. 49, No.5, May-June, 1970, pp. 827-852.

[10] H. C. Poon, "VT and Beyond", VLSI Modeling Workshop, Burlingame, California, March 29, 1979.

[11] W. Rosenzweig, private communication.

[12] H. K. Gummel, "A Self-Consistent Iterative Scheme for One-Dimensional Steady State Transistor Calculations", IEEE Trans. Elec. Dev., Vol. ED-11, October 1964, pp. 455-465.

[13] D. C. D'Avanzo, M. Vanzi, and R. W. Dutton, "One-Dimensional Semiconductor Device Analysis (SEDAN)", Stanford Electronics Labs., Tech. Rep. No. G-201-5, October 1979.

[14] W. L. Engl and H. Dirks, "Numerical Device Simulation Guided by Physical Approaches", Proc. of the NASECODE-I Conf., Dublin, Ireland, June 27-29, 1979, pp. 65-93. Also, "Functional Device Simulation by Merging Numerical Building Blocks", Proc. of the NASECODE II Conf., June 17-19, 1981, pp.34-62.

[15] W. Fichtner, D. J. Rose and R. E. Bank, "Semiconductor Device Simulation", Joint SIAM-IEEE Electron Devices Society Conference on Numerical Simulation of VLSI Devices, Boston, Mass., November 2-4, 1982.

[16] D. A. Antoniadis, S. E. Hansen, and R. W. Dutton, "SUPREM II - A Program for IC Process Modeling and Simulation", Technical Report No. 5019-2, Stanford Electronics Laboratories, June 1978.

[17] B. R. Penumalli, "Numerical Aspects of IC Process Simulation", Joint SIAM-IEEE Electron Devices Society Conference on

Numerical Simulation of VLSI Devices, Boston, Mass., November 2-4, 1982.

[18] A. R. Neureuther, "Simulating VLSI Wafer Topography", Tech. Dig. IEDM, December 1980, pp. 214-218.

[19] G. Persky, D. N. Deutsch, and D. G. Schweikert, "LTX - A System for the Directed Automatic Design of LSI Circuits", Proc. 13th Design Automation Conf., San Francisco, June 28-30, 1976, pp. 399-407. Also, G. Persky, D. N. Deutsch, and D. G. Schweikert, "LTX - A Minicomputer Based System for Automated LSI Layout", J. Des. Aut. & Fault-Tol. Comp., Vol 1, No.3, May 1977, pp. 217-255.

[20] D. N. Deutsch and P. Glick, "An Over-the-Cell Router", Proc. 17th Design Automation Conf., Minneapolis, Minn., June 23-25, 1980, pp. 32-39.

[21] C. H. Hao and S. V. Kartalopoulous, "VLSI Chip of the Time Compression Multiplexer for the Circuit Switched Digital Capability", Globecom '82, Vol. 1, pp. 386-389.

[22] D. N. Deutsch, "A Dogleg Channel router", Proc. of the 13th Design Automation Conf., San Francisco, CA, June 1976.

[23] B. W. Colbry and J. Soukup, "Layout Aspects of the VLSI Microprocessor Design", 1982 Proceedings of the International Symposium on Circuits and Systems, Rome, Italy, pp. 1214-1228.

[24] C. M. Lee, B. R. Chawla, and S. Just, "Automatic Generation and Characterization of CMOS Polycells", 18th Design Automation Conf., Nashville, Tenn., June 29-July 1, 1981, pp. 220-224.

[25] A. K. Lopez and H. F. Law, "A Dense Gate Matrix Layout Style for MOS LSI", 1980 IEEE ISCC, Vol 23, pp. 212-213. Also, D. E. Blahut, R. Krambeck, H-F. S. Law, H. Shichman and H. C. So, "Hierarchical Design Methodology for a Single Chip 32-Bit Microprocessor", Proc. ICCC 1982, pp.16-20.

[26] P. A. Swartz and H. K. Gummel, "Layout Characterization and Verification", CANDE Workshop, Lake Tahoe, Calif., Sept. 15-16, 1980.

[27] A. K. Bose, P. Kozak, C-Y. Lo, H. N. Nham, E. Pacas-Skewes, and K. Wu, "A Fault Simulator For MOS LSI Circuits", Proc. of the 19th Design Automation Conf., Las Vegas, Nevada, June 1982.

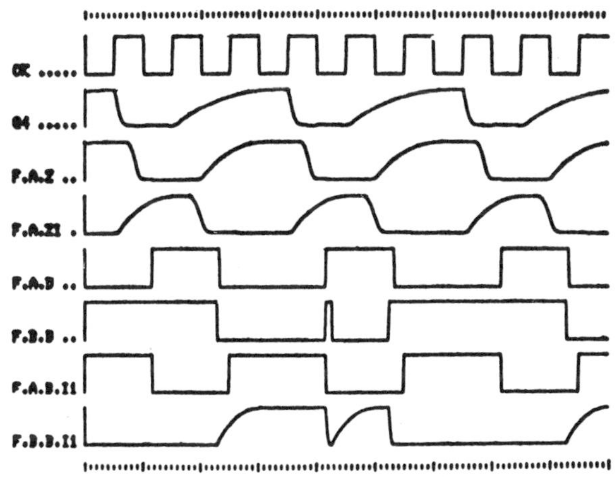

Figure 1 Example of MOTIS mixed-mode simulation.

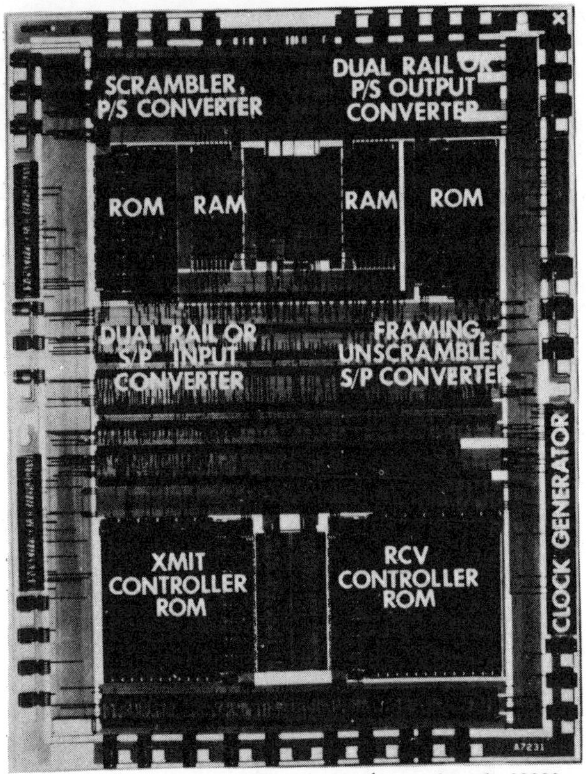

Figure 2 A custom CMOS chip (Approximately 32000 transistors).

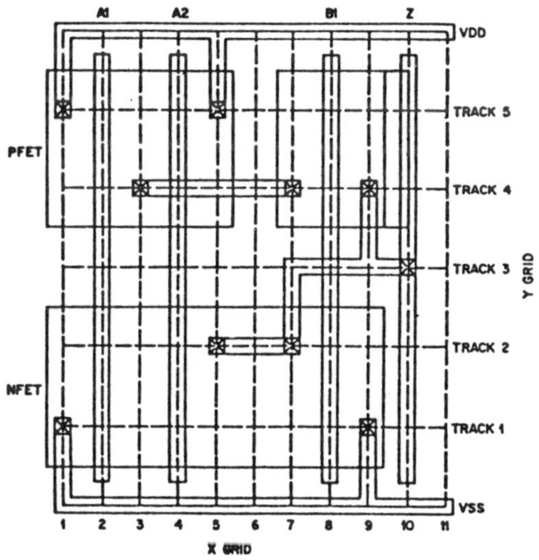

Figure 3 The symbolic layout of a CMOS polycell.

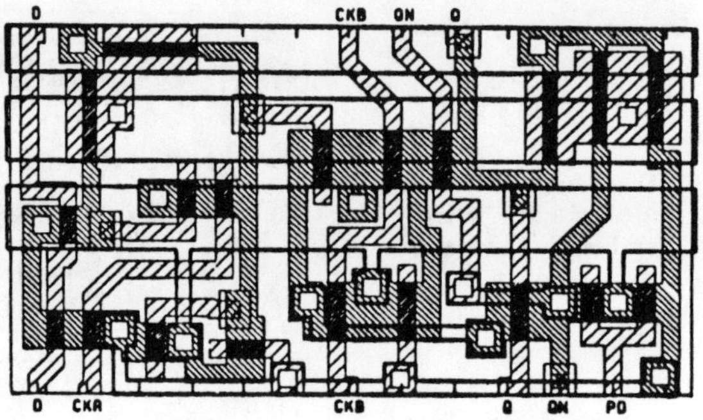

Figure 4 Shaded plot of mask produced by HCAP.

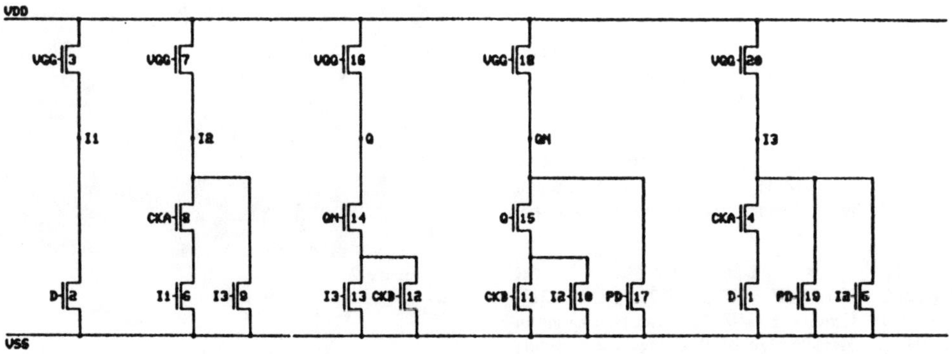

Figure 5 HCAP generated circuit diagram.

The VLSI Design Automation Assistant: From Algorithms to Silicon

T.J. Kowalski, D.J. Geiger,* W.H. Wolf, and W. Fichtner
AT&T Bell Laboratories

The complexity of integrated circuits has increased at an enormous rate, from a few hundred components on a chip in the 1960s to hundreds of thousands of components in the VLSI chips of the mid-1980s. This explosion in complexity was made possible by advances that continue to shrink the dimensions of devices in IC fabrication.

During the last few years, we have witnessed the realization of complete systems—32-bit microcomputers[1] and digital signal processors,[2] for example—on a single VLSI chip. While these designs were realized with conventional processes in two-micron design rules, considerably larger systems will be possible with state-of-the-art one-micron and submicron silicon technology.

Although we can build complete systems on a single silicon chip, the design of a VLSI chip is extremely expensive. It can take several years to go from the conception of an architecture to the completion of the final product. The complexity and expense of chip design makes it increasingly difficult to build cost-effective, low-volume, special-purpose VLSI systems.

We believe the only solution to this problem is to build new CAD tools that are capable of automatically performing more of the synthesis process. In the past few years, tools have been developed to act as expert assistants for stages in the VLSI design process, from architecture definition to layout. Our approach seeks to aid the designer with tools that automatically produce data paths and control sequences from an algorithmic system description within user-supplied constraints.

An automatic design aid offers several advantages. First—and probably most important—it drastically

Summary

Our ability to fabricate complex VLSI chips is outrunning our ability to design them. To reduce the design time for complex chips we are creating an expert design assistant—programs that automate as many of the design tasks as possible while allowing the human designer to control the synthesis process. A series of programs translates an algorithm into a chip layout. Some programs are expert systems while others use traditional algorithms.

Our early results lead us to believe that a knowledge-based expert system is an appropriate tool for bridging the gap between logical and physical design.

*Geiger is now with Carnegie-Mellon University.

Reprinted from *IEEE Design and Test of Computers*, pp. 33–43, August 1985.

decreases the time it takes to design a chip. Second, it allows the designer to investigate a wide variety of different strategies before deciding on a final design. This can even include the evaluation of different technologies in order to choose the one best suited to the particular problem. Third, it allows designers to use the most up-to-date fabrication technology available.

A cornerstone of our hardware-synthesis approach is the use of knowledge-based expert systems. Such systems make decisions based on knowledge, expressed as rules, obtained from expert designers. This approach to hardware synthesis has several advantages. First, much of the problem is too poorly understood, and the nature of the problem changes too rapidly to lend itself to solution by traditional algorithmic approaches. Second, a knowledge-based expert system can replace expensive backtracking with knowledge-directed search, thus producing good hardware designs in a reasonable amount of CPU time. Third, the organization of an expert system makes it easy to add, modify, and test new expert knowledge. Finally, the discipline of encoding expert knowledge into rules enriches our understanding of both the knowledge and the structure of the problem.

Therefore, while some stages of chip design—notably layout generation—can be handled well by algorithms, the most challenging areas require the application of knowledge-based expert systems.

While our research is not complete, we have promising results at both the high and low end of the chip synthesis problem. At the high end of the design process, we have demonstrated that a knowledge-based expert system can produce quality block-diagram designs from algorithmic descriptions of chips.[3] At the low end, we have built on the work done on symbolic layout[4-6] and on our work in compiling layout cells from simple descriptions.[7] This experience leads us to believe that a knowledge-based expert system floor planner can bridge the gap between the high and low ends of chip design.

In this article, we describe our approach to automatic design of VLSI silicon chips from algorithmic descriptions, briefly review related work, and present our plans for future work on the system.

Methodology for hardware synthesis

We view hardware synthesis as the creation of a detailed representation from a more abstract representation. The designer first specifies the chip's functionality as an algorithm. The algorithmic description is not tied to a particular implementation style, such as parallel or serial, sequential or pipelined. Similarly, it is not restricted to a fabrication technology, such as TTL, ECL, NMOS, or CMOS.

This algorithmic description is then transformed into a technology-independent network, built of modules of medium complexity—registers, adders, and multiplexers (MUXes)—and a symbolic control description. This network is used to synthesize a technology-dependent network that includes implementations of the technology-independent modules and the controller in the target technology. The next step is the construction of a floor plan that describes the placement and wiring of the chip in terms of the modules in the technology-dependent network. The floor plan guides the assembly of the final layout of the chip.

This design process is illustrated in detail in Figure 1. Typically, the chip proceeds through six different tools as it is transformed from an algorithm to a layout. Briefly, the steps in the synthesis process are:

• The design automation assistant (DAA)[8]: a knowledge-based expert system responsible for transforming the algorithmic description of the chip into a block diagram and a symbolic control specification. In addition to this algorithmic description, written in

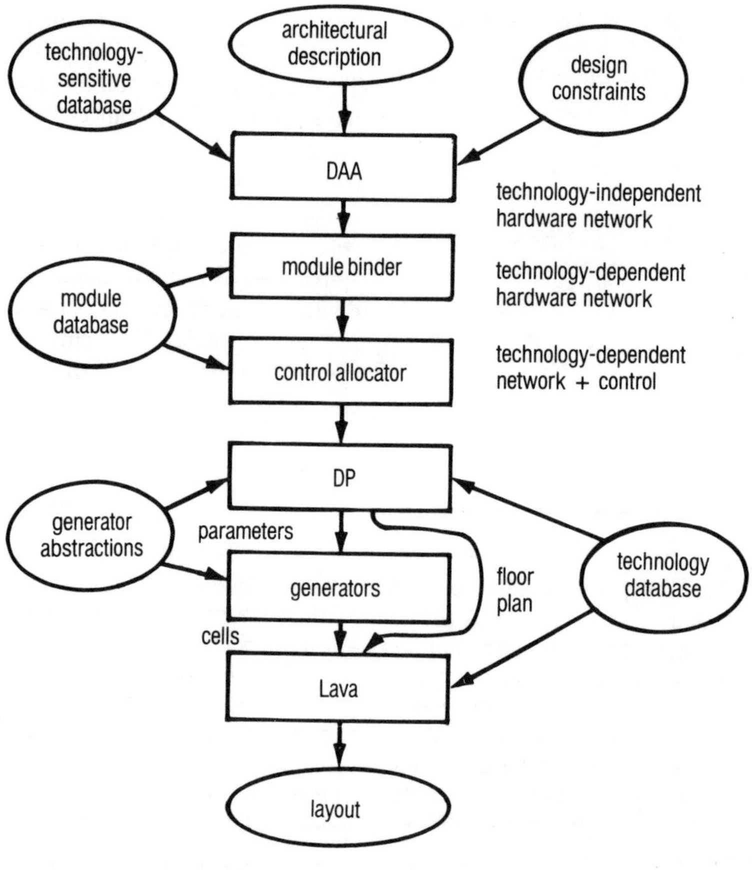

Figure 1. The VLSI design process.

ISPS,[9] the DAA uses a technology-sensitive database and a set of user-supplied constraints. The technology-sensitive database contains expert knowledge about the trade-offs particular to the target technology; the user-supplied constraints customize the design for the particular application. The DAA has been used to design several digital systems, including an IBM System/370. We are currently developing a way to create pipeline-style designs.

- The module binder[10]: an algorithmic program that builds the modules specified by the DAA from components in the target technology. The technology-dependent modules are either selected from the module database or fabricated from simpler, equivalent database entries. The selection process is guided by user-supplied constraints on area, power, and speed. The module binder is the master's thesis project of E. Dirkes of Carnegie-Mellon University.

- The control allocator[11]: an algorithmic program that designs the controller for the data-path logic designed by the DAA and the module binder. It can target the controller toward either a microprogram engine or programmable logic array (PLA) implementation. The control allocator was developed at Carnegie-Mellon University, and work is being done at AT&T Bell Laboratories to add pipeline-style designs.

- The data-path floor planner (DP): a knowledge-based expert system that is responsible for the complete physical design of the chip. It arranges the modules selected by the module binder and control allocator into a data-path floor plan and creates the cell-layout parameters. The floor planning process is constrained by user-supplied restrictions on area and speed and on knowledge supplied by two databases: the technology database and the generator-abstraction database. The technology database provides generic information such as unit-wire capacitance, while the generator-abstraction database provides estimates of the size and speed of particular cells. The DP is in the early stages of development.

- Generators: algorithmic programs that produce technology-specific custom cell layouts from parameters that describe logical, electrical, and layout requirements. Generators are not implemented as knowledge-based expert systems, but instead are written as algorithms by layout experts. Sample generators and programs exist to compile cells. Current work focuses on understanding the representation of generator knowledge, and expanding the library of generators.

- Lava: an algorithmic program that compacts a layout for each individual cell, then assembles the cells into a complete chip, stretching them when necessary to make the required connections. The result is a layout that can be used for fabrication. Lava, initially developed at Stanford University, has been used to design several chips.

While some stages of chip design can be handled well by algorithms, the most challenging areas require the application of knowledge-based expert systems.

Related research in synthesis design

The work done thus far in digital design synthesis is reminiscent of the early chess playing programs. While it is easy to get a computer to play a legal game of chess, it is exceedingly difficult to get it to play a good game of chess. Many design-synthesis projects have, like the early chess programs, produced a working result,[12] but the quality of the results leaves something to be desired. Some programs have produced designs of unacceptable quality, others are too expensive computationally for even small designs, and still others tackle only a small part of the VLSI design problem.

VLSI synthesis research can be divided into several major categories. One category is logical design—the synthesis of a technology-independent hardware description from an algorithmic specification. Research efforts in this category include a classical compiler,[13,14] an interactive optimization system,[15] a constraint language system,[16] a graph-theoretic system,[17] and an iterative heuristic approach.[18] These methods have various advantages and disadvantages, but in our opinion none of them produce acceptable results on realistic examples in a feasible amount of CPU time.

At the opposite end of the synthesis spectrum is physical design—the production of a layout from gate- or module-level hardware descriptions. Polycell systems like LTX II,[19] and standard-cell systems[20,21] can produce layouts with little or no human intervention by placing and routing predesigned cells that implement logic and memory functions. This category also includes module generators—programs that produce related cells according to the designer's specifications. Module generators include layout languages embedded in programming languages (Chisel),[22] compilation algorithms (such as Dumbo[7]), and knowledge-based expert systems.[23,24]

Between these two extremes is the silicon compiler. A silicon compiler takes as input an algorithmic description of a chip, and produces a layout. The compiler does this by using a target architecture for the logical design and a target floor plan for the layout design. The first silicon compiler was Bristle Blocks;[25] another well-known example is MacPitts.[26] In both systems, the logical and physical designs of the functional units are predefined by a human designer.

Design strategies

Our strategy is to break up synthesis into several programs. We believe our approach has some advantages for software engineering: it simplifies the implementation of the subsystems; it makes it easier for the designer to examine the state of the design and to modify it if necessary; and it allows the choice of different computing environments for different levels. While nonalgorithmic programs such as the DAA and DP are best implemented in the environment of a knowledge-based expert system, algorithmic programs such as Lava and the generator are best implemented in a programming-language environment.

We have two reasons for choosing algorithmic generators over a general knowledge-based expert system approach. First, we can write specialized generators corresponding to the hardware building blocks—ALUs and registers, for example—and make use of expert knowledge about their physical design. Second, we believe that knowledge-based expert-system layout generators cannot produce layouts for high-performance modules with the same success as has been achieved with algorithmic generators.[27]

DAA knowledge. The DAA is a knowledge-based expert system that uses a database of more than 500 rules to synthesize an architectural implementation from an algorithmic description with constraints. The description is written in ISPS; a sample description for a Digital Equipment Corp. PDP-8 is shown in Figure 2. The first part of the description lists the current-page and instruction words; the second part of the description tells how to calculate an effective address. The DAA actually works from a data-flow representation extracted from the ISPS description. This data-flow representation resembles the internal form used by most optimizing compilers, and it is less sensitive to the writing style of the algorithmic description.

The DAA produces a technology-independent hardware network. This network is composed of modules, ports, links, and symbolic microcode. The modules can be registers, operators, memories, and buses or MUXes with input, output, and bidirectional ports. The ports are connected by links and are controlled by the symbolic microcode.

The DAA uses a set of temporally ordered subtasks to perform the synthesis task. It begins by allocating the base-variable storage elements—constants, architectural registers, and memories with their input, output, and address registers—to hardware modules and ports. Then a data-flow BEGIN/END block is picked, and the synthesis operation assigns minimum delay information to develop a parallel design. Next, the DAA maps all data-flow operator outputs not bound to base-variable storage elements to register modules. Finally, it maps each data-flow operator with its inputs and outputs to modules, ports, and links. In doing so, the DAA avoids multiple assignments of hardware links; it supplies MUXes when needed. These last two mapping steps place the algorithmic description in a uniform notation for the expert analysis phase that follows.

The expert analysis subtask first removes registers from those data-flow outputs where the sources of the data-flow operator are stable. Operators are combined to create ALUs according to cost, partitioning, and pipeline estimators across the allocated design. The DAA also examines the possibility of sharing nonarchitectural registers. Where possible, it performs increment, decrement, and shift operations in existing registers. Where appropriate, it places registers, memories, and ALUs on buses. Throughout this subtask, constraint violations require trade-offs between the number of modules and the partitioning of control steps. The process is repeated for each data-flow BEGIN/END block.

The knowledge-based expert system approach followed by the DAA uses a weak method,[28] *match*, in place of extensive backtracking. The DAA uses match to explore the space of possible designs by extending a design from an initial state to a final state without any backtracking. The DAA proceeds through its major tasks in the same order for each problem; it never varies the order and it never backs up in any problem. This means that at any intermediate state, the DAA can determine how to extend the design to achieve an acceptable result.

DAA produces a technology-independent, but technology-sensitive implementation of the ISPS description.

```
Example :=
Begin

**Storage.Declaration**

cpage\current.page<0:4>,
i\instruction<0:11>,
  pb\page.0.bit< > := i<4>,
  pa\page.address<0:6> := i<5:11>

**Address.Calculation**

Global eadd\effective.address<0:11> :=
  Begin
  Decode pb =>
    Begin
    0 := eadd ='00000 @ pa,
    1 := eadd = cpage @ pa
    End
  End
End
```

Figure 2. Sample ISPS description.

The technology-independent network and associated control are described in the SCS language; a graphical representation is shown in Figure 3. The circles concatenate two sets of signals, the trapezoid is a two-input MUX, the small rectangles are registers, and the large rectangle is the controller. Because the initial ISPS was so simple, the final design includes only some wiring, the specified architectural registers, a MUX, and the controller. The control specification is also included in the SCS description. The control is described in terms of abstract operators—control steps and selects, for example—that can be implemented in a variety of controller technologies.

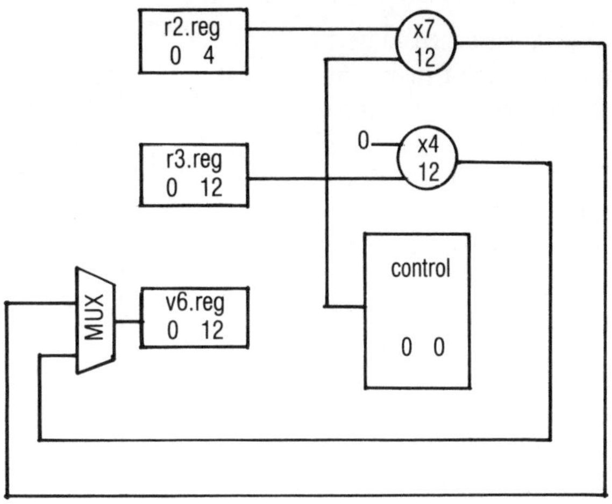

Figure 3. Sample structural specification.

Module binding. The module binder chooses physically realizable, technology-dependent cells to implement the modules in the DAA's design. The cells are chosen from the module database according to user-supplied constraints on functionality, power dissipation, delay, and area. The module binder may, for example, choose a ripple-carry adder when area is the major design constraint, or a carry-lookahead adder when performance is the major design constraint.

If the module database does not contain a cell with the right combination of functions, a combination of existing cells is used. For example, if the binder is unable to find a single module corresponding to some combination of an ADD and an OR, it will select separate ADD, OR and multiplexer cells to generate such a module. Similarly, if the module database does not contain a cell of the proper bit width, an existing cell may be replicated. Finally, if the DAA has specified the use of buses, the module binder will provide technology-dependent bus drivers.

Control allocator. The control allocator translates the technology-independent control specification, generated by the DAA, into a controller implementation in a chosen technology. A style-independent part determines the control signals that drive the data path and a style-dependent part designs the sequencer. This method is desirable because the style-independent part is the same regardless of the way the controller is implemented (PLA, microprogrammed controller, or random logic), but the style-dependent part changes with each different implementation.

The control allocator receives information from the technology-dependent hardware network and the module database to produce an output file for a specific style of controller. The structure-and-control-specification file[29] provides the control signals and the technology-dependent data path; the module database provides the required control line values. Currently, our control allocator can produce PLA-style output for the Berkeley[30] and the AT&T Bell Laboratories[31] PLA generation tools or a microprogrammed-style output for an AM2910 microengine.

Floor planning. The floor planner produces an abstract physical design for the technology-dependent network designed by the DAA and the module binder. The floor plan includes a placement and wiring plan for the

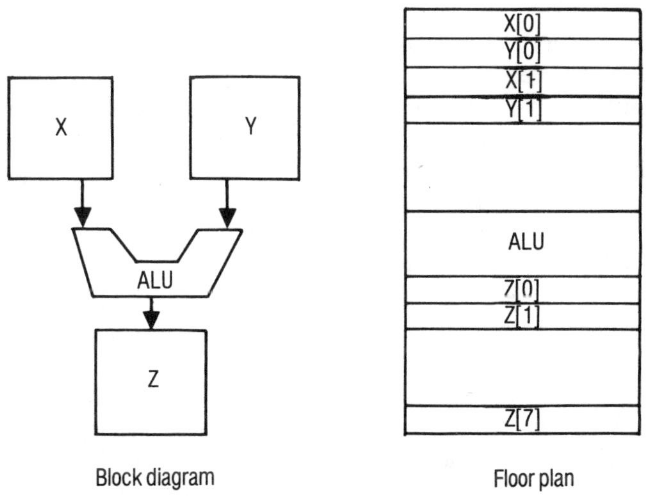

Figure 4. Sample netlist to floor plan transform.

modules and specifications for the design of the cells.

The DP knowledge-based expert system is driven by a set of rules that embody expert knowledge about floor planning. There are two categories of rules: those that build an initial floor plan for the chip based on the logical design and user constraints, and those that improve on the initial design.

The use of rules to build an initial floor plan can be illustrated using the technology-dependent design shown in Figure 4. The left side of the figure shows a block diagram for a system made of two input register arrays, X0-X7 and Y0-Y7; an ALU; and an output register array, Z0-Z7. The right side of the figure shows a common and efficient floor plan for this system: The input register arrays are interleaved above the ALU; the output register array is placed below the ALU.

This transformation from block diagram to floor plan suggests three rules: If word arrays of elements (such as X0-X7) are encountered, then rules are to be applied to each element independently; if elements are connected in parallel, then they may be interleaved; and if two elements are connected, then the source element should be placed above the sink element in the floor plan.

The second category of rules describes optimization methods, the type of knowledge we normally associate with expert designers. An example optimization transformation is shown in Figure 5. The critical timing path for the circuit flows from register B through the ALU and into the output latch. The initial floor plan, however, places register A closer to the ALU than register B. The transfer time from B to the ALU can be reduced by swapping A and B, thus reducing the capacitive load that B must drive.

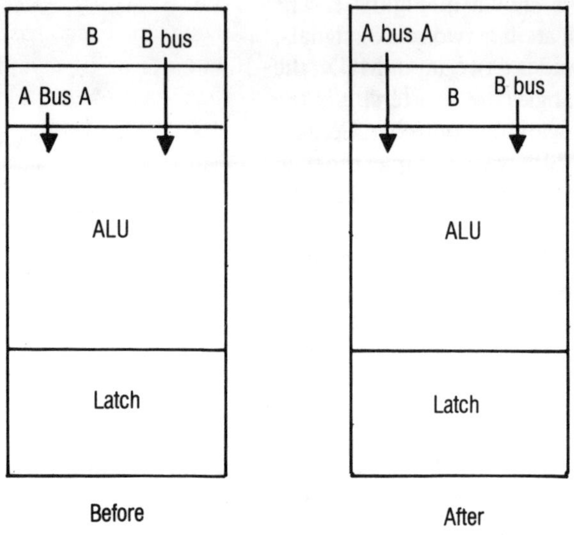

Figure 5. Sample optimization transformation.

A secondary source of expert knowledge, one not contained in the rule database, is the generator-abstraction database. Each generator is represented in the database by an abstraction that includes essential knowledge of the physical, electrical, and logical properties of the generated cells. This abstraction is necessary because the information is difficult to obtain directly from the generator. Much of the desired information—input load capacitance, aspect ratio, and drive current, for example—is included only implicitly in the generator code, so the floor planner has to make many decisions based on the information in the technology database. The desired cell

Using a series of algorithmic and non algorithmic techniques, our goal is to generate silicon from algorithms.

could be generated and then measured, but this would be too time-consuming for chip synthesis.

Some of the information in the abstraction is precise; other information is estimated from heuristics and particular knowledge about the generator. An abstraction may contain any mix of exact and estimated information, as specified by the abstraction's designer (usually, the generator designer). The database includes methods for estimating physical parameters (area and shape, for example) and timing parameters (delay and drive current) from partial descriptions of cells.

A floor plan is derived from the module binder's structural description by constructing an initial design and then improving the floor plan. Both initial placement and improvement are driven by the rule database and are guided by information provided by the generator abstractions. At the same time the floor plan is being built, the specifications for the modules—size, shape, delay, etc.—are being refined. These specifications will be passed to the module generators to guide the production of the cells.

The resulting floor plan is described in the Lava language. A sample floor plan for a bit-slice design is shown in Figure 6a, and a drawing of the floor plan is shown in Figure 6b. The floorplan header describes the chip's connections to the outside world. A signal can be a single wire, such as packet, or a bundle of wires, such as bitslicebot. (Actually, bitslicebot and bitslicetop are bundles of bundles.) The abut operator, |, asserts that two cells are to be abutted together and stretched if necessary to connect all wires between them. The keywords left, right, down and up show the relative placement of cells. Wire routing is contained in cells that are abutted to the cells implementing the modules.

Cell generation and assembly. A cell generator embodies expert knowledge of how to design a family of cells. The generator program produces a Lava description of a cell from parameters that define the requirements on the cell. At present we can generate modules from a database of predesigned modules or by compiling them using the Dumbo cell compiler. Cells can be added to the database via a Lisp-based symbolic layout editor.

Part of a cell-generator program is shown in Figure 7. This sample program produces an NMOS half-adder circuit in a transmission gate implementation.

The Dumbo compiler works from a predesigned cell, defined as gates and transistors, an external pin list, and hints about the form of the cell's layout. A preprocessor customizes the generic description based on the input parameters by selecting transistor sizes, optional components, etc. Dumbo then uses placement, routing, and component definition algorithms to

```
cell bitslice(bitslicebot, bitslicetop, packet, packetbar, word_out) abut
   bitslicebot : bitslice : down;
   bitslicetop : bitslice : up;
   packet, packetbar: diff: left;
   word_out: poly: right;
{
   shifter(packetbar, packet, bitslicebot.shift, bitslicetop.shift) |
leftof
   parity(databar, data, data_out, bitslicebot.par, bitslicetop.par) |
leftof
   iterate NWORDS | right queue(queueleft, queueright,
      bitslicebot.q, bitslicetop.q)
     {queueright # queueleft } |
leftof
   latchout(queueright[NWORDS].shiftsig, word_out, bitslicebot.outputs,
      bitslicetop.outputs);
data_out # queueleft[1].datasig;
}
```

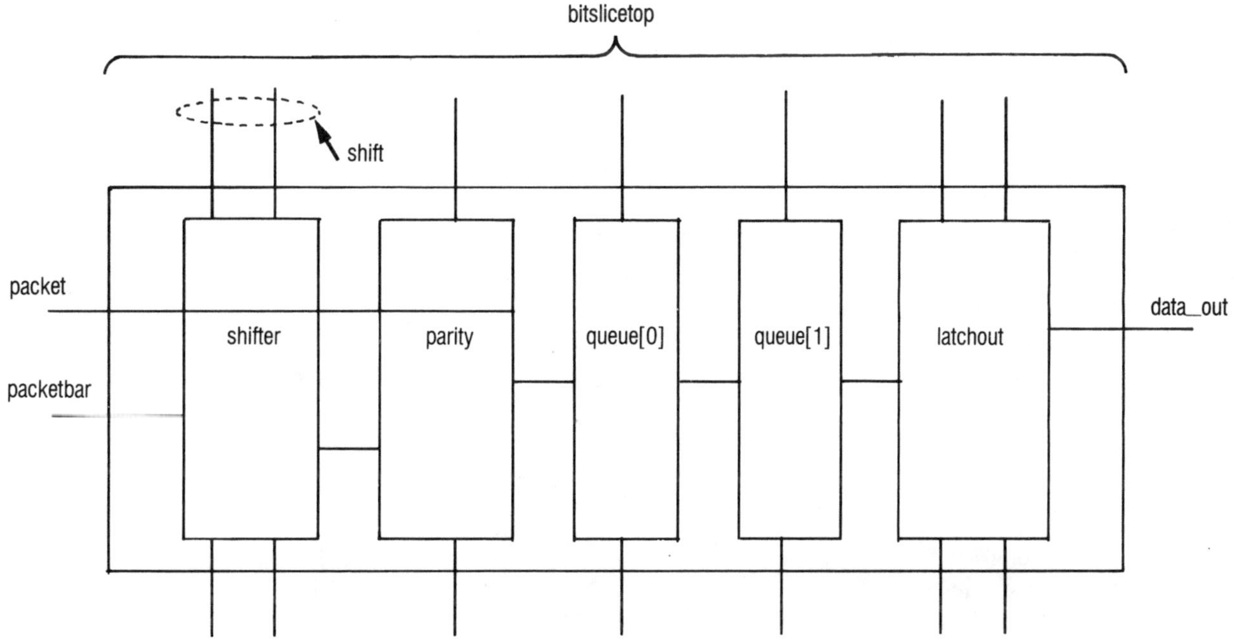

Figure 6a. Sample floor plan specification.

turn the generic cell into a symbolic layout.

Layout representation. The Lava chip compiler compacts the individual cells and assembles them into a chip according to the floor plan. Lava is responsible for stretching cells to make the electrical connections specified in the floor plan; it can also optimize the cells to reduce the layout's size. Lava produces a symbolic layout with its components placed at their exact positions in the final layout. This symbolic layout can be macro-expanded into a pure layout that can be used to fabricate the chip. Figure 8 shows the XYMASK version of the layout for the NMOS half-adder cell.

Future work

We are exploring the problem of synthesizing a working silicon chip from an algorithmic representation of a VLSI system. Using a series of algorithmic and nonalgorithmic techniques, our goal is to generate working chips from algorithms.

This article has outlined the work in progress on several tools. The first tool, the DAA, is still being developed, but is solidly in hand. Like human designers, the DAA becomes a better designer as its rule memory expands. Currently, DAA is being expanded to handle pipeline-style designs. This work is in its early development stages, but shows great promise.

```
/*A*/
cell tgha (vddtop, vddbot, vsstop, vssbot, xin, yin, sum, carry) leaf
vddtop : blue : up;
vddbot : blue : down;
vsstop : blue : up;
vssbot : blue : down;
xin : red : up;
yin : red : up;
sum : red : down;
carry : red : down;
{
      compact xy;
      (p0 = point blue [1,20]) #
            (width 4) vddtop;
      (p1 = point blue [22,20]) #
            (width 2) p0;
      :
      (t1 = load (2:2) rotate 6 [2,15]).source #
            (green) (p9 = via bg [15,15]);
      :
      p9 #
            (green) (t2 = tran enh (3:1) rotate 9 [15,13]).chant;
      :
      yin #
            (red) (p11 = point red [17,20]);
      :
      xin #
            (red) (p13 = via rg [29,19]);
      :
      carry #
            (red) t8.gate;
      :
      sum #
            (red) t11.gate;
}
```

Figure 7. Sample cell generator fragment.

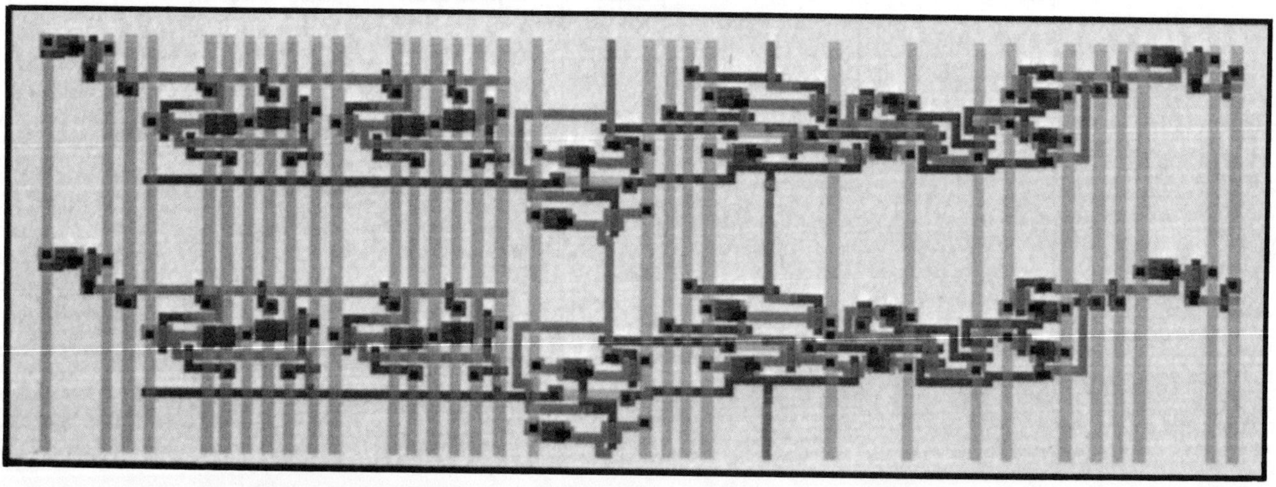

Figure 6b. Sample floor plan design.

The module binder is nearing completion at Carnegie-Mellon University.

The control allocator is complete, and has been used to design several PLA and microcoded controllers. The largest controller designed was for the RCA 1802 CMOS microprocessor. Changes will be needed in the control allocator if it is to design controllers for pipelined systems.

Development of the floor planning expert system is in early stages. We are building a floor planning knowledge base from interaction with chip designers and analysis of sample chip designs, such as the FLASH signal processor and other chips proprietary to AT&T Bell Laboratories. We are building the generator-abstraction database from interviews with layout designers and analysis of existing generator programs. Currently, we are redesigning several chips with our synthesis tools and comparing the automatic and hand designs.

After completing some working chip designs, we plan to experiment with high-level optimization based on layout analysis. Information about the physical design of the chip, gained during floor planning and cell generation, may be useful in optimizing the logic design or architecture of the chip. For example, if the floor planner is unable to produce a layout with an acceptable delay for a signal, the clocking scheme of the architecture may have to be redesigned. Automatic optimization is intriguing because the programs can generate and analyze much more information than can a human designer working alone, and because the new design can be reimplemented quickly. The problems are formidable, however, and our goals in this area are long-range.

This article has described our approach to automated VLSI design. We have discussed how we break the integrated circuit design process into stages and how to implement those stages as programs. We have aimed for a methodology that automates as much of the design work as possible, but still allows the designer to control the outcome of the synthesis. We hope to free the designer to solve the important problems in a chip design.

We strongly believe that only by automating more of the chip design process will we be able to exploit the possibilities of advanced VLSI fabrication technologies. We also believe that the combination of algorithms and knowledge-based expert systems offers significant benefits when building computer-aided design tools for VLSI. Some synthesis problems are solved best by algorithms; others have not yielded good results in reasonable amounts of computation time. Knowledge-based expert systems can solve these problems and yield good results in a reasonable amount of CPU time.

By combining expert systems with algorithmic approaches, we should be able to automate the entire custom chip design process. This development has profound implications. Automatically-produced designs can be guaranteed to be correct and can be produced far more quickly than if human intervention were required. The designer can either accept the automatic design or can use his knowledge to guide synthesis to a better result. Designers should be able to produce far better chip designs when aided by top-to-bottom synthesis aids than if they worked without CAD tools.

We still have much work to do, but we are encouraged by our results so far. We have demonstrated that the knowledge-based expert system methodology is an effective approach to high-level hardware synthesis. Our understanding of chip design leads us to believe that we can apply expert knowledge to the physical design of VLSI chips with similar success. □

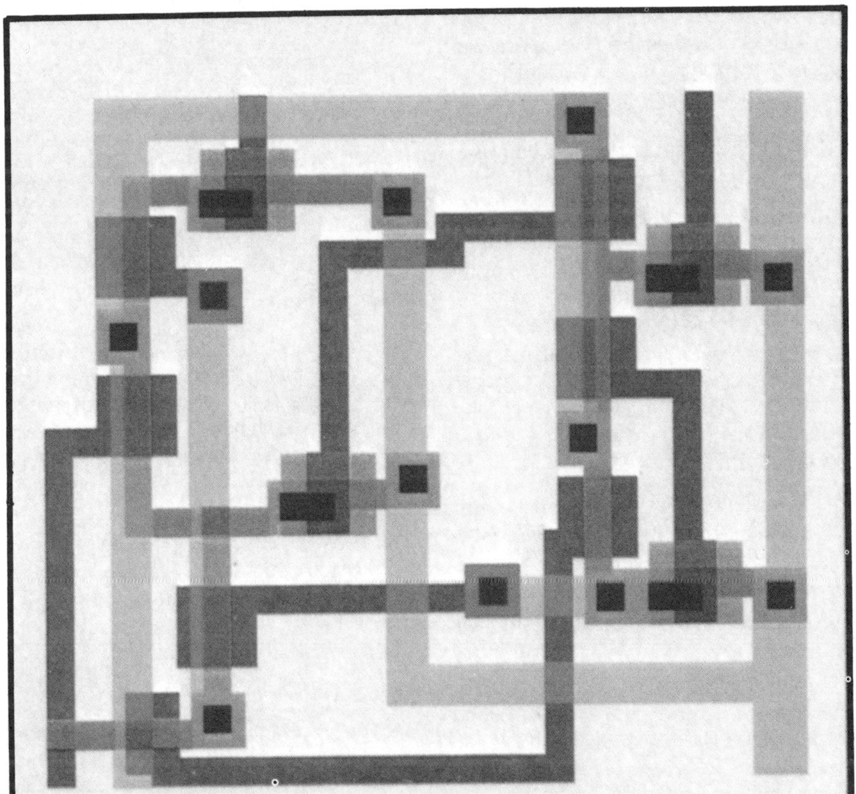

Figure 8. Layout in XYMASK form.

Acknowledgments

The work described here has built on work done by us and others at Carnegie-Mellon University and Stanford University. At Carnegie-Mellon we would particularly like to thank S. Director, E. Dirkes, C. Forgy, D. Gatenby, C. Hitchcock, J. Lertola, J. McDermott, M. McFarland, J. Nestor, A. Newell, J. Rajan, D. Siewiorek, D. Thomas, and R. Walker for many ideas, probing questions, and software support.

At Stanford we owe particular thanks to P. Eichenberger, D. Perkins, and T. Saxe, who designed and implemented Lava, and to R. Mathews, J. Newkirk, R. Dutton, and J. Hennessy for valuable discussions and support.

We would also like to thank J. Boddie, K. Chong, C. Davis, D. Ditzel, M. Maul, G. Mowery, A. Ross, C. Schneider, G. Williams, and A. Wilson for donating time to critique the designs. And we would like to thank J. Hartung, D. Hill, D. McCreary, and H. Moscovitz for taking time to discuss the problems of chip synthesis.

References

1. B.T. Murphy et al., "A CMOS 32b Single Chip Microprocessor," *Int'l Solid-State Circuits Conf.*, 1981, pp. 230-231.
2. R.N. Kershaw et al., "A Programmable Digital Signal Processor with 32b Floating Point Arithmetic," *32nd ISSCC,* Feb. 1985.
3. T.J. Kowalski and D.E. Thomas, "The VLSI Design Automation Assistant: An IBM System/370 Design," *IEEE Design and Test of Computers,* Vol. 1, No. 1, Feb. 1984, pp. 60-69.
4. R. Mathews, J. Newkirk, and P. Eichenberger, "A Target Language for Silicon Compilers," *Proc. Compcon*, Spring 1982, pp. 349-353.
5. N. Weste, "Virtual Grid Symbolic Layout," *18th Design Automation Conf.*, 1981, pp. 225-233.
6. W. Wolf, *Two-Dimensional Compaction Strategies,* PhD thesis, Stanford University, Stanford, Calif., Mar. 1984.
7. W. Wolf et al., "Dumbo, A Schematic-to-Layout Compiler," *3rd Caltech Conf. VLSI,* Mar. 1983, pp. 379-394.
8. T.J. Kowalski, *The VLSI Design Automation Assistant: A Knowledge-Based Expert System,* PhD thesis, Dept. Electrical and Computer Eng., Carnegie-Mellon University, Pittsburgh, Pa., Apr. 1984.
9. M.R. Barbacci et al., *The ISPS Computer Description Language,* Dept. of Computer Science, Carnegie-Mellon University, Pittsburgh, Pa., Aug. 1979.
10. E. Dirkes, *A Module Binder for the CMU-DA System*, masters thesis, Dept. Electrical and Computer Eng., Carnegie-Mellon University, Pittsburgh, Pa., Apr. 1985.
11. D.J. Geiger, *A Framework for the Automatic Design of Controllers,* masters thesis, Research Report No. CMUCAD-84-34, Carnegie-Mellon University, Pittsburgh, Pa., July 1984.
12. R. Di Russo, *A Design Implementation Using the CMU-DA System,* Masters thesis, Dept. Electrical Eng., Carnegie-Mellon University, Pittsburgh, Pa., Oct. 1981.
13. L. Hafer, *Data-Memory Allocation in the Distributed Logic Design Style,* masters thesis, Carnegie-Mellon University, Pittsburgh, Pa., Dec. 1977.
14. A.C. Parker et al., "The CMU Design Automation System: An Example of Automated Data Path Design," *Proc. 16th Design Automation Conf.*, June 1979, pp. 73-80.
15. P. Marwedel and G. Zimmermann, *MIMOLA Software System User Manual,* 1, Institute Fur Informatik und Praktische Mathematik, Christian-Albrechts-Universitat Kiel, May 1979.
16. L.J. Hafer, *Automated Data-Memory Synthesis: A Format Method for the Specification, Analysis, and Design of Register-Transfer Level Digital Logic,* PhD thesis, Dept. Electrical Eng., Carnegie-Mellon University, Pittsburgh, Pa., June 1981.
17. C.J. Tseng and D.P. Siewiorek, "Facet: A Procedure for the Automated Synthesis of Digital Systems," *Proc. 20th Design Automation Conf.*, June 1983, pp. 490-496.
18. C.Y. Hitchcock, "Automated Synthesis of Data Paths," CMUCAD-83-4, SRC-CMU Center for CAD, Carnegie-Mellon University, Pittsburgh, Pa., Jan. 1983.
19. A.E. Dunlop, "Automatic Layout of Gate Arrays," *Proc. Int'l Symp. Circuits and Systems,* 1983, pp. 1245-1248.
20. B.T. Preas, *Placement and Routing Algorithms for Hierarchical Integrated Circuit Layout,* PhD thesis, Stanford University, Stanford, Calif., Aug. 1979.
21. C. Horng and M. Lie, "An Automatic/Interactive Layout Planning System for Arbitrarily Sized Rectangular Building Blocks," *Proc. 18th Design Automation Conf.*, 1981, pp. 293-300.
22. S.C. Johnson and S.A. Browning, *The LSI Design Language i,* AT&T Bell Labs internal memorandum, Murray Hill, N.J., 1980.
23. P.W. Kollaritsch and N.H.E. Weste, "A Rule-Based Symbolic Layout Expert," *VLSI Design,* Aug. 1981, pp. 62-66.
24. J. Kim and J. McDermott, "TALIB: An IC Layout Design Assistant," *Proc. AAAI,* 1983, pp. 197-201.
25. D. Johannsen, "Bristle Blocks: A Silicon Compiler," *Conf. VLSI,* Jan. 1979, pp. 303-310.
26. J.M. Siskind, J.R. Southard, and K.W. Crouch, "Generating Custom High Performance VLSI Designs from Succinct Algorithmic Descriptions," *Proc. Conf. Advanced Research in VLSI,* Jan. 1982, pp. 28-40.
27. K. Chu and S. Ramautar, "A Technology-Independent MOS Multiplier Generator," *Proc. 21st Design Automation Conf.*, 1984, pp. 90-97.
28. A. Newell, "Heuristic Programming: Ill-Structured Problems," in *Progress in Operations Research 3,* J. Aronofsky, ed., Wiley, New York, 1969, pp. 360-414.
29. J. Vasantharajan, *Design and Implementation of a VT-Based Multilevel Representation,* masters thesis, Dept. Electrical Eng., Carnegie-Mellon University, Pittsburgh, Pa., Feb. 1982.
30. G. De Micheli and A. Sangiovanni-Vincentelli, "KISS: A Program For Optimal State Assignment of Finite State Machines," *Int'l Conf. Computer-Aided Design,* Nov. 1984, pp. 209-211.
31. M.J. Meyer, P. Agrawal, and R.G. Phister, "A VLSI FSM Design System," *Proc. 21st Design Automation Conference,* June 1984, pp. 434-440.

VLSI CAD Research at MCC

Tom Rhyne, MCC

The Microelectronics and Computer Technology Corporation, MCC, began its Austin-based research operations in October, 1983. Prior to this starting date, however, extensive planning for MCC's technical programs had defined the technical and financial scope of each program. These detailed program definitions were made by technical task teams assembled from the staffs of the companies involved in the initial organization of MCC.

The "CAD/CAM Task Force" that planned what has now become MCC's Computer-Aided Design Program, completed its initial program plan in late 1982. In answering the question "Why is this program needed?" the Task Force wrote:

> CAD tools available to VLSI designers have been lagging behind the [semiconductor] industries' ability to build dense chips. Major advancements are needed in CAD capabilities if the industry is to take advantage of the opportunities created by the state of the art of IC processing. The normal evolutionary progress being made by individual companies is insufficient to keep pace, let alone catch up with silicon density advancement.

To meet this challenge, the Task Force defined an eight-year research and development effort, "to substantially advance the state-of-the-art in CAD/CAM tools and systems so that MCC participants have access to design automation capabilities second to none." The current CAD Program research plan is based almost entirely on the research objectives outlined by the Task Force.

Multi-company support for MCC's CAD Program is based on the expectation that the companies, and ultimately US industry in general, will benefit from sharing the large expenses and scarce human resources involved in CAD research and development.

The basic concept put forward by the Task Force was for researchers at MCC to develop new CAD concepts and algorithms and to develop prototypes from the most promising ideas. After benchmarking, these prototypes would be transferred to the participating companies under appropriate licensing agreements. The companies would then incorporate the tools into their own operations or develop commercial products based upon them.

MCC's CAD Program began in November 1983 with the hiring of John Hanne as its vice president and program director. Before joining MCC, Hanne spent 17 years with Texas Instruments, serving as design automation strategy manager as well as design automation department manager from 1965 to 1976.

In this article we present a summary of the CAD Program's progress during its first year of operations and an overview of its major research plans for the period from 1984 to 1986.

Staffing

At present, 12 of the 19 MCC shareholding companies participate in the CAD Program. The Program is about 60 percent staffed, with full staffing planned for 1986. The current staff includes research specialists in a wide variety of CAD-related disciplines, plus a number of experienced VLSI designers.

Mission and objectives

The mission of the CAD Program, as currently stated, is

> To advance substantially the state-of-the-art in VLSI CAD tools and systems so that MCC participants will have access to robust, integrated, state-of-the-art CAD capabilities spanning all major design approaches. These include (but are not limited to) design with highly regular structures, semi-custom, and fully handcrafted designs.

To accomplish this mission, four objectives have been established.

(1) To develop a hierarchical, integrated, state-of-the-art CAD system capable of supporting chip- and system-level designs with mixed semiconductor technologies and/or technology upgrades.

(2) To reduce significantly the time of the VLSI design process and to improve design quality and engineering productivity at reduced cost.

(3) To provide to Program participants the technology to integrate the MCC CAD system (or portions of it) within their companies.

(4) To ensure CAD system capability to support device density levels and complexity projected for the 1990's.

To meet Objectives (1) and (2), CAD research and development efforts are currently aimed at creating, by 1986, a vertically integrated CAD system supporting the design of multi-chip systems of significant complexity. This, the MCC CAD System, will thereafter evolve over the duration of the CAD Program to incorporate additional CAD capabilities and to support in-

Reprinted from *IEEE Design and Test of Computers*, pp. 104–107, November 1984.

creases in device complexity. Objective (3) will be met through the technology-transfer procedures being developed by the liaisons. The specific density and complexity targets of Objective (4) are not publicly stated, although the intent is to achieve aggressively-set goals.

Strategy and major deliverables

The major 1984 Program activity, in addition to recruitment and capital equipment acquisition, has been the development of a detailed research plan covering the 1984 to 1986 period. This research plan attempts to map a clear development path so the participants can integrate future deliverables. The plan also seeks to maintain, as much as possible, the delivery schedules of the original Task Force plan.

The deliverables required from the CAD Program by the original Task Force plan fall into three categories.

(1) The detailed research plan mentioned above, including strategies for research on hardware description languages; design databases; simulation techniques; design verification; timing analysis techniques; device/system test and fault analysis; layout methodologies; technology tracking; design synthesis; and a variety of other key CAD research topics.

(2) The definition of the initial hardware and software configuration of the MCC CAD System that is to serve as the test site for all benchmarking activities, the verification site for evaluating new research developments, and the transfer system for MCC-developed CAD tools. The forecast evolution of the MCC CAD System is also a deliverable.

(3) The definition and publication of key interfaces to the MCC CAD System and the software developed for it, thereby allowing participants and CAD vendors to develop CAD tools that can be easily integrated into the MCC CAD System environment. These published interfaces are to be kept as stable as possible, thereby allowing evolution of the hardware and software of the MCC CAD System without invalidating existing capabilities. Published interfaces for user-to-tool interaction, tool-to-data access, hardware description languages, design data interchange, and other areas of system operation are also called for.

The detailed CAD Program research plan has been completed, subject to final approval by MCC management and the participants during November and December, 1984. The plan outlines the design methodology to be supported by the MCC CAD System and the research products needed to support that methodology. The principal design approach is the use of top-down design refinement leading to the eventual selection, placement, and routing of very powerful parameterized modules. Tools for supporting high-level, hierarchical design capture in both functional and structural form are under development, along with a variety of simulators, verifiers, planning/estimation tools, module-selection tools, placement tools, and routing tools. The provision for incorporating other layout approaches into the chip (gate arrays or hand-crafted blocks, for example) is also supported under the planned design methodology.

All CAD tools needed to support this methodology will be operated within a consistent system environment that presents the user with a standard interface throughout all design operations. A "buy-before-build" approach has been chosen, so that tools meeting CAD Program requirements and available from vendors or universities can be incorporated into the MCC CAD System in lieu of internal development of tools with similar capabilities. This allows the resources of the CAD Program to be concentrated on areas where conventional CAD approaches are not likely to meet the Program's needs.

Benchmarking of the initial MCC CAD System is scheduled for 1986. The System must support the design of multi-chip systems of significant per-chip complexity, including hierarchical layout and automatic placement and routing with interactive cleanup. The System should use a consistent hardware description language, user interface, and database, and it should support the layout of a given chip in a few weeks.

Most of the tools available to the 1986 System will be human-directed. Thereafter, however, research plans call for the incorporation of more and more artificial-intelligence technology. The use of AI techniques is planned for design session management; automated design synthesis; early estimation of design parameters; automatic placement and routing; and many other areas of the design process.

The initial and second-year characteristics of the MCC CAD System have been defined (Table 1). The plans strongly emphasize the use of industry-wide standards for operating systems, network protocols, and non-proprietary workstation hardware, seeking thereby to maximize future opportunities to incorporate improvements in hardware or software into the MCC CAD System environment.

The development of the MCC CAD System serves both the research and technology-transfer components of the CAD Program. By serving as an *in situ* test bed, it allows researchers to quickly evaluate new algorithms or tool concepts against the current "best" tools installed in the System. A body of benchmark designs will be kept within the System for this purpose. For technology transfer, the System provides a stable, standardized environment through which newly developed MCC CAD technology will pass.

The Lisp environment within the MCC CAD System is being used to develop the high-level, hierarchical design capture system mentioned above, along with new tools for design session management, chip planning, performance estimation, and user interfacing. Further, early incorporation of Lisp-based workstations will provide a base for bringing appropriate artificial intelligence techniques (rule-based design, for example) into the MCC CAD System. The Unix environment provides a variety of useful utilities (file management, electronic mail, etc.), plus computational power for tools written in C and Pascal, as needed.

The key interfaces applicable to the MCC CAD System will be published during 1985. These include the publication of standard data access methodologies, stan-

Table 1.
The MCC CAD System for 1985-86.

ASPECT	1985	1986
System Concept	Lisp- and Unix-based workstations in a high-speed local-area network	Same equipment plus a possible higher-power Unix engine and/or lower-cost Lisp/Unix workstations
Operating System	4.2 BSD Unix plus Lisp environment	System V Unix plus Lisp environment
Application Programs (Tools)	Selected vendor or university tools plus some prototype MCC tools	New non-MCC tools plus MCC-developed tools
Communications	TCP/IP	TCP/IP plus possible MCC-developed enhancements

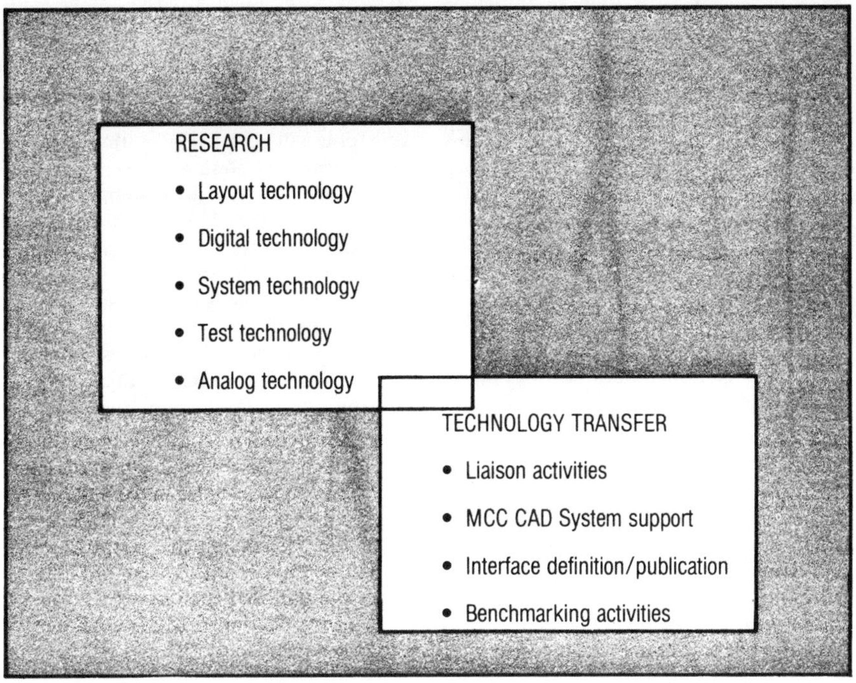

Figure 1. CAD Program organization.

dard user-interface methodologies, and the multi-level hardware description language selected by MCC. In the latter regard, the CAD Program has followed the VHDL and EDIF standards efforts very closely, and will decide whether to adopt these two standards during early 1985. The requirement for HDL standards includes both the need for a standard design description language covering design data from the highest level (commonly referred to as the architectural, functional, or behavioral level) down to the geometric level, plus the need for a design data interchange standard. EDIF, if selected, can very likely satisfy the lower parts of the first requirement and all of the second. VHDL, if selected, can fill much of the upper levels of the first requirement.

Organization

The CAD Program is organized into five research and development laboratories and one group dealing with technology transfer. These efforts overlap, as shown in Figure 1, assuring that research activities are closely linked to the CAD requirements of the participating companies. Each of these six activities has its own director, as shown in Table 2.

The management mix reflects the nature of the CAD Program's research strategy—blending an academic-like research effort with the application orientation typical of an industrial development program, a blending made possible when the two academic researchers offered their services to the CAD Program.

Analog technology. Marvin Daniel, most recently director of ECAD for Sandia Laboratories, directs the Analog Technology Laboratory within the CAD Program. The research activities of this laboratory center on the analog characteristics of the digital devices used within VLSI chips, rather than upon the use of analog devices within VLSI chips. Specific research efforts within Analog Technology include the development of very powerful hierarchical circuit simulators; improved models for circuit elements and the processes used to fabricate them; improved tools for extracting circuit parameters from chip layout data; improved tools for optimizing circuit performance and process yields; improved tools for timing performance analysis; and other related issues.

Digital technology. The Digital Technology Laboratory is directed by Harvey Garner, who is on sabbatical from the University of Pennsylvania to fill this position until late 1985. Digital Technology is responsible for research into design synthesis, analysis, verification, and simulation. Their 1984 to 1986 plans include the specification of the CAD HDL at and above the logic level; the development of powerful multi-level simulation tools; the development of improved timing analysis tools to support system performance evaluation and critical path identification; the development of new verification tools; and other efforts.

Layout technology. The Layout Technology Laboratory is directed by Ed Lien, who previously served as head of the Computing Technology Research Department at Bell Laboratories. Layout Technology research efforts during 1984 to 1986 are concentrated on the design, description, and use of highly complex parameterized modules as the basic building blocks of the MCC CAD System. Their research activities include both automated and interactive tools for designing these modules, and for placing and routing groups of them to form a composite chip.

System technology. The Systems Technology Laboratory is directed by the author of this article, Tom Rhyne, most recently of Texas A&M University where he served as a professor of electrical engineering and director of their Digital Systems Laboratory. Systems Technology is responsible for integrating the research products from the other Laboratories into a consistent environment; for developing standards to support long-term hardware and software evolution; for developing utility programs to support CAD tool developers and VLSI designers; and for developing new tools for capturing, trans-

Table 2.
CAD Program management.

LABORATORY/GROUP	DIRECTOR	BACKGROUND
Analog Technology	Marvin Daniel	Industrial
Digital Technology	Harvey Garner	Academic
Layout Technology	Ed Lien	Industrial
System Technology	Tom Rhyne	Academic
Test Technology	Open	
Technical Liaison	Jerry Werner	Industrial

ferring, and managing information during the design process.

Test technology. A director for the Test Technology Laboratory has not yet been identified, and recruitment of a well qualified individual for this position is a high-priority task within the CAD Program. A detailed test technology research plan was developed jointly by CAD Program staff members and a group of test specialists from the participating companies. This plan addresses the development of appropriate test methodologies, including design for testability; automated test pattern generation for detecting high level faults; test grading, including advanced fault-simulation tools; and new tools for fault diagnosis. The ready availability of pooled expertise from the participating companies has proved to be a major asset to the CAD Program.

Technical liaison. Jerry Werner, most recently editor of *VLSI Design* magazine, is the director of Technical Liaison, a group that includes both the liaisons from the participating companies and a small technology-transfer support group. The development and execution of the technology transfer strategy are its responsibility. The director of Technical Liaison also spends time on the CAD system definition requirements. This will allow immediate interaction with both MCC and the participants regarding the system and its requirements.

Interactions with other MCC programs

A major asset to each of the research programs within MCC is the presence of the other programs. From the CAD point of view, MCC's programs in artificial intelligence, database, human interface, software technology, parallel processing, and packaging provide on-site expertise for supporting research and development. The internal MCC environment facilitates program-to-program cooperation, allowing researchers in different programs to combine their talents in support of specific research projects. Areas of joint research now underway or planned by the CAD Program include:

Artificial intelligence. Automated design synthesis; automated verification; rule-based design optimization; intelligent session management; and other expert systems applications to CAD.

Database. Advanced design databases; networked design databases; and configuration management strategies.

Human interface. Improved graphics methodologies; natural-language user input; and possible use of speech.

Software technology. Improvements in productivity of CAD tool developers and interprocess communications methodologies.

Parallel processing. Design and prototyping of specialized high-performance processors for CAD applications.

Packaging. Packaging and interconnection strategies for multi-chip systems.

Technology-transfer methodology

The fundamental concept behind MCC is the successful transfer of jointly supported research into the companies that provide the support. Planning for successful transfer is an MCC-wide activity. Within the CAD Program, planning for transfer has defined the concept of *transferable tools,* which are prototype software packages that have been tested and documented and are ready for release to participants. A specialized Systems Integration Group, serving as staff to the Technical Liaison Group, has been defined to support this transfer. Transferable tools will be delivered with test cases, user documentation, and user training. Following transfer, the Systems Integration Group and the liaisons will serve as a collection point for user problems with the tools, and will call upon the researchers who developed the tools for assistance in resolving problems.

Close working relationships between the MCC CAD Program and its participating companies are essential to the success of the transfer process. To create and maintain these relationships, the Program has formed the CAD Advisory Committee, asking each company to send appropriate representatives to frequent CADAC meetings at MCC. CADAC has met seven times during 1984, reviewing the CAD Program's planning at each meeting. These meetings have provided early direct feedback from the companies.

An early comment concerning the formation of MCC's CAD Program was that it would be impossible to assemble sufficient talent to meet the goals put forward by the original CAD/CAM Task Force. After almost one year of Program operation, this concern has not arisen. The Program, in the opinion of all observers, is well staffed, well planned, and well into successful operation.

For additional information on MCC, write to MCC at 9430 Research Boulevard, Austin, TX 78759. For additional information on the CAD Program, write to John Hanne at the above address.

CHIPBUSTER VLSI DESIGN SYSTEM*

Bill McCalla, Beatriz Infante, Dennis Brzezinski, Joe Beyers
Hewlett-Packard Company, Circuit Technology Group

1. A SYSTEM FOR VLSI DESIGN

ChipBuster is an integrated, UNIX workstation-based IC design system. It provides a high degree of configurability which allows it to be customized to support a spectrum of integrated circuit design and process technologies. Currently, it is configured to support the design of CMOS VLSI full custom IC's as depicted by the following representative floorplan:

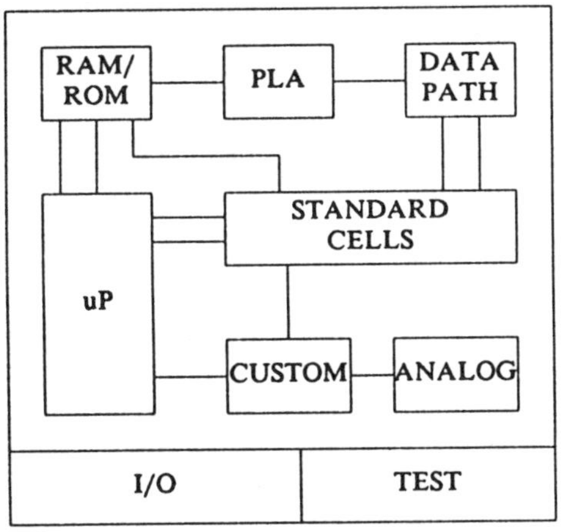

ChipBuster Floorplan

ChipBuster was conceived and designed by tool developers and IC designers working together. This cooperative effort has resulted in an integrated circuit design system that recognizes the following:

- Chips must be designed and represented hierarchically - there is simply too much data associated with a chip having over 500,000 transistors to be understood or designed "flat."
- Design complexity management is best handled by a top-down definition, bottom-up implementation approach.
- The design tools should complement and support the design methodologies - if cells are assembled hierarchically, they should be verified hierarchically.
- Individual tools, no matter how powerful, are rarely useful unless well integrated into a system. The sheer complexity of VLSI problems makes moving from tool to tool more complicated and time-consuming than actually using the tool.
- Integrated circuit design has evolved from a sequential to a parallel process. Multiple designers working simultaneously on a single chip necessitates proper management of shared libraries and concurrent access to designs.

* Disclaimer: The ChipBuster VLSI Design System is being developed only for use within Hewlett-Packard Company. It is not intended for sale or distribution either commercially or to universities.

2. UNIFIED DESIGN ENVIRONMENTS

The design of an integrated circuit can be characterized by three major phases:

- Chip definition and decomposition
- Leaf block design
- Chip recomposition and release

Each of these phases requires a different set of tools, with different interactions among those tools. One of **ChipBuster's** greatest contributions to IC design productivity is its integration of tools into coordinated design environments. For each of the three design phases, **ChipBuster** provides design environments which present the engineer with sets of closely coupled tools. This coupling goes beyond the ability to invoke programs from a parent shell - the various tools run and communicate with each other **concurrently**. A design change made using one tool is recognized by the other environment tools, so that each representation of the design remains current.

2.1 Chip Definition and Decomposition Phase

Throughout the definition and decomposition phase, **ChipBuster** provides continuous access to the tools necessary to

- Create a behavioral description of the design
- Enter circuit structure, either graphically or textually
- Create simulation vectors
- Perform simulations
- View the results directly on the schematic or in a text window without ever having to leave the design environment

Advanced capabilities include on-line library searching utilities to find blocks which have already been designed, module generator definition tools, the ability to use the same I/O vectors for behavioral, functional, gate, FET (switch), and SPICE level simulation, and the ability to create and format test vectors from simulation vectors.

2.2 Custom Block Design Phase

While designing low level cells, the engineer is able to simultaneously display the layout, schematic and simulation waveforms, thereby providing the ability to

- Concurrently edit more than one block and/or block representation
- Probe a node on either a layout or schematic and see simulation results on that node
- Highlight inconsistencies between the artwork and schematic after running a database block comparison
- Receive immediate feedback on design rule violations during artwork editing
- Directly overlay the artwork with DRC and encapsulation information
- Retain the same simulation capabilities as provided in the decomposition environment

2.3 Chip Recomposition and Release Phase

In the chip recomposition environment, **ChipBuster** provides the ability to place and route standard cell and custom blocks

Reprinted from *IEEE Int. Conf. CAD*, pp. 20–23, 1986.

including chips, to compare schematics and artwork, extract artwork connectivity and parasitics, and run simulations. Extremely tight coupling between the artwork, schematic, and simulation tools is maintained. Additional capabilities include:

- The ability to push and pop through various representations in the hierarchy (e.g. select the schematic symbol of a register cell and push into the layout representation of that cell)
- Assertion checking of expected versus actual interconnect capacitances, port directionality, and simulation results

3. CHIPBUSTER DESIGN FLOW

For each design phase the design environment itself must be altered according to the current activity of emphasis even though the same tools and/or design windows are needed. For example, during schematic capture a large window is usually needed for graphical editing. As the activity of emphasis evolves to simulation, a smaller window would be used to display the schematic in order to provide a large window for waveform display and editing. The configurability designed into **ChipBuster** allows adaptation of the design environment to support multiple activities of emphasis within each design phase.

Driven by a commitment to provide such a task-oriented as opposed to a tool-oriented environment, the **ChipBuster** system has been structured as shown in the picture below. The analogy is that of a software card-cage into which individual tools are inserted. Virtually all tools -- the behavioral simulator **BEST**, the **SPICE** circuit simulator, the **TRANTOR** graphic editor and design verification package, the module generators, etc. -- share a common user interface and are supported by a common design data management package.

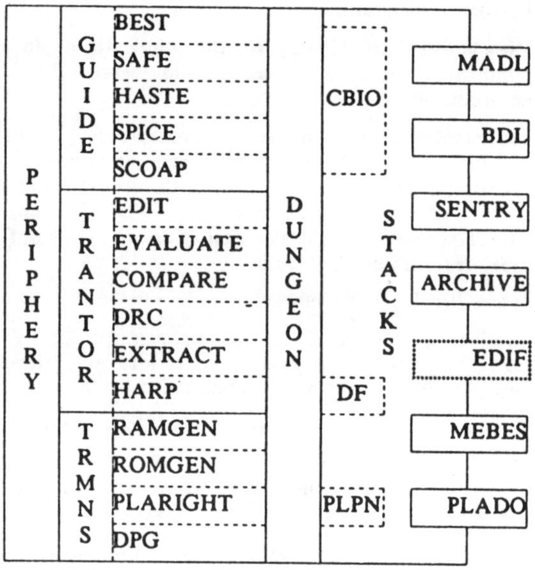

ChipBuster System Architecture

This section addresses in more detail the specific capabilities of most of the tools indicated above as they relate to the three design phases.

3.1 Block Decomposition

Much time is spent by engineers during the decomposition phase in defining the structure of a block which previously had been described behaviorally from the hierarchy above and in simulating and comparing the results against the previously specified behavior. The **GUIDE** simulation shell in concert with the **TRANTOR** design shell provides the major part of this design environment. Both are linked individually as single programs but communicate continuously via command pipes. The **GUIDE** shell accesses and manages the simulator kernels:

- **BEST** (BEhavioral and Specified Timing simulator)
- **SAFE** (SoftwAre FEt simulator)
- **HASTE** (HardwAre Switch Timing Engine)
- **SPICE** (Simulation Program with Integrated Circuit Emphasis)
- **SCOAP** (Sandia Controllability and Observability Analysis Program)

BEST, which supports behavioral, functional and logical simulation, is primarily used at the system, chip and functional block level. Both behavioral and structural models can be described using **MADL**, a hardware description language. Structure can also be inferred from schematic and block diagrams. For lower-level sub-blocks where more accurate simulation of blocks is required, the same **GUIDE** environment with the software and hardware switch-level simulator kernels, **SAFE** and **HASTE**, respectively, and **SPICE** are provided.

GUIDE initiates editing sessions of the **MADL** source used for describing behavior and circuit structure. **GUIDE** is used to invoke the **MADL** compiler which (a) translates any **MADL** structural definition into the **TRANTOR** structural database and (b) creates compiled behavioral models for execution by **BEST**. **GUIDE** provides commands, usually organized into scripts, for simulation control (such as setting breakpoints), for generating input vectors to drive simulation, and for querying the simulation results. Synchronous and asynchronous test vector generation capabilities are provided.

Changes may be made to a block's structural description either in the **TRANTOR** schematic or in the **MADL** text, whichever was used for original entry. Changes can also be made to the model descriptions of child blocks in the **MADL** source. The new version of the block can be loaded and a new simulation begun. Design data management provides controls for minimizing the amount of recompilation, reexpansion, and reloading between circuit changes.

3.2 Leaf Block Design

Leaf block design begins with a primitive-level "decomposition" phase that is analogous to the previously described block decomposition process, only here the "sub-blocks" to be interconnected are IC primitives -- transistors. Simulation is performed on the transistor circuit in the form of both functional and transient analysis. As artwork is composed, the electrical and parasitic qualities of the extracted artwork are frequently compared against the specified structural description, and the extracted data is simulated for performance comparisons as well.

The schematic for the leaf block is entered in the **TRANTOR** schematic editor, and its structure is determined using **TRANTOR** evaluation scripts. Properties of the schematic can include transistor specifications as well as capacitance (and often resistance) values of the interconnect. The structure and the properties together are expanded into a **SPICE** circuit description or are used directly by **SAFE/HASTE** for switch level simulation.

The switch-level software simulator **SAFE** and/or hardware engine **HASTE** operate in the **GUIDE** environment with command scripts providing input vector stimulus and simulation control. **GUIDE** provides a screen-oriented output and control environment or can display simulation outputs for nets/signals selected from the **TRANTOR** schematics viewport.

Once the design is conceptually correct, the designer begins to compose the artwork of the block in the **TRANTOR** artwork editor. As the layout progresses, design-rule correctness can be monitored with **TRANTOR** DRC scripts. Extraction scripts identify and characterize transistors, determine the connectivity (structure) of the block, and compute parasitic capacitance of the block's layout.

When the block is completed, scripts encapsulate the block's artwork and layout structure and store that information in the database for the block's hierarchical role in the recomposition process of higher-level blocks.

The extracted version of the block can be compared against the schematic version and/or structural **MADL** description both by using **DBCOMPARE** and also by running additional simulations on this artwork version of its circuitry. Differences discovered by **DBCOMPARE** are noted in the **TRANTOR** database for display on the schematic and/or the artwork representations of the block.

3.3 Chip Recomposition

The design task to be carried out during the recomposition phase is fundamentally very similar to the latter stages of leaf block design where the circuit has been realized in artwork. Thus, the tool requirements and their relationships exhibit considerable overlap with what has already been described. Two major differences are the addition of routing tools and the requirements for hierarchical support.

A suite of placement and routing modules collectively known as **HARP** can be accessed in a batch mode or pseudo-interactively at the block level. Pseudo-interactive mode supports manual block placement and orientation, spatial and hierarchical repartitioning, and complete routing at the block level. Results are returned automatically to the binary **TRANTOR** database which holds structure, artwork, and parametric data.

Hierarchical design support for the recomposition phase resides in **TRANTOR**. DRC and circuit extraction in **TRANTOR** are performed on a per-block basis with hierarchy accounted for at the script customization level. Data management utilities provide functionality directly embedded into **TRANTOR** to insure cross-hierarchy design integrity. The flattener module expands binary **TRANTOR** structure data and produces a binary flat connectivity schema for **GUIDE/HASTE/SAFE**. Expansion control and design management utilities allow efficient and flexible transformation from the hierarchical to the flat description.

4. CHIPBUSTER TOOL ARCHITECTURE

The following figure shows the architecture common to all interactive **ChipBuster** tools. The analogy is that of a software cardcage where new or modified tools may be conveniently added to the overall framework. Exceptions to this general architecture are batch-style processes which have no need of the interactive user interface module. Such programs may still use the database access module.

The capabilities of any **ChipBuster** tool are inherent in its command modules. There are many common functions required by most, if not all, tools. For example, the commands for display setup and handling are useful to any tool which uses the graphics capabilities of the **ChipBuster** workstation. The functions which are unique to an application can also be loaded into the program as command modules. Within **ChipBuster** related command modules have been grouped into three shells, two of which **GUIDE** and **TRANTOR** have already been described implicitly. The third, **TERMINUS**, incorporates module generators. All three shells are built upon a common user interface, **PERIPHERY**, by appropriate extensions to the base command table.

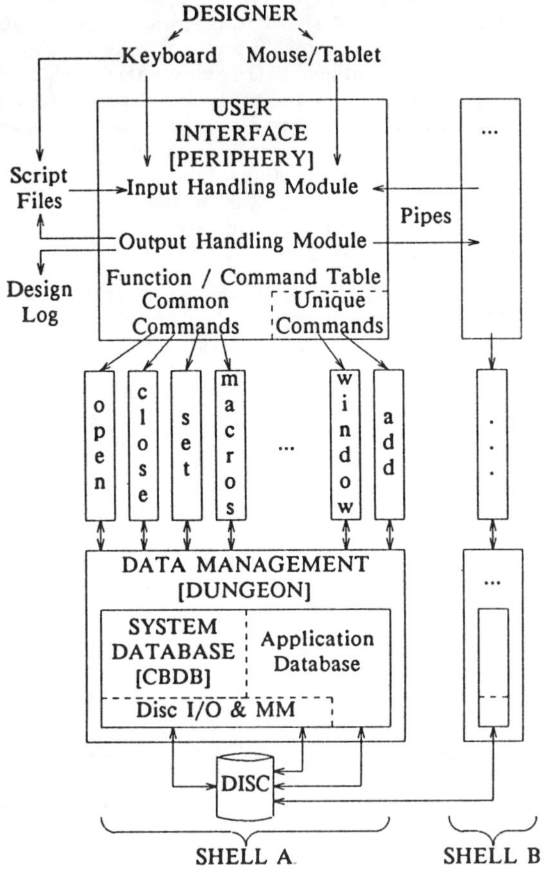

ChipBuster Shell Architecture

Each command module may read, modify, and write objects of the coherent **ChipBuster** database referred to as **CbDb**. This access is handled by database procedures. For performance efficiency, data objects are available in the program's allocated memory for the duration of the active process rather than on disc. Fast disc I/O is part of the **CbDb** database technology as well as database and process memory management. Arenas of data required for a given operation in the program which are not currently in the program's memory are demand-loaded by **CbDb** while arenas of inactive data may be freed up to make room for newly required data.

An application may extend **CbDb** by defining its own arenas of specialized data within the **CbDb** technology. Such new arenas benefit from the fast disc I/O and memory management modules of **CbDb**. Still easier methods of extending **CbDb** are available for application data which does not require the permanence of disc storage but only exists for the duration of the active process.

The **PERIPHERY** module provides the program's user with access to the tool's capabilities. The syntax of these commands is similar to HP-UX commands in that a combination of flags and arguments are specified on the command line. Flags modify the exact functionality of the command while arguments typically specify the data objects to be treated. Commands used to create, modify, or select displayed objects in the database are passed coordinates from the graphics input device attached to the workstation.

Sequences of commands can be stored in a text file and invoked by the user as a "script" of operations to be performed by the tool. Scripts are invoked by typing their HP-UX file name, and they also accept user arguments. Command scripts vary from an elaborate hierarchical DRC script for the

TRANTOR artwork tool to a simple sequence of commands written by an individual designer to perform a task he repeats frequently. Any script can invoke another "nested" script, and a script can request coordinate data from the the graphics input device or information from the keyboard. The **PERIPHERY** command interpreter understands programming "flow of control" statements, and this capability is used especially in scripts to allow the operation of the entire **ChipBuster** tool to be "microprogrammed." Individual commands may be typed or picked from a graphic menu. Menus are configurable via the **PERIPHERY** menu commands.

Additional configurability of a **ChipBuster** program is possible by defining substitution "macros." The command interpreter identifies any defined macro and replaces it with a user-defined substitution string. Although it is probably not the most productive use of this capability, with the **PERIPHERY** macro system, the entire command set of a tool can be mapped to an alternative set of commands.

Finally, **PERIPHERY** is responsible for handling any graphical display associated with a **ChipBuster** application. Common commands for the system interact with the **PERIPHERY** module to set up viewports associated with an active process.

Interprocess communication between **ChipBuster** tools is essential to provide designers with the interactive design environments described above. The HP-UX "pipe" mechanism provides a safe and acceptable initial communication channel. Two types of information may be passed among shells. First, actual commands handled by the other's **PERIPHERY** to control major actions of the parallel process can be passed. Second, data to be shared by the parallel processes might also be passed. For example, the designer might send a **TRANTOR** command from the simulation environment to the schematic editor environment to display a particular bus port in the circuit's hierarchy. Then the simulated vector value for that bus would be passed directly as data (rather than as a **TRANTOR** command) to the schematic environment for display on the schematic.

In summary, **ChipBuster** is tightly coupled yet highly configurable and may be customized via environment variables, macros, and scripts. Further, new commands, programs and/or shells may be written by other developers and linked into the system.

5. CHIPBUSTER FILE SYSTEM

Since the data storage needs of many tools must be accommodated, a simple, flexible scheme for using the **HP-UX** file system greatly eases the tool integration process. Accordingly, **ChipBuster's** disc-based data is organized in a three-tiered hierarchy of similar **HP-UX** directories. The tiers of the hierarchy are shown below:

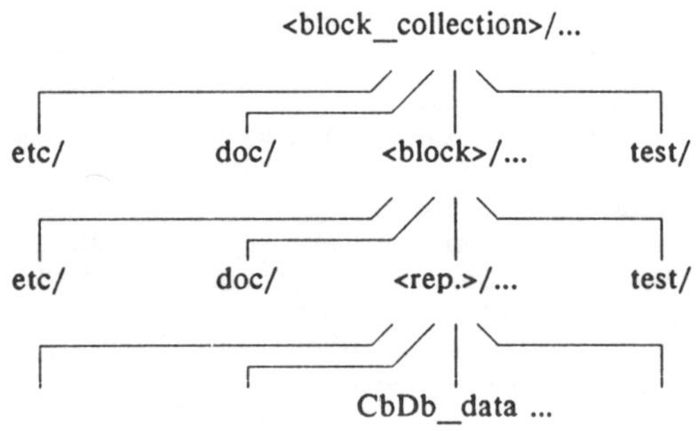

ChipBuster File Organization

The block collection contains any number of block directories; the block can hold any number of representation directories; and the representation, any number of **ChipBuster data files**. In addition, at each level the directories **doc**, **etc**, and **test** may occur. Nothing precludes the use of additional directory types at any level, if any particular tool or **ChipBuster** configuration deems it necessary.

6. CHIPBUSTER DESIGN DATA MANAGEMENT

The design data management system in the **ChipBuster** system provides several different capabilities. In prioritized order, they include:

1. database comparison of different block representations
2. specification and maintenance of data dependencies
3. concurrent access of blocks and a block's representations
4. access to distributed (including LAN) block collections
5. creation, distribution, and maintenance of libraries
6. archive storage and retrieval of multiple block collections (i.e designs)

Database compare is done across representations of a given block. The amount of comparison that is done can be configured based upon the given methodology. The database entities which can be compared include ports, nets, instances, devices, and properties. Comparisons may be made pairwise between different block representations including: **MADL**, schematic, block symbol, layout, encapsulated layout symbol, etc.

A very important aspect of data management is the ability to describe data dependencies and control what actions occur to "update" a dependency which is "out of date". Dependency management in the **ChipBuster** system is modeled after the "make" capabilities available for source code control and revision for software development on HP-UX. This ability is available in all environments of **ChipBuster**.

Concurrent access of blocks and block representations is important both in terms of coordination among tools used within the same design session as well as among different sessions. In particular, synchronization when writing data is essential so that no loss of data occurs. Data management software modules provides such synchronization.

Access to blocks stored in libraries and/or other chips is accomplished in the **ChipBuster** system by use of search paths and by providing a means to globally change a given search path in a particular block or block collection.

Another important function of data management is the creation, maintenance, and updating of libraries. Libraries are used to leverage past development work. Librarian capabilities provide the ability to:

- control access to the data, both storage and query
- maintain version information of included blocks
- allow updating of individual blocks in the library
- generate specifications for building a data sheet

Archive storage/retrieval of multiple block collections is required to put a chip into long term storage and be able to retrieve it in order to update it. It is important to get all data items necessary for changing and verifying the chip into the archive. These would include not only the database files but also any scripts, tools or even the particular operating system. Data management provides the ability to determine. what a block uses or depends upon.

ACKNOWLEDGMENT

The authors extend their thanks and appreciation to the many engineers, project managers, and section managers throughout Hewlett-Packard Company who have made **ChipBuster** a reality.

CAD Systems for VLSI in Japan

TSUNETA SUDO, TATSUO OHTSUKI, SENIOR MEMBER, IEEE, AND SATOSHI GOTO, MEMBER, IEEE

Invited Paper

***Abstract*—With the advent of the VLSI era, computer-aided design (CAD) is increasing its importance, and much effort is now being expended on CAD by many IC manufacturers and laboratories in Japan. This paper reviews the historical aspect of the CAD systems developed in this field, and describes the current status of VLSI CAD systems and technologies, from device to system levels, in Japan. The CAD development activities for IC's were initiated in the late 1960's. At present, VLSI CAD systems and related CAD technologies in Japan seem to be in the adolescent stage—partly capable of practical use and partly still in the immature state.**

I. Introduction

At the beginning of the 1960's, the development of computer-aided design (CAD) systems started in the electronic industry in Japan. CAD programs were recognized as essential to support the design and manufacture of electronic equipment for 1) design turn-around time reduction, 2) elimination of human errors, and, as a result, 3) reduction in design and manufacturing costs. At the earliest stage, efforts were expended mainly toward developing logic simulators for large electronic circuits and physical design programs for printed-circuit or back-wiring boards, which were used in mainframe computers or electronic switching systems. These CAD systems, as well as the fundamental algorithms, were developed originally by Japanese people.

As soon as the large-scale integrated circuit (LSI) concept appeared in the late 1960's, CAD tools for LSI design were also recognized to be essential for efficient LSI design, and manufacturing or development efforts have been made since then. In developing CAD programs for LSI design, LSI CAD designers have made use of numerous technologies and extensive know-how derived from their predecessors. The LSI placement and routing problem was considered to be one of the main subjects in the LSI CAD area, and various kinds of layout programs have been developed by many IC manufacturers and laboratories. In the early LSI development stage, numerous kinds of LSI layout design approaches were proposed such as planar layout, masterslice, building-block, multichip LSI, etc. The planar layout approach, which features a single metal layer for routing, offered interesting themes to theorists, but soon proved to be impractical because of the rapid progress in LSI technologies which has made possible the use of two or more layers for signal wiring. The masterslice and building-block LSI's have shown themselves to be more practical approaches in conjunction with the development of layout CAD programs, and, at present, have become the most popular LSI's in Japan. The multichip or hybrid LSI is also in practical use in mainframe computers. At the beginning, the masterslice approach was dedicated to bipolar technologies, while building-block approaches were used only for MOS LSI's. However, such a distinction disappeared as both the LSI fabrication and CAD technologies progressed, and, currently, either approach is being applied to both bipolar and MOS technologies. For these layout design systems, Japan was always a technical leader in the world, as the first papers on these topics were published by Japanese researchers.

Entering the 1970's, research and development on LSI CAD programs fluorished over a wider range, including device, circuit and logic simulations, placement and routing, artwork verification, and test generation. Although strenuous efforts have been made to develop LSI CAD programs, CAD designers always faced a great many difficulties in putting their programs into practical use because the automatically designed results were often of lower quality than manually designed ones. Among the layout programs, CAD programs for masterslice LSI's containing several hundred gates were successfully applied to the design of "semicustom" LSI's for mainframe computers or electronic switching systems at that time.

With in-house CAD system development efforts being made, the so-called stand-alone interactive design systems were introduced in many IC manufacturing plants and laboratories, and mass-produced LSI's, such as memories, calculators, and microprocessors, which have severe requirements for high packing density to hold the fabrication cost down as low as possible, have mainly been designed manually with the aid of these interactive graphic systems. In most cases, these stand-alone systems have magnetic tape interfaces with mainframe computers to share design tasks. According to a report [1] by the Research Committee on Electronic Equipment Design Technology of the Information Processing Society of Japan (IPSJ) regarding the status of design automation in Japan, about 40 percent of the CAD systems used for LSI layout design are such stand-alone interactive design systems purchased from outside vendors, while the remaining 60 percent of the CAD systems were developed in house. Along with the introduction of interactive stand-alone graphic systems, CAD application programs have also been introduced and applied for practical use, in combination with in-house originally developed CAD programs. SPICE [2], SLIC [3], and ASTAP [4] have been used for circuit analysis of LSI circuits, and TEGAS [5] for logic verification and test generation in several companies.

At present, the greatest concern is for the development of CAD systems, covering from device to system levels, for VLSI, which is always increasing in scale and complexity and creating numerous kinds of design difficulties. These technical problems are summarized as follows: 1) enlargement of design ob-

Manuscript received April 15, 1982; revised October 5, 1982.
T. Sudo is with the Nippon Telegraph and Telephone Public Corporation, Tokyo, Japan.
T. Ohtsuki is with the Department of Electronics and Communication Engineering, Waseda University, Tokyo, Japan.
S. Goto is with C & C Systems Research Laboratories, Nippon Electric Co., Ltd., 1-1 Miyazaki Yinchome, Miyamae-ku, Kawasaki, Japan 213.

Reprinted from *Proc. IEEE*, vol. 71, no. 1, January 1983.

TABLE I
DESIGN STAGES AND DESIGN AUTOMATION (DA) RESOURCES

	Process Design	Device Design	Circuit Design	Logic Design	Layout	Test Pattern Generation
Verification	Process Siml.	o Device Siml.	o Circuit Siml.	o Logic Siml. Function Siml.	o Design Rule Checker o Connectivity Checker o Timing Siml.	o Fault Siml.
Automatic Design	Process Optimizer	Device Optimizer	Circuit Optimizer	RTL → gate translator o μ program assembler PLA minimizer	o Placer o Router	o Test Pattern Generator
DA hardware	—	—	—	o TSS terminal Graphic System Logic Diagram Reader	o Graphic System	—
DA-oriented LSI Structure	—	—	o DA-oriented Cell	o PLA	o Building Block o Master Slice o PLA	o Scan Path

o --- most popular

jectives from device to system level; 2) increase of the amount of design data contained on a single chip; 3) overall optimization requirements with respect to logical and physical design factors; 4) strict requirements regarding design result quality; and 5) ever-progressing LSI technologies.

A hierarchical design concept is considered to be indispensable to alleviating these technical problems. Extensive efforts are being made toward the establishment of hierarchical CAD systems for VLSI design in many IC manufacturing companies and laboratories. As an example of hierarchically oriented CAD systems for VLSI design, an outline of the VLSI DA system developed by NTT is described in Section VI. It seems that VLSI CAD systems and related CAD technologies in Japan have reached the "adolescent" stage—partly capable of practical use and partly still in the immature state. Table I shows the present status regarding VLSI CAD–VLSI design stages and available CAD resources for each design phase.

Surveys on CAD systems in the electronic industry in Japan have been presented several times at various international conferences [6]–[14]. They will complement this paper, furnishing more information on CAD development in various fields of the electronic industries in Japan.

II. LOGIC DESIGN AND TEST GENERATION TOOLS

Logic design aids are mainly divided into three parts—synthesis, verification, and design methodology. Logic synthesis is finding an automatic way to realize in logic a set of Boolean expressions or a representation of a sequential machine. However, there is currently a great deal of human intervention in the synthesis process and models used in the synthesis process are not complete. At this moment, verification plays an important role to guarantee the logic function before the chip is built (simulation) and after (testing).

A. Logic Design

Fig. 1 shows the logic design tools which are used in Japan (Figs. 1–5 are taken from [1] and [11]). Logic simulators,

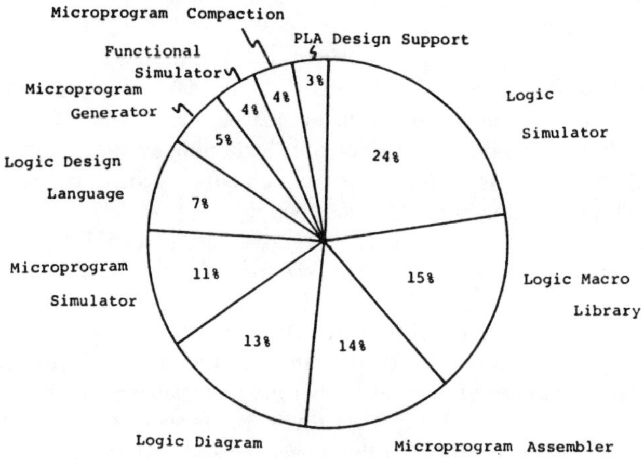

Fig. 1. Logic design tool.

hardware design languages, and microprogram support tools are now mainly used.

Logic simulators are the most popular logic design tools and have been used by many organizations. Most logic simulators used in Japan have been developed individually by the organizations which use them and improved year by year. Some are imported from the United States. Among these TEGAS [5] is a typical program. The input data for logic elements are shown in Fig. 2. Table II summarizes the logic simulators developed in Japan.

Gate-level simulators are the most commonly used in logic design. LOGOS [15], developed by the Nippon Electric Company (NEC), is a typical gate-level simulator, based on an event-driven method with mapping mechanism. It can handle unit, maximum, minimum, rise, fall, or wire delay time with 3 values. Recently, it was extended for simulating complex MOS circuits, including analog or bidirectional circuits with 4 values [16]. Toshiba first developed a 7-value logic simulator for MOS LSI circuits, which can treat transmission gates, clocked gates, and tri-state buffers successfully [17].

In the past few years, functional-level simulators have been

TABLE II
LOGIC SIMULATORS DEVELOPED IN JAPAN

System Name	Developer	Year	Logic Hierarchy	Function
LOGOS [15]	NEC	1977	Gate	3-value Simulator
ALEPH [18]	Hitachi	1979	RTL, Gate	Simulator
DDL [21]	Fujitsu	1979	RTL → Gate	Translator
ALS-4 [17]	Toshiba	1980	Gate	7-value Simulator for MOS
BINARY [19]	Oki	1980	RTL, Gate, Timing	Simulator, Verifier
MIXS [20]	NEC	1980	RTL, Gate	Mixed Level Simulator
FOREST [35]	NTT	1981	RTL, Gate	Simulator

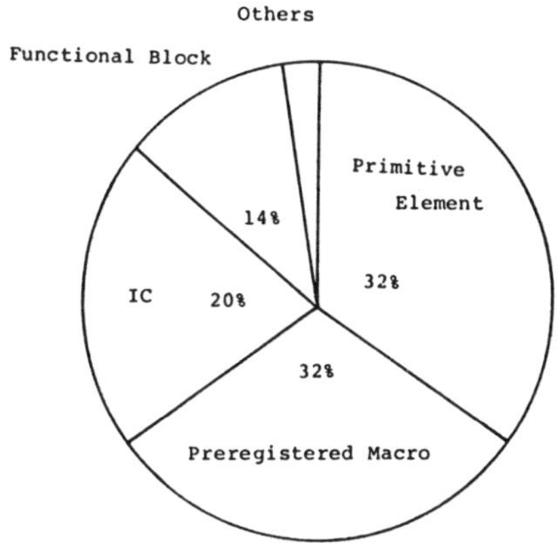

Fig. 2. Logic simulator input data.

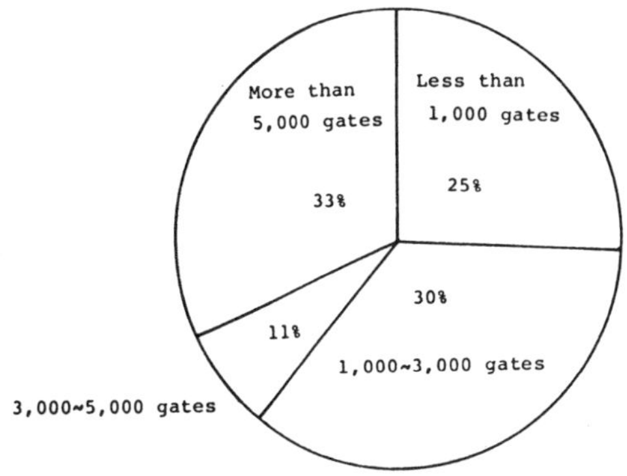

Fig. 3. Test application circuit size.

developed by computer manufacturers. Hitachi has developed a logic simulator [18], called ALEPH, which can handle register-transfer-level logic descriptions in addition to gate-level ones with capability of 750 000 gates. Oki's logic simulator [19], BINARY, can simulate logic circuits with three levels, functional, logic, and timing, in a hierarchical structure.

Functional-level, or register-transfer-level simulators can handle a logic circuit almost 10 times larger in half the computation time, compared to gate-level simulators. On the other hand, gate-level simulators are able to simulate logic circuits with more accurate delay time. Recently, mixed-level simulators have been developed to incorporate the features of the functional and gate-level simulators. Recently, NEC has developed a mixed-level simulator [20], called MIXS, suitable for a hierarchical or top-down support of large digital systems. It has realized a unified simulation mechanism for both gate- and functional- or register-transfer-level models, based on time-wheel and selective-trace algorithms.

In order to integrate the design and the verification stages, Fujitsu Labs has developed a DDL system [21] which can support register-transfer-level simulation, translation from DDL description to gate-level logic, and automatic logic verification. Also, an automatic input and interactive editing system for logic circuit diagrams has been developed [22].

"Array logic" was conceived to reduce design time and cost in comparison to random logic, so the popularity of the PLA is steadily increasing. NEC [23], [24] and Nippon Telegraph and Telephone (NTT) [25] have made a large effort in this area. They established the CAD system which first performs Boolean minimization and then automatically maps the logic function into the PLA personality matrix, together with the logic verifier. As we enter the VLSI era, the advantage of regularity is increasingly emphasized, and as device sizes scale down and chip area may be plentiful, the PLA or regular structural design will be more popular and practical in use.

Universities are active in the research and development of logic design [7], [26], [27]. The University of Tokyo has developed a logic simulation system using a data flow machine, which achieves a high-speed parallel logic simulation [28].

B. Test Generation

Testing and diagnosis of digital logic circuits (IC cards) have been studied actively for the past decades in Japan [29]. Recently, with the appearance of LSI's or VLSI's, a greater effort in research and development has enthusiastically been made by universities, public corporations, and manufacturers. The bulk of digital testing can be summarized as using single stuck-at fault models, generating test patterns, and using fault simulation to find all the faults which can be detected by these test patterns. This procedure is iterated until an acceptable percentage of all faults is covered.

The logic circuit size, to which testing tools can be applied, is shown in Fig. 3. Most tools are applied to combinatorial circuits but are rarely applied to sequential ones. Also, more than half of them can treat asynchronous circuits.

The test generation methods used are shown in Fig. 4. Full automatic generation is quite rare, but most methods are combined with manual and automatic methods. The generation algorithm is either a D-algorithm or path-sensitizing method. For fault simulation, the parallel method is most popular, but concurrent or deductive methods are also adopted (see Fig. 5). In 1968, NEC first proposed a method to convert sequential circuits to combinatorial circuits by evaluating circuits clock by clock, which facilitated the easy generation of test patterns for complicated sequential circuits [30].

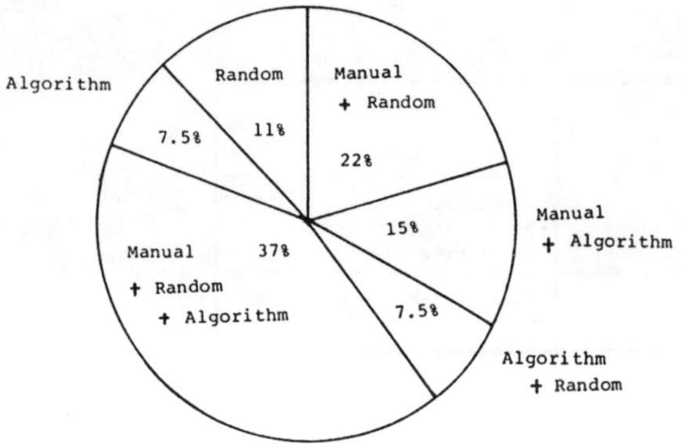

Fig. 4. Test generation method.

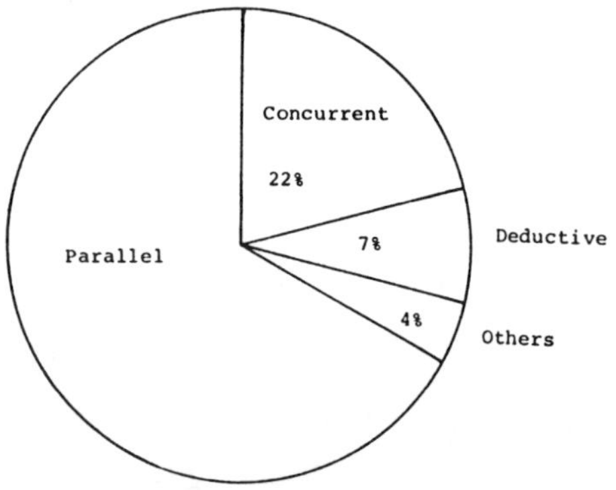

Fig. 5. Fault simulation method.

Now, let us look at the test systems in more detail. Oki has developed an automatic test generation system called FUNTASY by using Muth's 9-value method [19]. Hitachi has also developed a diagnostic system [31], consisting of an automated test generator, a fault simulator, and a fault locator. Its technique is based on a 9-value D-algorithm and a 6-value concurrent fault simulator with functional modeling. NEC has developed a test generation system [32], called TESTA, whose newly proposed algorithm is particularly efficient for large asynchronous and highly sequential circuits. It can handle up to 2000 gate sequential circuits, and can also be applied to MOS dynamic circuits.

Basic research efforst have been made in universities. Hiroshima and Saga Universities have jointly developed a test generator for combinatorial circuits using the structured description function and Boolean difference approach [33]. Recently, Osaka University devised an efficient generating algorithm for combinatorial circuits by introducing the testability measure [34].

The testing problems increase exponentially with the circuit chip complexity. Partitioning and hierarchical structure of VLSI design help to keep the testing problem within control. The major techniques used for testability are circuit partitioning or module design, the scan-path method, and built-in self-testing.

NTT has developed a test generation system [35] called FOREST which generates test patterns block by block. This system can treat a very large logic circuit efficiently by using a block-sensitizing algorithm in a hierarchical manner.

The scan-path method has been used practically by NEC for the first time [36] to improve the design testability (controllability and observability), while avoiding situations which cannot be tested easily. The method uses a serial shift register approach, which converts sequential circuits to combinatorial circuits. It has been applied to system-level test generation and testing, in addition to chip and printed-circuit assembly levels. FTT-700 has been developed by NEC and can handle up to 100K block digital systems [37].

Built-in self-testing research is now going on in several organizations.

III. Circuit and Timing Simulators

An electrical circuit simulator performs simulation wherein detailed electrical waveforms are taken into account. It comprises two main simulator categories, a "general circuit simulator" (Section III-A), and a "MOS timing simulator" (Section III-B). A considerable number of papers have been written by researchers in NTT and electronics manufacturers. However, there is very little activity in this field by the universities. This seems to be a notable difference when Japan is compared with the United States.

A. General Circuit Simulator

General circuits simulators, or circuit analysis programs, were developed to simulate any kind of electrical network and are capable of detailed modeling of electrical behavior. However, the simulation often becomes prohibitively expensive, particularly in this VLSI era.

Circuit analysis programs, developed in the United States, such as ASTAP [4] and SPICE [2], have been widely used in Japan. Also, NTT and electronics manufacturers have modified the programs originally developed in the United States or developed new programs for their own use. NTT's ECSS [38], Fujitsu's FNAP [39], NEC's ANAP (formerly called "NECTAR") [40], etc., summarized in Table III, are the programs developed in Japan. These programs are either unpublished or are available only to users who have installed computers produced by the program developers.

Among these programs, some of them have remarkable features, which those developed in the United States do not have. For example, NEC's COSMOS is a program dedicated to only MOS logic circuits. Although the class of applicable circuits is limited, COSMOS is reported to run much faster than any other existing circuit analysis program. NEC's ANAP [40] has another such feature in that convergence of DC analysis is guaranteed by exploiting the piecewise-linear technique [41], unlike the programs using Newton–Raphson type iteration.

Even in this VLSI era general circuit simulators are still used for verification of circuit design. However, the present research efforts focus primarily on development of timing simulators, which are capable of analyzing larger scale circuits.

B. MOS Timing Simulator

MOS timing simulators, such as MOTIS [42] and MOTIS-C [43], attempt to compromise between "speed," as in logic simulation, and "accuracy," as in general circuit simulation. An MOS device is essentially unidirectional, which leads to a decoupling of successive stages in MOS combinatorial logic. Also, the physics of the MOS device is much simpler than that of a bipolar device. Taking advantage of the simplicity of MOS circuits, together with pertinent decomposition and approximation techniques, MOS timing simulators are reported to perform transient analysis faster than general circuit simulators by a factor of one hundred. To maintain accuracy, however, the class of applicable circuits is reduced considerably.

TABLE III
CIRCUIT AND TIMING SIMULATORS DEVELOPED IN JAPAN

System Name	Developer	Year	Function			
			DC	AC	Transient	Timing
ECSS [38]	NTT	1973	o	o	o	
LILAS	Mitsubishi	1973.	o		o	
FNAP [39]	Fujitsu	1973	o	o	o	
ANAP6 [40]	NEC	1975	o	o	o	
ECAP	Hitachi	1976	o	o	o	
ECAP 6	Toshiba	1976	o		o	
COSMOS 2	NEC	1976	o		o	
NAP II	Oki	1977	o	o	o	
MATIS [44]	Toshiba	1978				o
FNAP (Extended) [46]	Fujitsu	1979				o
LOTAS [47]	NTT, NEC	1979				o
LNAP	NTT	1980	o		o	
MOSTAP [48]	NEC	1980				o

Japanese LSI designers were stimulated by the development of MOTIS [42] and MOTIS-C [43]. The basic ideas included in these programs have affected the direction of their research and development activities. As it turned out, however, developers were not satisfied with the accuracy and class of applicable MOTIS or MOTIS-C circuits. NTT and electronics manufacturers have developed or are now developing their own timing simulators [44]–[50], which are summarized in Table III. These activities include fundamental research on high-speed simulation techniques, such as a bypass technique for latent or inactive gates, macro modeling for logic gates, circuit decomposition, and a direct table lookup technique for the calculation of MOS device equations.

Japanese LSI designers have started to use their MOS timing simulators for transient analysis of large-scale circuits, including hundreds or thousands of transistors, which are beyond the capability of general circuit simulators. It is reported that degradation of accuracy compared with general circuit simulators can be restrained to within a few percent by careful device modeling. For example, a cascade circuit including ten inverters was simulated by both a general circuit simulator (COSMOS) and a timing simulator (MOSTAP). Results are compared in Fig. 6.

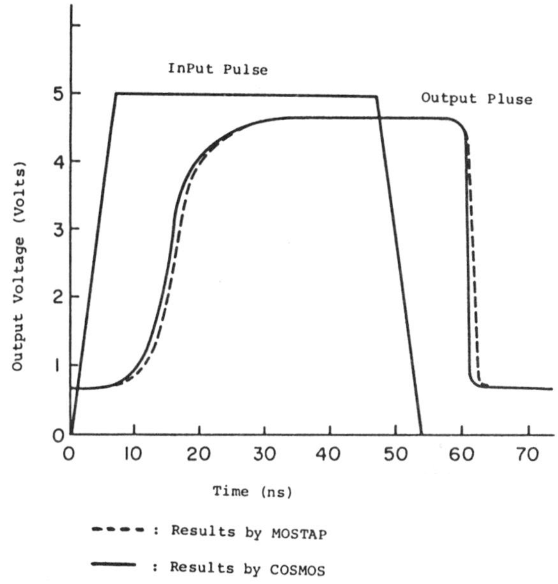

Fig. 6. Transient analysis by a general circuit simulator (COSMOS) and an MOS timing simulator (MOSTAP).

At present, some engineers are still reluctant to use MOS timing simulators, because the technologies are not yet as mature as those in general circuit simulation. However, the authors expect that timing simulators will be further improved and used by many more engineers in the near future.

IV. LAYOUT SYSTEM AND ARTWORK CHECKING TOOLS

In the layout design automation field, several important contributions have been made by Japanese researchers. The line search method [57] proposed by Mikami and Tabuchi greatly improved the efficiency of the Lee-type maze routing algorithm for Printed-Circuit Board (PCB) and LSI, and the algorithm has been widely used in many layout systems. Aiming at efficiency and quality improvement of routing for PCB, Hashimoto and Stevens' two-phase routing method [52] was proposed, and its concept has been implemented in various routing systems for VLSI as well as PCB. NEC people have proposed an algorithm to pack a set of rectangles into a rectangle for a planar electrical circuit as small as possible [53]. This approach, again, is looked at as a useful technique for building-block LSI's.

The routing method reported in COMPAS [54], a routing system for multichip or hybrid LSI, adopted almost the same method as the Hashimoto and Stevens method.

A. Building-Block Layout Systems

In the VLSI layout design process, chip mask patterns are generally constructed by using predefined cells. These cells correspond to gates or more complex logic blocks. They are placed in several rows on a chip, and the interconnections between cells are wired within the channel regions. A channel router is generally applied to interconnections between two rows of cells. This layout design style is called the building-

TABLE IV
AUTOMATIC LAYOUT SYSTEMS DEVELOPED IN JAPAN

System Name	Developer	Years	Layout Style	Application	Routing Method	Placement Method
LILAC [55]	Hitachi	1972	Building block	MOS	Channel router 2-layer	Clustering
Advanced-LILAC [56]	Hitachi	1974	Building block	MOS	Channel router 2-layer	
— [69]	Oki	1974	Masterslice		Line search method Maze router 2-layer	Partitioning by a seeding algorithm Iterative improvement
ROBIN [57]	NEC	1976	Building block	MOS	Global router Channel router 2-layer	Manual placement
CAD75 [70]	Hitachi	1976	Masterslice	ECL	Maze router 2-layer	Random method Pair interchange
MARC [72]	Musashino ECL NTT	1978	Masterslice	ECL MOS	Global router Line search method 2-layer	Attractive and repulsive (AR) method
MIRAGE [61]	Mitsubish	1979	Hierarchical layout		Global router Channel router 2-layer	Manual placement
MASTER [77]	NEC	1979	Masterslice	Bipolar	Global router Line search method 2-layer/3-layer	SORG GFDR
— [60]	Osaka University	1980		MOS	Left edge method	Clustering Iterative improvement
— [76]	Fujitsu	1980	Masterslice	CMOS	Global router Channel 2-layer	Min-cut placement One-dimensional improvement
PLASMA/COSMIC — [67]	Musashino ECL NTT	1980	Hierarchical layout	CMOS Bipolar	Channel router 3-layer	Linear placement Pseudo 2-dimensionalization
MARS-M3 [74]	Mitsubishi	1981	Masterslice	ECL	Global router Channel router 2-layer	Top-down placement FDPR
MILD [63]	Mitsubishi	1981	Hierarchical layout	NMOS CMOS	Global router Grid-free Channel router 2-layer	2-dimensional placement Linear placement
SHARPS [65]	Sharp	1981	Hierarchical layout		Global router Channel router 2-layer	Clustering Linear placement FDR
ALPHA [68]	Musashino ECL NTT	1982	Hierarchical layout		Global router Channel router 3-layer	Linear placement Pairwise exchange Shift insertion
LAMBDA [80]	NEC	1982	Masterslice		Global router Maze router Dynamic router	SORG GFDR Two level hierachical placement

SORG = Sub - Optimum - Random - Generation FDPR = Force - Directed - Pairwise - Relaxation
GFDR = Generalized - Force - Directed - Relaxation FDR = Force - Directed - Relaxation

block or polycell layout approach, and has mainly been applied to MOS LSI's.

Automatic layout systems for building-block LSI's were developed at a relatively early stage in Japan. LILAC [55], Advanced LILAC [56], and ROBIN [57] are layout systems of this type, as shown in Table IV. In the LILAC system [55], an algorithm for minimizing the number of tracks needed in practical routing is used. The LILAC system is considered to be the first polycell-type layout design system in the world. In the Advanced LILAC system [56], in addition to standard cells, nonstandard cells like PLA's are allowed. A wiring model is proposed where channel routing is always guaranteed in a predefined sequence. In addition, the strategy is implemented which can eliminate cyclic loops in the constraint graph without the rearrangement of cells.

ROBIN [57] is a layout system applicable to more general building-block LSI's. A chip structure used in ROBIN is shown in Fig. 7. A group, which usually consists of a row of cells, is used as a layout unit. Besides these usual groups, some large predefined blocks, such as ROM's, RAM's, and PLA's, are also treated as a group. The determination of interconnection routes between these groups is performed according to the following procedure [58], [59]: 1) for a given topological layout, a mapping process sequence is determined, which can be applied to each street without contradiction; 2) by a heuristic algorithm, the interconnection nets are assigned to streets, aiming to minimize the chip area (routing process); and 3) using a heuristic approach, the physically precise positions of wire segments are also determined in order to minimize the width of streets (mapping process). The graph-

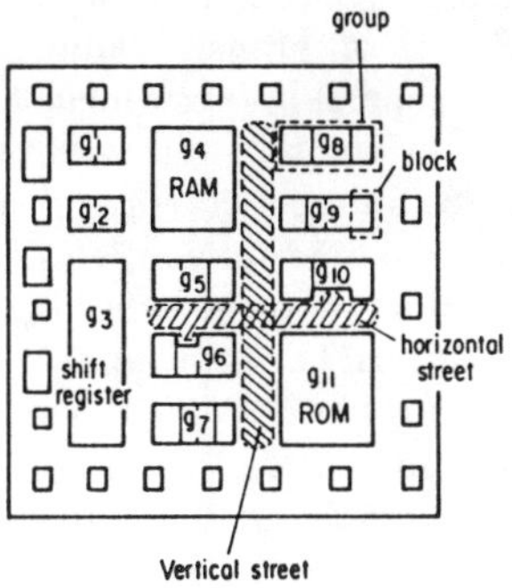

Fig. 7. Chip structure of ROBIN [57].

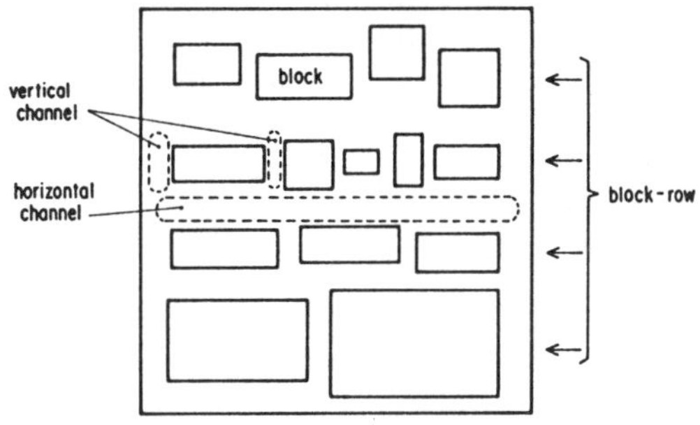

Fig. 8. Layout model of MIRAGE [61].

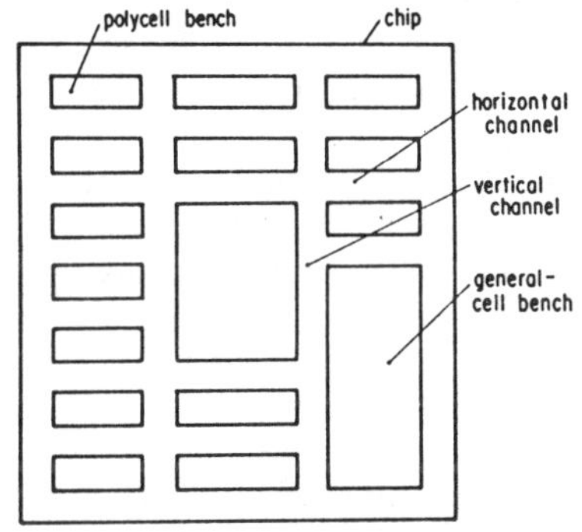

Fig. 9. Chip structure of MILD [63].

theoretical model proposed in ROBIN is the first such publication [58] in the world. Since its publication, many papers on the general building-block or hierarchical layout problems which are based on the ROBIN model have appeared.

A layout system [60] developed by Osaka University is used for the layout of the random logic portion of MOS LSI, which is based on the one-dimensional MOS complex gate-array ordering technique.

B. Hierarchical Layout Systems

MIRAGE [61] is the first hierarchical layout system developed in Japan. Fig. 8 shows a layout model of the MIRAGE system. In this system, in order to simplify the placement and routing process, the following restrictions are imposed [62]: 1) blocks are placed in rows called "block rows"; and 2) pin location of each block is allowed only at the upper and lower edges of the block. The vertical channels are used only for the vertical feedthrough lines.

MILD [63] is an automatic layout system for custom-made LSI's. This system can treat general cells, such as memory blocks (i.e., RAM's, PLA's, and ROM's) as well as standard cells. The chip structure for MILD is shown in Fig. 9. "Bench" is defined as an area in which cells are placed. The placement

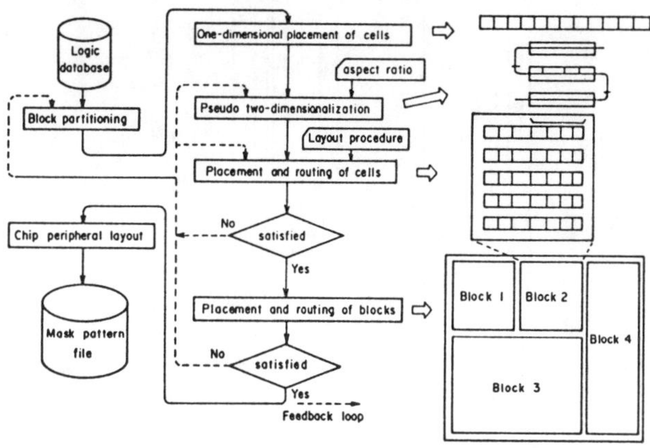

Fig. 10. Layout process flow in PLASMA/COSMIC [67].

process consists of the following three steps: 1) element assignments in benches; 2) determination of the intrabench positions of elements; and 3) improvements in the results in 2) within individual benches. The routing process is achieved by the global routing and grid-free channel router [64].

SHARPS [65] is a hierarchical layout system for fully custom-made LSI's. This system has the following specific features: 1) rectangular functional blocks of arbitrary size and shape can be treated; 2) contradictions in the channel processing order can be automatically dissolved by a new shrinking operation; and 3) in the track assignment process, relative positions of blocks adjacent to the current channel are determined so as to minimize the channel area, and a modified dogleg channel router [66] is adopted. The system also has interactive utilities, such as maze-running router and initial-placement program, based on a constructive approach.

PLASMA/COMIC [67] is an automatic layout system recently announced. A hierarchical layout procedure is also provided with this system. The chip structure in each hierarchical layout stage is simplified to the polycell layout style. This restriction does not cause serious damage in respect to chip density. Fig. 10 shows the layout process flow in PLASMA/COSMIC. In this system, the layout process is divided into four stages, namely, one-dimensional placement by COSMIC; pseudo two-dimensionalization, placement, and routing of functional blocks by PLASMA; and chip peripheral routing. In each step, a designer can select various kinds of placement and routing strategies aiming toward higher density. An appropriate hierarchical structure for optimal layout can be selected in each step. Also, not only higher density but also better electrical performance can be realized in the placement and routing procedures.

A new layout system, ALPHA [68], presented recently uses the hierarchical top-down layout design method. A global route assignment process reduces the redundant interblock wiring area. Routing in a single path over the whole chip enables one to use chip area more efficiently.

C. Masterslice or Gate Array Layout Systems

Automatic layout systems for masterslice or gate array LSI's in Japan are shown in Table IV. A layout design system [69] is the first announced system for masterslice LSI in Japan. The automatic layout system reported in [70] is also one of these systems. In the routing process, a modified Lee algorithm [71] is used, and rerouting is also tried.

MARC [72] places functional blocks on a chip and wires between these blocks for masterslice LSI's. It is impossible to

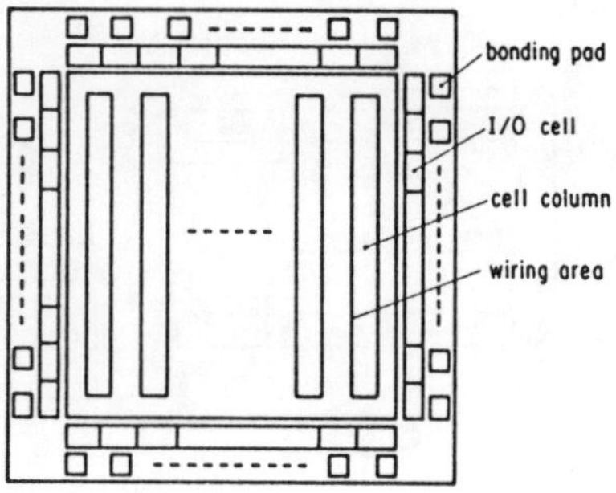

Fig. 11. Masterslice structure [76].

guarantee 100 percent wiring completion in the masterslice LSI, in which the interconnection area is fixed and limited. Therefore, in the MARC system, manual modification can be performed through the stand-alone graphic system interfaced with the main computer.

The three afore mentioned automatic layout systems use the Lee algorithm [71] and the line search algorithm [51] as routing algorithms. However, to deal with VLSI chip complexity, channel routing algorithms have been widely adopted recently.

MARS-M3 [73], [74] is an automatic layout system for ECL and MOS gate array LSI's using two metal layer levels. In the placement phase, hierarchical placement and top-down initial placement techniques are utilized [75]. The routing is also hierarchically achieved in two phases. The first phase is a global routing called channel assignment. The second is a local routing, which assigns tracks by modified dogleg channel router.

Similarly, an automatic layout program [76] also uses channel routing algorithm. The system has been applied to the layout design for CMOS masterslice LSI's. Fig. 11 shows a masterslice structure treated by this system. The placement process is performed as follows. First, the initial placement is determined by the min-cut placement algorithm. Next, in order to avoid local congestion, a one-dimensional placement improvement process is iterated for the initial placement. The routing process consists of two phases, i.e., topological routing and channel routing.

MASTER [77] is also an automatic layout system for masterslice LSI's. In this system, a new placement algorithm [78] is adopted, which is considered to be the most efficient iterative improvement method in the world. The routing process consists of two phases–global routing and detailed routing–in order to avoid local wiring congestion. Detailed routing is performed by using a line search method [51] with a few modifications, which can be applied to three-layer wiring.

LAMBDA [79], [80] is the newest automatic/interactive layout design system for masterslice LSI's developed in Japan. In this system, automatic and interactive functions are completely integrated and, therefore, can achieve complete net connectivity in a quick design time. The placement algorithm consists of two kinds of heuristic methods, a constructive one called SORG and an iterative improvement one called GFDR [78]. A maze algorithm is adopted with some modifications to find the exact wiring pattern. And the dynamic router iterates by removing routes and putting them back to improve the connection rate.

D. Design Rule, Connection, and Electrical Performance Checker

Many artwork verifiers have been developed in the last half decade, and development efforts are being continued in Japan. Although automatic layout programs which produce error-free layout patterns from logic descriptions have been developed, an artwork verifier is of primary importance because most memory LSI's and general-purpose logic LSI's are still being designed manually, in combination with automatic programs or with the so-called interactive graphic systems. Generally speaking, there are two kinds of verifiers. One is a checker for geometrical design rules, and the other is a checker for interconnections and electrical performances.

1) Design Rule Checker: A design rule checker has been developed in many companies. Many simple function programs reside in interactive graphic systems and are used for checking small-size blocks. An elaborated checker was reported from Toshiba in 1977 [81] which can perform a rather sophisticated function like a Boolean operation between layout patterns, a high-speed expansion/shrinking operation, and a design-rule violation checking. Since then, intensive investigations have been made in many companies [82], [83] about how to easily and speedily handle a great deal of geometrical data. New technologies and data structures have been applied not only to checking but also to more general data-handling purposes.

2) Interconnection and Electrical Performance Checker: Interconnection and electrical performance check verifiers can be classified into several groups. The most primitive verifier checks only the consistency of interconnections between block terminals identified with names. More general checking capability for connections is provided as a part of the other kinds of verifiers presented in the following.

Electrical performance checkers, which derive various circuit parameters from layout patterns, have been widely developed for MOS LSI's [84], [85] because the MOS circuit performance is very sensitive to parasitic capacitances, which directly depends on the layout patterns. Similar function programs for an analog bipolar LSI have also been developed in Matsushita [86] and Toshiba. These programs derive circuit-level descriptions from layout patterns, draw circuit diagrams, and calculate circuit parameters such as transistor sizes and capacitance values. The circuit parameters obtained by these programs can be directly input into circuit simulators, and a detailed circuit performance analysis can be achieved.

Resistance calculation is another function requested especially for the estimation of long wirings or analog circuits. One of the difficulties in this calculation lies in the fact that it is necessary to recognize networks which are only implicitly expressed in layout patterns. Multiterminal interconnections have to be broken up into many two-terminal networks because resistance calculation is essentially performed between a couple of terminals. For this purpose, a center line extraction algorithm for any geometrical pattern shape has been reported by Matsushita [86]. Another problem for resistance calculation is in regard to accuracy. A complex potential method has been proposed recently by NTT and Hitachi which can offer an accurate resistance calculation with a comparatively small memory. The finite element method reported by Toshiba also seems promising. Practically, especially for MOS LSI's, resistance calculation is needed only for a few long wirings on critical paths. Therefore, a simple and interactive program developed by NTT [87], which can calculate resistances within a limited condition, is also very useful from the practical point of view.

A logic function consistency verifier has become one of the indispensable tools for logic LSI design. The most popular and widely used method is one that takes advantage of identification marks given on each gate and terminal in advance [88]. In order to eliminate the possibility of giving marks incorrectly

TABLE V
ARTWORK VERIFIERS DEVELOPED IN JAPAN

Program	Developer	Year	Design Rule Check	Connection Check	Electrical Performance Check	Application
MAICAP [87]	NTT	1978			o	MOS
PAS [84]	NTT	1978		o	o	MOS
PALLEQ	Mitsubishi	1979		o		MOS
PALMS [88]	NEC	1979		o		MOS
IVS	Hitachi	1979		o		MOS
EMAP [85]	NTIS	1980	o	o	o	MOS
PANAMAP [86]	Matsushita	1980		o	o	Bipolar
LAS	Hitachi, NTT	1980		o	o	Bopolar/MOS
LIVES [91]	NTT	1981		o		MOS
MIAS	Oki	1981		o	o	MOS

and to remove such laborious work, the program developed by NTIS makes use of a logic simulation technology [85], [89]. For the same purpose, employment of the graph isomorphism to this problem has been attempted by NEC [88] and Osaka University [90]. A new verifier for a logic consistency check using this graph isomorphism technology has been reported recently by NTT [91].

As with other CAD tools, it is also desirable for verifier programs to be free from device change as well as fabrication technology changes. If they are not free from changes, continuous version upgrading efforts become inevitably necessary due to technological changes with rapid LSI progress. In order to avoid this problem, a library concept has been implemented into several programs [81], [85]. In these programs, a user defines the relationship between device technologies and desired operations in the form of a library with the prepared definition language.

Artwork verifiers, which have been developed in Japan, are summarized in Table V. Though there are many well-designed programs already, continuous efforts are being made to develop new verifiers having enough capabilities to deal with much larger LSI data in less computing time. They will be closely combined to the other CAD programs and will become an indispensable part of total CAD systems for future VLSI design.

V. DEVICE AND PROCESS SIMULATORS

The device simulator has become an indispensable tool for device design. Also, a process simulator, such as SUPREM [92], has been widely used in process evaluation prior to fabrication. Furthermore, a device simulator coupled to a process simulator [93] has also played a significant role as a method for predicting device parameter variations from process specifications.

In Japan, much effort has been made to develop device simulators applicable to various devices, including bipolar [94]-[96] and unipolar devices [97]-[101]. Two-dimensional device simulators are widely used for device design.

Three-dimensional simulators for very small device analysis [102], [103] have now been reported. A two-dimensional simulator, applicable to transient conditions, has also been investigated [104]. Process simulator developments have also become active [105], [106]. As a basic approach, a process/device optimization simulator [107], which can obtain optimum process conditions for realizing desired device performance, has been developed. Device and process simulators developed in Japan are summarized in Table VI.

A. General-Purpose Device Simulators and FET Device Simulators

1) General-Purpose Device Simulators: General-purpose device simulators are capable of analyzing a wide variety of bipolar and unipolar devices, wherein Gummel's iteration method [108] is applied to linearized Poisson's equation and the continuity equations. These partial-differential equations are formulated by using the finite difference or finite element method suitable for a nonrectangular grid. To solve the resulting matrix, the relaxation [109], [110] or direct method is adopted, according to its matrix structure.

Physical models and boundary conditions are implemented, corresponding to device material, such as Si or GaAs, and corresponding to device structure, such as MOSFET, MESFET, or bipolar devices, respectively. Application examples of these simulators for bipolar devices are presented as follows.

Current-voltage characteristics can easily be obtained, corresponding to impurity profiles. In addition, the junction capacitance and cutoff frequency are numerically calculated. Therefore, the transistor parameters are precisely calculated and served for the Gummel-Poon model [111] or for a table lookup model, instead of a test element device for the model parameter determination. Device performance can be evaluated through a circuit simulator, based on the above model. This application example for IIL is reported, including the experiments [96]. However, phenomena such as current spreading due to the extrinsic base region in bipolar devices can be analyzed with only a three-dimensional simulator. Fig. 12 shows the current-voltage characteristics calculated by a three-dimensional simulator [102]. These results suggest that the three-dimensional effects are dominated in the higher current region, compared with two-dimensional results.

As another example of a bipolar device simulator, IIL analysis with multistream function was investigated [94]. In this simulator, in order to solve the continuity equations with multistream function, the currents are divided into recombination current component and electron/hole current components injected into each electrode.

2) FET Device Simulators: FET device simulators are capable of analyzing only FET devices based on unipolar approximation with shorter CPU time. These simulators use

TABLE VI
DEVICE AND PROCESS SIMULATORS DEVELOPED IN JAPAN

Program	Developer	Year	Function	Feature	Application	Algorithm
TRANAL [95],[96],[102],[104]	NTT	1977	D.S.	2D, 2D-transient, 3D	General	FDM, FEM
FEDAS [99],[103]	NTT	1979	D.S.	2D, 3D	FET	FDM
POSEIDON [107]	NTT	1981	D.P.O.	SUPREM+FEDAS	FET	Flecher-Powell
CADDET [94],[97],[98]	Hitachi	1978	D.S.	2D	FET, Bipolar	FDM
——	Hitachi	1981	D.O.	SUPREM+CADDET	FET	
MODEST [101]	Toshiba	1981	D.S.	2D	FET	FDM
TOPICS [106]	Toshiba	1980	P.S.	2D	FET	FDM
SUBMODAN [113]	Toshiba	1981	D.O.	SUPREM+MODEST	FET	
PIEECHPIE [105]	NEC	1980	P.S.	1D	FET	
—— [115]	NEC	1981	D.O.	PIEECHPIE+D.S.(2D)	FET	
—— [100]	Fujitsu	1979	D.S.	2D	FET	FDM

D.S. --- Device Simulator D.O. --- Device Optimizing simulator

P.S. --- Process Simulator D.P.O. --- Device and Process Optimizing simulator

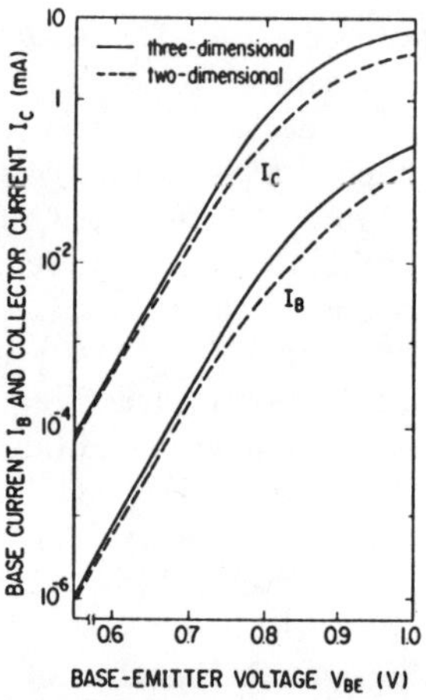

Fig. 12. Forward base and collector current as a function of base-to-emitter bias using TRANAL [102].

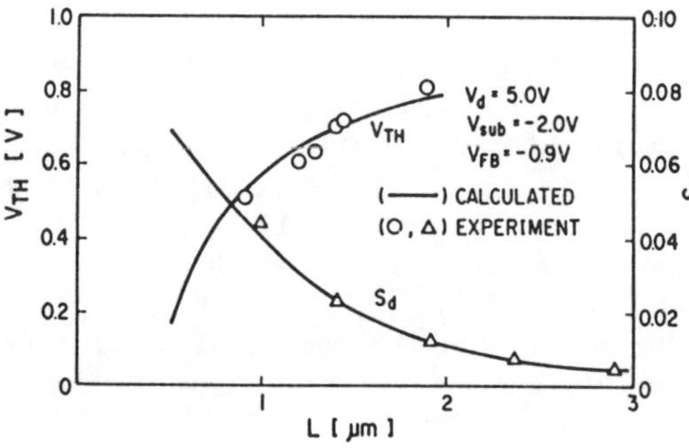

Fig. 13. Threshold voltage V_{TH} and its sensitivity for drain voltage S_d versus channel length L characteristics. Open circles and triangles denote experimental results for V_{TH} and S_d, respectively. Solid lines indicate calculated results [99].

Mock's algorithm [112] to solve the continuity equations with stream function and Gummel's method, which linearizes Poisson's equation. The difference equations are solved by the relaxation method.

Device parameters such as voltage–current, g_m-V_G and C_G-V_G characteristics, can be easily deduced in a manner similar to a bipolar device and serve for circuit analysis on device parameters. The other parameters—for example, threshold sensitivities based upon profiles—are estimated numerically, as shown in Fig. 13 [99]. Moreover, avalanche breakdown analysis for short channel NMOS was also investigated by a two-dimensional simulator [98]. A two-dimensional simulator cannot easily handle performance variations in a small-size MOSFET whose width is similar to that of the channel length. A three-dimensional simulator was successfully applied to examination of short and narrow channel effects [103].

B. Device and Process Optimization

Process-oriented device design is an inevitable state in order to reduce fabrication costs. Table VI also shows the process simulators developed in Japan, including those coupled with the device simulators.

There are two trends for process modeling. One is a one-dimensional simulator, which has advantages in regard to wide application and savings in computational time. For example, PIEECHPIE [105] and modified SUPREM II [92] in oxidation-rate and depth-dependent diffusion coefficient, etc., are commonly used. However, these methods have limitations in accuracy under the very small device dimensions. The other one is two-dimensional process modeling, taking into account lateral spreading. TOPICS [106], for example, has been developed for MOSFET process design and reported two-dimensional implanted Boron distribution.

In either case, if the device structure is closely related to the fabrication process simulator, it would be possible to design desired device performance through an optimized fabrication process and device structure. Therefore, the integration of the device simulator and the process simulator is considered to make overall device and process optimization, as shown in Table VI. Much work has been accomplished in MOSFET's synthesizing pseudo-two-dimensional impurity profile. Then process sensitivities for device characteristics can be analyzed using a coupled simulator. Although problems involving the

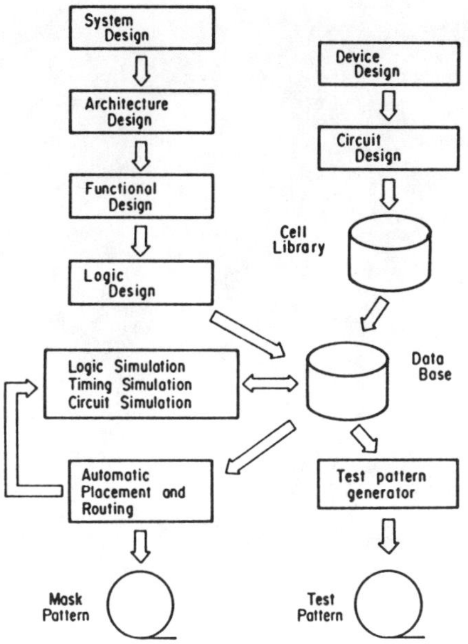

Fig. 14. VLSI design process flow.

short-channel effects remain to be solved, these simulators can be applicable to submicron devices [113]. Coupling the device simulator with a two-dimensional process simulator, instead of a one-dimensional simulator, enables the design of a device with shorter channel length.

The afore mentioned optimization procedure depends upon the engineer's capability or work load. On the contrary, the computer-aided process/device optimization method POSEIDON has been reported [107] owing to the progress in computer facilities. The Flecher–Powell method [114], commonly used in circuit design, is applicable to calculating the process sensitivity for device characteristics.

The optimization simulator is expected to be a very effective tool for process/device optimum design through the use of an array processor.

VI. Integrated VLSI Design System

A. Overview

In this section, a VLSI CAD system [116], [117] developed by NTT is described as an example of hierarchical-oriented CAD systems. Several main features of this system are as follows:

1) A hierarchical and structural design method is adopted throughout the design phases from LSI specification to mask pattern data.

2) The LSI design language (HSL) is extensively used as common input data.

3) A hierarchical design process is mainly achieved by the iterative interactiveness of the placer and the router (COSMIC/PLASMA).

4) A design database controller (HIDEMAP) administrates logical and physical data.

5) In addition to highly automatic processes, man–machine interfaces are provided to facilitate an interactive design.

B. General VLSI Design Flow

The VLSI design process flow is shown in Fig. 14. After the VLSI system architectural design is accomplished, functional specifications for each functional block are fixed. Then the logical design for each block follows, and the resultant logical design data are registered into the VLSI design database. On the other hand, standard function cells are designed, confirmed through circuit simulations, and provided within a cell library in the database.

After the logic expansion, the logical data is verified using logic and circuit simulations. Then automatic placement and routing programs are executed. After the layout process, the resultant patterns are evaluated through critical path check and timing simulation. If the critical path delay exceeds a specified value, the routing parameters are changed and the layout programs are rerun. Placement and routing are repeated until the given specifications are satisfied. Finally, the resultant data are converted to a format file for mask fabrication. In addition, test patterns are also generated, using a test pattern generator from the verified logical design data.

C. Hierarchical Specification Language: HSL

The Hierarchical specification language (HSL) is used as a common input language to the VLSI design system. HSL has been developed, based on the structure description language (SDL) [118] developed by vanCleemput of Stanford University, which is appropriate for hierarchical LSI design methodology. In HSL, the language specification is extended with respect to the following items:

1) For verification purposes, module function specification items are added.

2) Gate and media delay parameters are added.

3) Pin function and attribute can be specified.

4) Schematic information is added.

5) Database management information is provided.

The VLSI cannot be designed by using only VLSI structural information since the verification and evaluation processes require accurate design data. For example, in the case of circuits with bidirectional buses or wired gates, the physical structure of the network becomes different from the logical one. Therefore, the design language has to be provided with appropriate functions to deal with these problems.

In HSL, the DEFAULT statement is introduced to overcome these problems. If the DEFAULT statement is specified, a circuit with dotting function is automatically converted to a circuit with no dotting function. The stack values of unconnected pins are also specified by the DEFAULT statement.

D. Hierarchical Structure

The VLSI CAD system consists of three major subsystems, as shown in Fig. 15, i.e., language processing, verification, and placement and routing subsystems. The language processing subsystem, especially, has a unique feature. It is integrated into the program called HIDEMAP (Hierarchical design database manipulator), which has the following functions: 1) HSL compiler and decompiler; 2) macro expander; 3) scissor (block divider); 4) wired function (dot) and bidirectional bus processor; 5) database manager; and 6) interface between database and application programs.

1) Compiler: The HSL compiler checks for a syntax error in input data described by HSL and converts it to the object file which consists of a collection of modules directly accessible by module names or its data items. The object file can be directly registered into the database, and broken down and converted back to the HSL source file or list.

2) Database Manager and Macro Expander: In the database, modules are managed by name and management items. The

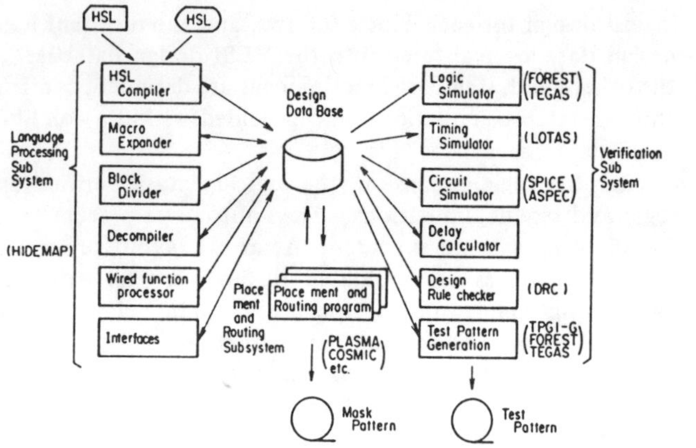

Fig. 15. VLSI design system configuration [116], [117].

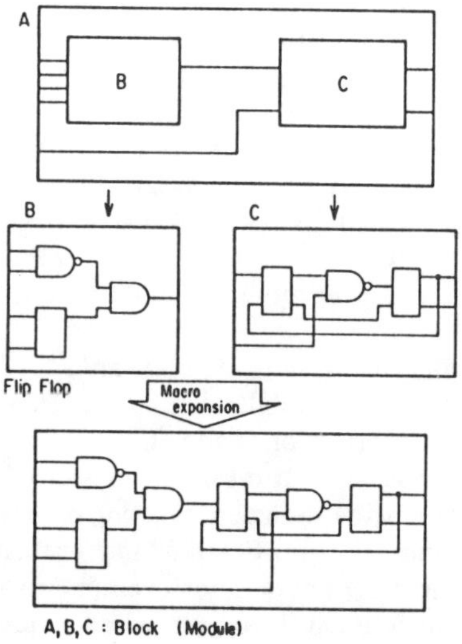

Fig. 16. Macro expansion.

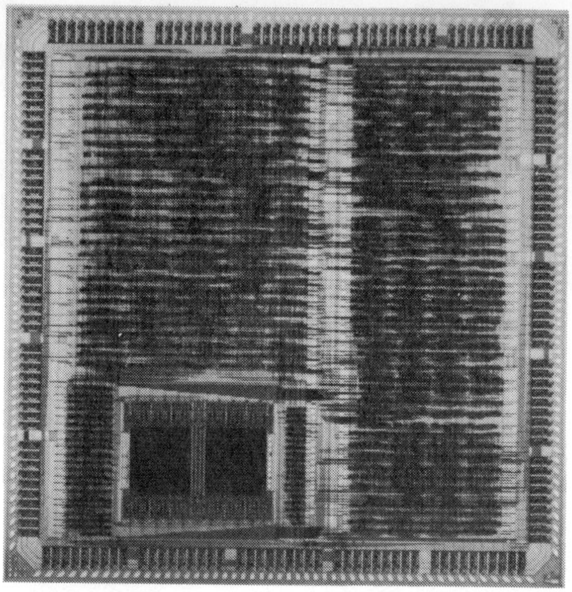

Fig. 17. Microphotograph of an automatically designed 32-bit VLSI processor chip (containing 20K gates) [119].

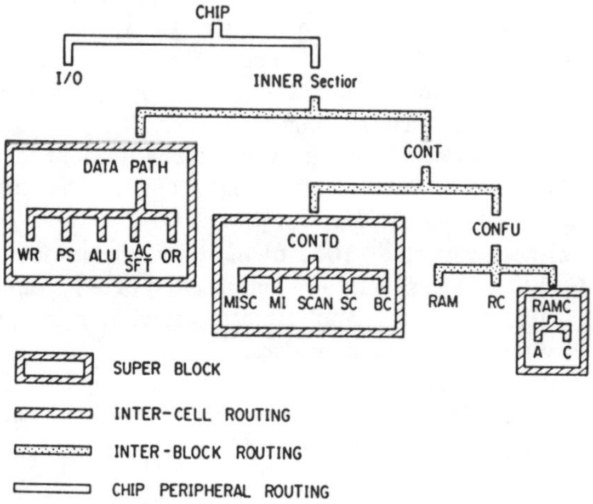

Fig. 18. Layout hierarchy of the 32-bit VLSI processor chip [119].

database consists of library, project, and user file. The library file which keeps cell libraries is accessible by any user who has access permission. The design data can be shared among several users working on the same project. Database control commands are provided to handle the collection of modules and get statistical information easily.

The macro expander program transforms higher level modules to lower level modules. An example is shown in Fig. 16. Macro expansion is also controlled by HIDEMAP commands specifying top module, and stop level name or stop module names. By using this function, designers can set up network data which are necessary for the hierarchical layout program. Input data to the macro expander can be retrieved from the database by specifying only top module name or management item priority.

3) Block Divider: The block divider is also a useful tool for the hierarchical VLSI design approach. In the hierarchical-oriented system, since the design is made in a top-down manner, it often occurs that some blocks are too large and other blocks are too small to be processed in a single run of the layout program. Therefore, it is desirable that the block structure is reorganized into a proper size for the routing process. The block divider (or scissor) is provided with such a function, which can automatically change structures by simple commands.

4) Layout Program: After the hierarchical structure is reorganized, a layout process takes place. At first, functional cells are automatically routed. Then intercells are routed to build functional blocks. These functional blocks are then routed and compose the whole chip. In the hierarchical layout method, the layout quality is controlled by: i) block partitioning; ii) block shape and location; iii) interblock wiring channel location; and iv) external pin locations within each block.

E. System Application Example of VLSI CAD System

The VLSI CAD system has been applied to a number of VLSI designs [119], [120] since 1979. Fig. 17 shows a microphotograph of a 32-bit CMOS VLSI processor chip containing a 17K gate random logic and a 2304-bit RAM, which has been designed by using the above CAD system. In this particular circuit, the logic hierarchy consists of three levels—cell, block, and chip. The cell size is 3 gates on an average, and each block contains about 100 to 300 cells or 300 to 1000 gates.

The original three-level hierarchy was extended to seven levels, as shown in Fig. 18, such that it becomes well fitted to the automatic layout process. The RAM block was designed manually, while RC block, DATA PATH, CONTD and RAMC superblocks were automatically designed by using the cell placer and intercell router. Layout design efforts were within two man-months for the particular circuit which contains 78K transistors and 17K interconnection nets. The CAD system has achieved a high packing density of 542 Trs/mm^2 for random logic circuits. This result is considered equal to higher layout quality, compared to a manual design.

VI. Conclusion

The developmental aspect and the present status of CAD systems for VLSI design in Japan have been presented. With the explosive increase in VLSI scale and complexity, CAD will play an unprecedentedly important role. To meet future VLSI design requirements, great efforts must be continued for the development of CAD systems and technologies. In order to effectively aid 50K-gate or larger VLSI design, in addition to improvements in traditional CAD systems, technological breakthroughs are keenly needed. In the research and development on VLSI CAD hereafter, the following subjects will be the principal themes: 1) more intelligent CAD algorithms and methodologies; 2) CAD-dedicated machine or usage of supercomputer in CAD area; 3) highly interactive design environment; and 4) vertically integrated CAD system covering system concept to the device fabrication process.

With the remarkable progress in LSI technologies, the LSI world is widening its boundary and the population of LSI designers, i.e., CAD users, is rapidly increasing as a result. Under these circumstances, standardization becomes indispensable for the efficient usage of CAD resources among numerous users at various levels and with different backgrounds. Consequently, it is earnestly desirable that strenuous efforts are made toward standardization as well as accomplishment of CAD systems for future VLSI design.

Acknowledgment

The authors wish to thank K. Ueda, T. Watanabe, T. Adachi, M. Nagatani, M. Tomisawa, T. Hoshino, H. Miyashita, and K. Yokoyama of Musashino ECL, NTT for preparing the manuscript. We would also like to thank A. Yamada, T. Kurobe, and N. Tanabe of Nippon Electric Co., Ltd. for their valuable suggestions. Finally, the authors are greatly indebted to the Research Committee on the Design Technology of Electronics Equipment of IPSJ for providing the data on design automation status in Japan.

References

[1] Research Committee on the Design Technology of Electronics Equipment, "Computer aided design of electronic equipment (4)—the CAD system in use or under development," *J. Inform. Process. Soc. Japan*, vol. 21, no. 2, pp. 165–167, Feb. 1980 (Japanese).

[2] L. W. Nagel and C. O. Pederson, "Simulation program with integrated circuit emphasis," in *Proc. 16th Midwest Symp. Circuit Theory* (Waterloo, Ont., Canada, Apr. 1973).

[3] T. E. Idleman, F. S. Jenkins, W. J. McCalla, and D. O. Pederson, "SLIC—a simulator for linear integrated circuits," *IEEE J. Solid State Circuits*, vol. SC-6, pp. 188–204, Aug. 1971.

[4] W. T. Weeks, A. J. Jimenez, G. W. Mahoney, D. Mehta, H. Qassemzadeh, and T. R. Scott, "Algorithms for ASTAP—a network analysis program," *IEEE Trans. Circuit Theory*, vol. CT-20, pp. 628–634, Nov. 1973.

[5] S. A. Szygenda and E. W. Thompson, "Digital logic simulation in a time based table driven environment," *Computer*, Mar. 1975.

Survey on CAD Systems in Japan

[6] K. Kusunoki, "DEX-design automation system cooperation development," in *Proc. 10th Design Automation Workshop*, 1973, pp. 187–192.

[7] T. Moto-Oka, T. Kurachi, T. Shiino, and M. Sugimoto, "Logic design system in Japan," in *Proc. 12th Design Automation Conf.*, June 1975, pp. 241–250.

[8] K. Ueda, "Computer-aided layout design for LSI; state of the art," *Proc. 2nd USA-Japan Computer Conf.*, 1975, pp. 556–561.

[9] K. Kani, A. Yamada, and M. Teramoto, "CAD in the Japanese electronics industry," presented at the Symp. CAD of Digital Electronic Circuits and Systems, Brussels, Belgium, Nov. 1978.

[10] M. Watanabe, "CAD tools for designing VLSI in Japan," in *Proc. ISSCC 79*, Feb. 1979, pp. 242–243.

[11] A. Yamada, "Design automation status in Japan," in *Proc. 18th Design Automation Conf.*, 1981, pp. 43–50.

[12] T. Hosaka, K. Ueda, and H. Matsuura, "A design automation system for electronic switching systems," in *Proc. 18th Design Automation Conf.*, 1981, pp. 51–58.

[13] T. Sudo, "VLSI and CAD," *Proc. IPSJ*, vol. 22, no. 8, 1981 (Japanese).

[14] T. Sudo, "Present status of CAD technologies in Japan," to be published.

Logic Design

[15] T. Kurobe, S. Nemoto, Y. Shikata, and K. Kani, "Logic simulation system; LOGOS 2," in *Monograph of Technical Group on DA of Inform. Process. Soc. Japan*, DA31-2, 1977 (Japanese).

[16] Y. Fujinami, N. Yasuda, H. Sakuma, and T. Kurobe, "A logic simulation system for MOS LSI circuits with bidirectional elements," in *Monograph of Technical Group on Circuits and Systems of Inst. Electron. Commun. Eng. Japan*, Feb. 1982.

[17] J. Watanabe, J. Miura, T. Kurachi, and I. Suetsugu, "Seven value logic simulation for MOS LSI circuits," in *Proc. IEEE Inter. Conf. Circuits and Computers 80*, Oct. 1980, pp. 941–944.

[18] Y. Ohno, M. Miyoshi, and K. Sato, "Logic verification system for very large computers using LSI's," in *Proc. 16th Design Automation Conf.*, June 1979, pp. 367–374.

[19] M. Murakami, N. Shiraki, and K. Hirakawa, "Logic verification and test generation for LSI circuits," in *Proc. Test Conf.*, Nov. 1980, pp. 467–472.

[20] T. Sasaki, A. Yamada, S. Kato, T. Nakazawa, K. Tomita, and N. Nomizu, "MIXS: a mixed level simulator for large digital system logic verification," in *Proc. 17th Design Automation Conf.*, June 1980, pp. 626–633.

[21] N. Kawato, T. Saito, F. Maruyama, and T. Uehara, "Design and verification of large-scale computers by using DDL," in *Proc. 16th Design Automation Conf.*, June 1979, pp. 360–366.

[22] M. Ishii, Y. Ito, M. Iwasaki, M. Yamamoto, and S. Kodama, "Automatic input and interactive editing systems of logic circuit diagrams," in *Proc. 18th Design Automation Conf.*, June 1981, pp. 639–645.

[23] K. Hayamizu and T. Nanya, "An automatic design system for PLA," in *Proc. Nat. Conf. Inform. Process. Soc. Japan*, 5G-4, 1978 (Japanese).

[24] M. Tsuboya, T. Takagi, M. Teramoto, and T. Nanya, "PLA design support system," in *Proc. Nat. Conf. Inform. Process. Soc. Japan*, 5G-5, 1978 (Japanese).

[25] H. Miyashita and Y. Sugiyama, H. Sakuma, and H. Ishizuka, "PLACAD: PLA design automation system," in *Proc. Nat. Conf. Inst. Electron. Commun. Eng. Japan*, 410, 1981 (Japanese).

[26] M. Tokoro, M. Sato, M. Ishigami, E. Tamura, T. Ishimitsu, and H. Ohara, "A module level simulation technique for systems composed of LSIs and MSIs," in *Proc. 15th Design Automation Conf.*, June 1978, pp. 418–427.

[27] M. Itoh, G. Kohno, T. Makita, and R. Tadenuma, "A logic design automation system," in *Monograph of Technical Group on Design Technology of Electronics Equipment of Inform. Process. Soc. Japan*, 4-3, Mar. 1980 (Japanese).

[28] T. Fukazawa, K. Kurihara, T. Suzuki, H. Tanaka, and T. Moto-Oka, "Logic simulator by data flow machine," in *Monograph of Technical Group on Design Technology of Electronics Equipment of Inform. Process. Soc. Japan*, 6-3, Oct. 1980 (Japanese).

Test Generation

[29] S. Funatsu, N. Wakatsuki, and T. Arima, "Test generation system in Japan," in *Proc. 12th Design Automation Conf.*, June 1975, pp. 112–114.

[30] H. Kubo, "A procedure for generating test sequence to detect sequential circuit failures," *NEC Res. Develop.*, no. 12, pp. 69–78, 1968.

[31] S. Goshima, T. Kozawa, Y. Oka, T. Mori, Y. Takeguchi, and Y. Ohno, "Diagnostic system for large scale logic cards and LSIs," in *Proc. 18th Design Automation Conf.*, June 1981, pp. 256–259.

[32] T. Yamada, M. Saisho, and Y. Kasuya, "TESTA-1: test generation system using state diagram tracing approach," *NEC Res. Develop.*, no. 65, pp. 68–73, Apr. 1982.

[33] K. Kinoshita, Y. Takamatsu, and M. Shibata, "Test generation for combinatorial circuits by structure description functions," *FTCS-10*, pp. 152–154, Oct. 1980.

[34] T. Shimono, H. Fujiwara, and H. Ozaki, "An efficient test generating algorithm based on testability measurements," in *Monograph of Technical Group on Design Technology of Electronics Equipment of Inform. Process. Soc. Japan*, 13-4, Feb. 1982 (Japanese).

[35] T. Hoshino, Y. Wada, and T. Sudo, "A test pattern generation system by using a block sensitizing method," in *Monograph of Technical Group on Design Technology of Electronics Equipment of Inform. Process Soc. Japan*, 9-3, June 1981 (Japanese).

[36] A. Kobayashi, S. Matsue, and H. Shiba, "Flip-flop circuit with FLT capability," in *Proc. Nat. Conf. Inst. Electron. Commun. Eng. Japan*, no. 892, 1968.

[37] A. Yamada, N. Wakatsuki, T. Fukui, and S. Funatsu, "Automatic system level test generation and fault location for large digital systems," in *Proc. 15th Design Automation Conf.*, June 1978, pp. 347-352.

Circuit and Timing Simulators

[38] M. Sugimori, "DEMOS E circuit analysis program: ECSS," in *Proc. Nat. Conf. Inst. Electron. Commun. Eng. Japan*, 1974, p. 1784 (Japanese).

[39] T. Kojima and K. Watanabe, "CARD (computer assisted research and development) for electrical circuit," *Fujitsu*, vol. 24, no. 7, pp. 175-189, 1973 (Japanese).

[40] K. Kawakita and T. Ohtsuki, "NECTAR-2–circuit analysis program based on piecewise linear approach," in *Proc. IEEE Inter. Conf. Circuits and Systems*, 1975, pp. 46-51.

[41] T. Ohtsuki, T. Fujisawa, and S. Kumagai, "Existence theorems and a solution algorithm for piecewise-linear resistor networks," *SIAM J. Math. Anal.*, vol. 8, pp. 69-99, 1977.

[42] B. R. Chawla, H. K. Gummel, and P. Kozak, "MOTIS–an MOS timing simulator," *IEEE Trans. Circuits Syst.*, CAS-22, vol. 12, pp. 901-910, 1975.

[43] S. P. Fan, M. Y. Hsueh, A. R. Newton, and D. O. Pederson, "MOTIS-C: a new circuit simulator for MOS LSI circuits," in *Proc. IEEE Inter. Conf. Circuits and Systems*, 1977, pp. 700-703.

[44] K. Hirabayashi and J. Watanabe, "MATIS–macromodel timing simulator for large scale integrated MOS circuits," in *Proc. 3rd USA-Japan Computer Conf.*, 26-1-1, 1978.

[45] T. Shikage and H. Sato, "A circuit simulation method based on macro-modeling of circuit blocks," in *Proc. Nat. Conf. Semiconductors of Inst. Electron. Commun. Eng. Japan*, 1979, p. 62 (Japanese).

[46] J. Watanabe, "Macro-model timing simulator–MATIS," in *Proc. Nat. Conf. Semiconductors of Inst. Electron. Commun. Eng. Japan*, 1979, p. 63 (Japanese).

[47] J. Yamada, T. Takeya, N. Miyahara, and H. Nakamura, "A large-scale circuit simulator (LOTAS)," in *Proc. Nat. Conf. Semiconductors of Inst. Electron. Commun. Eng. Japan*, 1979, p. 65 (Japanese).

[48] N. Tanabe, H. Nakamura, and K. Kawakita, "MOSTAP: an MOS circuit simulator for LSI circuits," in *Proc. IEEE Inter. Conf. on Circuits and Systems*, 1980, pp. 1035-1038.

[49] N. Tanabe, M. Endo, and N. Miyahara, "Timing simulation: a fast analysis method," in *Proc. Nat. Conf. Inst. Electron. Commun. Eng. Japan*, 450, 1981, p. 215 (Japanese).

[50] N. Miyahara, M. Endo, N. Tanabe, and H. Nakamura, "Timing simulator (LOTAS)," in *Monograph of Technical Group on Electron Devices of Inst. Electron. Commun. Eng. Japan*, EDD82-13, 1982, pp. 61-69 (Japanese).

Automatic Layout Design

[51] K. Mikami and K. Tabuchi, "A computer program for optimal routing of printed circuit conductors," in *Proc. IFIP Cong. 68*, 1968, pp. 1475-1478.

[52] A. Hashimoto and J. Stevens, "Wire routing by optimal channel assignment within large apertures," in *Proc. 8th Design Automation Workshop*, 1971, pp. 158-169.

[53] N. Sugiyama, T. Ohtsuki, S. Nemoto, K. Kani, and H. Watanabe, "An integrated circuit layout design program based on graph-theoretical approach," *ISSCC*, 1970.

[54] Y. Sugiyama, K. Ueda, K. Kani, and M. Teramoto, "Routing program for multichip LSIs," in *Proc. USA-Japan Design Automation '75*, Aug. 1975, pp. 87-94.

[55] T. Kozawa, H. Horino, K. Watanabe, M. Nagata, and H. Fukuda, "Block and track method for automated layout generation of MOS LSI array," *ISSCC*, 1972, pp. 62-63.

[56] T. Kozawa, H. Horino, T. Ishiga, and S. Sato, "Advanced LILAC–an automated layout generation system for MOS/LSIs," in *Proc. 11th Design Automation Workshop*, 1974, pp. 26-46.

[57] K. Kani, H. Kawanishi, and A. Kishimoto, "ROBIN: a building block LSI routing program," in *Proc. 1976 IEEE Inter. Conf. on Circuits and Systems*, 1976, pp. 658-661.

[58] H. Kawanishi, S. Goto, T. Oyamada, H. Kato, and K. Kani, "A routing method of building block LSI," in *Proc. 7th Asilomar Conf. on Circuits, Systems and Computers*, 1973, pp. 119-124.

[59] H. Kato, H. Kawanishi, and S. Goto, "On automated wire routing for building-block LSI," in *Proc. 1974 IEEE Inter. Conf. on Circuits and Systems*, 1974, pp. 309-313.

[60] I. Shirakawa, N. Okuda, T. Harada, S. Tani, and H. Ozaki, "A layout system for the random logic portion of MOS LSI," in *Proc. 17th Design Automation Conf.*, 1980, pp. 92-99.

[61] K. Sato, T. Nagai, H. Shimoyama, and T. Yahara, "MIRAGE–a simple model routing program for the hierarchical layout design of IC masks," in *Proc. 16th Design Automation Conf.*, 1979, pp. 297-304.

[62] K. Sato and T. Nagai, "A method of specifying the relative locations between blocks in a routing program for building block LSI," in *Proc. 1979 IEEE Inter. Conf. on Circuits and Systems*, 1979, pp. 673-676.

[63] K. Sato, T. Nagai, M. Tachibana, H. Shimoyama, M. Ozaki, and T. Yahara, "MILD–A cell based layout system for MOS LSI," in *Proc. 18th Design Automation Conf.*, 1981, pp. 828-836.

[64] K. Sato, H. Shimoyama, T. Nagai, M. Ozaki, and T. Yahara, "A grid-free channel router," in *Proc. 17th Design Automation Conf.*, 1980, pp. 22-31.

[65] T. Chiba, N. Okuda, T. Kambe, I. Nishioka, T. Inufushi, and S. Kimura, "SHARPS: a hierarchical layout system for VLSI," in *Proc. 18th Design Automation Conf.*, 1980, pp. 820-827.

[66] D. N. Deutsch, "A dogleg channel router," in *Proc. 13th Design Automation Conf.*, 1976, pp. 425-433.

[67] M. Nagatani, H. Miyashita, H. Okamoto, K. Tansho, and Y. Sugiyama, "An automated layout system for LSI functional blocks: PLASMA," in *Monograph of Technical Group on Design Technology of Electronics Equipment of Inform. Process. Soc. Japan*, 41, 1980, pp. 1-10.

[68] T. Adachi, H. Kitazawa, M. Nagatani, and T. Sudo, "Hierarchical top-down layout design method for VLSI chip," presented at 19th Design Automation Conf., 1982.

[69] Y. Ozawa, M. Murakami, and K. Suzuki, "Master slice LSI computer aided design system," in *Proc. 11th Design Automation Conf.*, 1974, pp. 19-25.

[70] R. Kamikawai, K. Kishida, A. Osawa, I. Yasuda, and T. Chiba, "Placement and routing program for masterslice LSIs," in *Proc. 13th Design Automation Conf.*, 1976, pp. 245-250.

[71] C. Y. Lee, "An algorithm for path connections and its applications," *IRE Trans. Electron. Comput.*, vol. EC-10, no. 3, pp. 346-364, 1961.

[72] K. Ueda, Y. Sugiyama, and K. Wada, "An automated layout system for masterslice LSI: MARC," *IEEE J. Solid State Circuits*, vol. SC-13, no. 5, pp. 716-721, 1978.

[73] C. Tanaka, S. Murai, S. Nakamura, T. Ogihara, M. Terai, and K. Kinoshita, "An integrated computer aided design system for gate array masterslices: part 1 logic reorganization system LORES-2," in *Proc. 18th Design Automation Conf.*, 1981, pp. 59-65.

[74] C. Tanaka, Su. Murai, H. Tsuji, T. Yahara, K. Okazaki, M. Terai, R. Katoh, and M. Tachibana, "An integrated computer aided design system for gate array masterslices: part 2 the layout design system for master-slice LSIs," presented at *19th Design Automation Conf.*, 1982.

[75] S. Murai, M. Kakinuma, M. Imai, and H. Tsuji, "The effects of initial placement techniques on the final placement results constructive vs. top-down techniques," in *Proc. 1980 IEEE Inter. Conf. on Circuits and Computers*, 1980, pp. 80-82.

[76] H. Shiraishi and F. Hirose, "Efficient placement and routing techniques for master slice LSI," in *Proc. 17th Design Automation Conf.*, 1980, pp. 458-464.

[77] H. Yoshizawa, H. Kawanishi, S. Goto, A. Kishimoto, Y. Fujinami, and K. Kani, "Automatic layout algorithms for master slice LSI," in *Proc. 1979 IEEE Inter. Conf. on Circuits and Systems*, 1979, pp. 470-473.

[78] S. Goto, "An efficient algorithm for two dimensional placement problem in electrical circuit layout," *IEEE Trans. Circuits and Syst.*, vol. CAS-28, no. 1, pp. 12-18, Jan. 1980.

[79] S. Goto, "An automated layout design system for masterslice LSI chip," in *Proc. Europ. Conf. on Circuit Theory and Design '81*, 1981, pp. 631-635.

[80] T. Matsuda, T. Fujita, K. Takamizawa, H. Mizumura, H. Nakamura, F. Kitajima, and S. Goto, "LAMBDA: a quick, low cost layout design system for master-slice LSIs," presented at 19th Design Automation Conf., 1982.

Artwork Verifier

[81] K. Yoshida, T. Mitsuhashi, Y. Nakada, T. Chiba, H. Ogita, and S. Nakatsuka, "A layout checking system for large scale integrated circuit," in *Proc. 14th Design Automation Conf.*, 1977, pp. 322-330.

[82] T. Kozawa, A. Tsukizoe, J. Sakemi, C. Miura, and T. Ishii, "A

concurrent pattern operation algorithm for VLSI mask data," in *Proc. 18th Design Automation Conf.*, 1981, pp. 563–570.
[83] H. Kawanishi, T. Nishide, and T. Masuda, "An algorithm on graphical manipulation of LSI mask pattern," *Trans. Inst. Electron. Commun. Eng. Japan*, vol. J65-A, pp. 611–618, 1980 (Japanese).
[84] S. Yamada and T. Watanabe, "A mask pattern analysis system for LSI (PAS-1)," in *Proc. IEEE Inter. Conf. on Circuits and Systems*, 1979, pp. 858–861.
[85] T. Mitsuhashi, T. Chiba, M. Takashima, and K. Yoshida, "An integrated mask artwork analysis system," in *Proc. 17th Design Automation Conf.*, 1980, pp. 277–284.
[86] J. Yoshida, T. Ozaki, and Y. Goto, "PANAMAP-B: a mask verification system for bipolar IC," in *Proc. 18th Design Automation Conf.*, 1981, pp. 690–695.
[87] H. Yoshimura, K. Tansho, N. Ohwada, and T. Nishide, "An algorithm for resistance calculation from IC mask pattern information," in *Proc. IEEE Inter. Conf. on Circuits and Systems*, pp. 478–481, 1979.
[88] A. Kishimoto, H. Kawanishi, H. Yoshizawa, H. Ohno, Y. Fujinami, and K. Kani, "An interconnection check algorithm for mask pattern," in *Proc. IEEE Inter. Conf. on Circuits and Systems*, 1978, pp. 669–672.
[89] K. Hirabayashi and M. Kawamura, "MACLOS–mask checking logic simulator," *IEEE J. Solid State Circuits*, vol. SC-15, no. 3, pp. 368–370, 1980.
[90] N. Kubo, I. Shirakawa, and H. Ozaki, "Algorithm for testing graph isomorphism, " in *Proc. IEEE Inter. Conf. on Circuits and Systems*, pp. 641–644, 1979.
[91] N. Miyahara, T. Watanabe, M. Endo, and S. Yamada, "A new CAD system for automatic logic interconnection verification," in *Proc. IEEE Inter. Conf. on Circuits and Systems*, 1981, pp. 114–117.

Device and Process Simulators

[92] D. A. Antoniadis, S. E. Hanson, and R. W. Dutton, "SUPREM–a program for IC process modeling and simulation," Stanford Electronics Labs, Stanford Univ., Stanford, CA, Tech. Rep., SEL78-020, 1978.
[93] R. W. Dutton, D. A. Divekar, A. G. Gonzales, S. E. Hansen, and D. A. Antoniadis, "Correlation of fabrication process and electrical device parameter variations," *IEEE J. Solid-State Circuits*, vol. SC-12, pp. 349–355, Aug. 1977.
[94] T. Toyabe, M. S. Mock, T. Okabe, K. Ujiie, and M. Nagata, "A two-dimensional analysis of I^2L with multi-stream function technique," in *Proc. NASECODE 1st Conf.*, 1979, pp. 290–292.
[95] T. Adachi, A. Yoshii, and T. Sudo, "Two-dimensional semiconductor analysis using finite-element method," *IEEE Trans. Electron Devices*, vol. ED-26, pp. 1026–1031, July 1979.
[96] M. Tomizawa, H. Kitazawa, A. Yoshii, S. Horiguchi, and T. Sudo, "An accurate design method of bipolar device using a two-dimensional device simulator," *IEEE Trans. Electron. Devices*, vol. ED-28, pp. 1148–1153, Oct. 1981.
[97] K. Yamaguchi, S. Asai, and H. Kodera, "Two-dimensional analysis of stability criteria of GaAs FET's," *IEEE Trans. Electron Devices*, vol. ED-23, pp. 1283–1290, Dec. 1976.
[98] T. Toyabe and S. Asai, "Analytical models of threshold voltage of short-channel MOSFET's derived from two-dimensional analysis," *IEEE J. Solid-State Circuits*, vol. SC-14, pp. 375–383, Apr. 1979.
[99] K. Yokoyama, A. Yoshii, and S. Horiguchi, "Threshold sensitivity minimization of short-channel MOSFET's by computer simulation," *IEEE Trans. Electron Devices*, vol. ED-27, pp. 1509–1514, Aug. 1980.
[100] H. Oka, K. Nishiuchi, T. Nakamura, and H. Ishikawa, "Computer analysis of a short-channel BC MOSFET," *IEEE Trans. Electron Devices*, vol. ED-27, pp. 1514–1520, Aug. 1980.
[101] M. Konaka, K. Natori, and R. Dang, "Development of an MOS device simulator (MODEST) and some related problems in the two-dimensional numerical calculation," *Inst. Electron. Commun. Eng. Japan*, Tech. Rep. SSD81-10, 1981 (Japanese).
[102] A. Yoshii, H. Kitazawa, M. Tomizawa, S. Horiguchi, and T. Sudo, "A three-dimensional analysis of semiconductor devices," *IEEE Trans. Electron Devices*, vol. ED-29, pp. 184–189, Feb. 1982.
[103] R. Kasai, K. Yokoyama, A. Yoshii, and T. Sudo, "Threshold voltage analysis of short and narrow channel MOSFET's by three-dimensional computer simulation," *IEEE Trans. Electron Devices*, to be published.
[104] M. Tomizawa, K. Yokoyama, A. Yoshii, and T. Sudo, "Two-dimensional device simulator for gate level characterization," *Solid-State Electron.*, to be published.
[105] Y. Ohno and Y. Okuto, "Computer aided Si-MOSFET process designing," in *Proc. 11th Conf. Solid State Devices* (Tokyo, Japan), vol. 19, sup. 19-1, 1980, pp. 65–69.
[106] K. Taniguichi, M. Kashiwagi, and H. Iwai, "Two-dimensional computer simulation models for MOSLSI fabrication process," *IEEE Trans. Electron Devices*, vol. ED-28, May 1981.
[107] K. Yokoyama, A. Yoshii, T. Adachi, and R. Kasai, "Application of Fletcher-Powell's optimization method to process/device simulation of MOSFET characteristics," *Solid-State Electron.*, vol. 25, pp. 201–203, Mar. 1982.
[108] H. K. Gummel, "A self-consistent iterative scheme for one-dimensional steady-state transistor calcualtion," *IEEE Trans. Electron Devices*, vol. ED-11, pp. 455–465, Oct. 1964.
[109] H. L. Stone, "Iterative solution of implicit approximations of multidimensional partial differential equations," *SIAM J. Numer. Anal.*, vol. 5, pp. 530–558, Sept. 1968. . . . .
[110] J. A. Meijerink and H. A. van der Vorst, "An interactive solution method for linear systems of which the coefficient matrix is a symmetrical M-matrix," *Math. Comput.*, vol. 31, pp. 148–162, Jan. 1977.
[111] H. K. Gummel and H. C. Poon, "An integral charge control model of bipolar transistors," *Bell Syst. Tech. J.*, vol. 49, pp. 827–852, May–June 1970.
[112] M. S. Mock, "A two-dimensional mathematical model of the insulated-gate field effect transistor," *Solid-State Electron.*, vol. 16, pp. 601–609.
[113] R. Dang, T. Shibata, M. Konaka, S. Onga, and M. Kashiwagi, "Complete process/device simulation system and its application to the design of sub-micron MOSFET," in *Proc. 1981 Symp. VLSI Technology*, 1981, pp. 86–87.
[114] R. Flecher and M.J.D. Powell, "A rapid convergence descent method for minimization," *Comput. J.*, vol. 6, pp. 163–168, June 1963.
[115] Y. Okuto, Y. Ohno, and M. Fukuma, "Device process designing aids for MOS LSIs," in *Proc. 2nd NASECODE Conf.*, 1981.

Integrated VLSI CAD System

[116] T. Sudo, "VLSI Design Methodology," *Monograph of Technical Group on Design Technology of Electronics Equipment of Inform. Process. Soc. Japan*, 1-1, 1979 (Japanese).
[117] T. Sudo and Y. Sugiyama, "Custom VLSI design system," in *Proc. Inform. Process. Soc. Japan*, vol. 22, no. 8, pp. 791–796, 1981 (Japanese).
[118] W. M. vanCleemput, "An hierarchical language for the structural description of digital systems," *Proc. 14th Design Automation Conf.*, pp. 378–385, June 1977.
[119] S. Horiguchi, H. Yoshimura, R. Kasai, and T. Sudo, "An automatic designed 32b CMOS VLSI processor," *ISSCC82*, pp. 54–55, 1982.
[120] Y. Sugiyama, M. Suzuki, Y. Kobayashi, and T. Sudo, "A sub-nano second 12K gate bipolar 32 bit VLSI processor," in *Proc. of Custom Integrated Circuit Conf.*, pp. 73–77, 1981.

An Integrated Design Automation System for VLSI Circuits

O. Karatsu, T. Hoshino, M. Endo,
H. Kitazawa, T. Adachi, and K. Ueda
Atsugi Electrical Communication Laboratories

Our Integrated Design Automation System consists of an integrated design database, automated design processors, verification tools, and an interactive capture system. The automatic logic synthesis program, Angel, and the hierarchical layout system Champ/Alpha, have been particularly important in reducing the total design effort. A unified design language, HSL-FX, has been developed to broaden LSI design system coverage and to obtain better results from automatic logic synthesis.

Using this system, a 10K-gate CPU has been designed in two man-months—from function-level description to layout pattern.

With the advances of VLSI technology, the number of gates integrated into chips has increased, causing several new design problems: Design objectives have expanded from device to system level; the design data contained on a single chip have increased; logical and physical design optimization requirements have increased; design quality requirements are more strict; and LSI technologies continue to progress.

To solve these problems, the Integrated Design Automation System has been developed at NTT's Atsugi Electrical Communication Laboratories. VLSI circuits with up to 100K gates can be automatically designed using this system, which covers all steps, from functional design to layout design.

Evolution of the system

The evolution of the DA system and the resulting reduction in design effort is illustrated in Figure 1. More than five years ago, the design of a typical 20K-gate VLSI circuit took 170 man-months to complete, using primitive CAD tools such as a logic simulator and a commercially available graphic editing system.

Reprinted from *IEEE Design and Test of Computers*, pp. 17–26, October 1985.

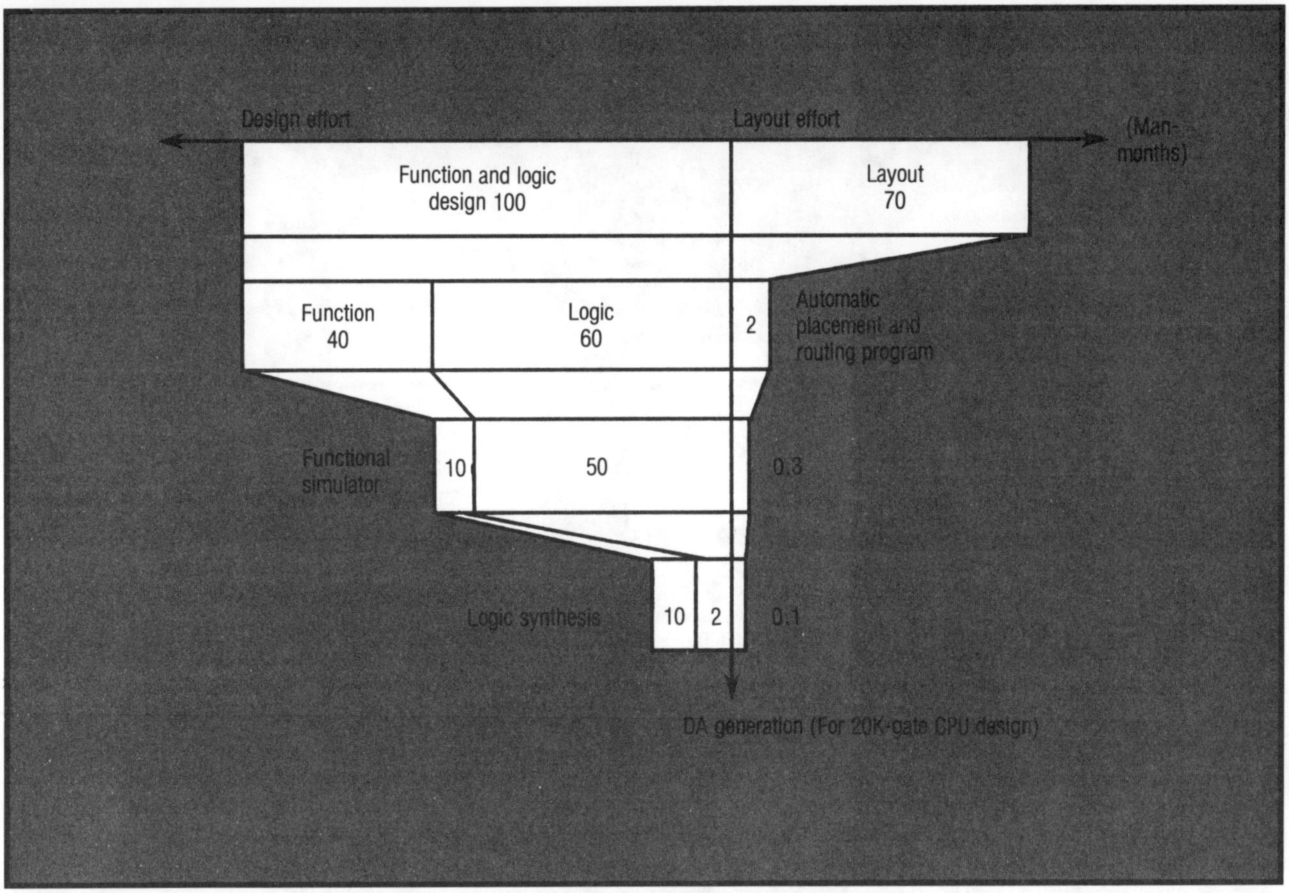

Figure 1. Evolution of the DA system and the resulting reduction in design effort.

Our major DA development effort was first concentrated in a layout database, a common design language processing system, and a design database. When the automatic placement and routing program and environment system were finished, design effort for layout had been reduced from 70 man-months to two man-months. Further development has since reduced layout design effort to less than one man-month.

Our next major target for DA development was the function and logic design. A function-level design system has been developed and there has been a big increase in design verification capability. One hundred man-months of function and logic design effort was reduced to 60 man-months. An automatic logic synthesis program, which can synthesize logic circuits from function-level design data, has further reduced the design effort from 60 to 12 man-months.

Using this integrated design system, the total design period has been greatly shortened. The first-stage development of this system has already been introduced in a previous publication.[1] This article will present the second-stage development of this system – function-level design support, logic synthesis program, user interface improvement, and a more sophisticated layout system.

System overview

Figure 2 shows an overview of the Integrated Design Automation System. The DA system is organized into four tightly-coupled components:

(1) The design database stores function, logic, circuit, mask, and test-pattern data. These design data can be handled by routines such as the hierarchical expander, the hierarchical reorganizing program (Scissor), and any one of several format converters.

(2) The design processors, including the automatic logic synthesizer, Angel,[2] the chip floor-plan program, Champ,[3] and the layout program, Alpha.[4] These are top-down design tools.

(3) The verifiers such as function simulators, logic simulators, timing simulators, circuit simulators, and several checkers. These are bottom-up design tools.

(4) The user interfaces, which allow users to interactively specify the VLSI circuit's function and structure; the interfaces can also examine the DA programs' process outputs.

VLSI design language

Formerly, two languages were used in the DA system: HSL, for structural descriptions, and FDL, for functional descriptions. Currently, a new language, HSL-FX,[5] which can be used for both structural and functional de-

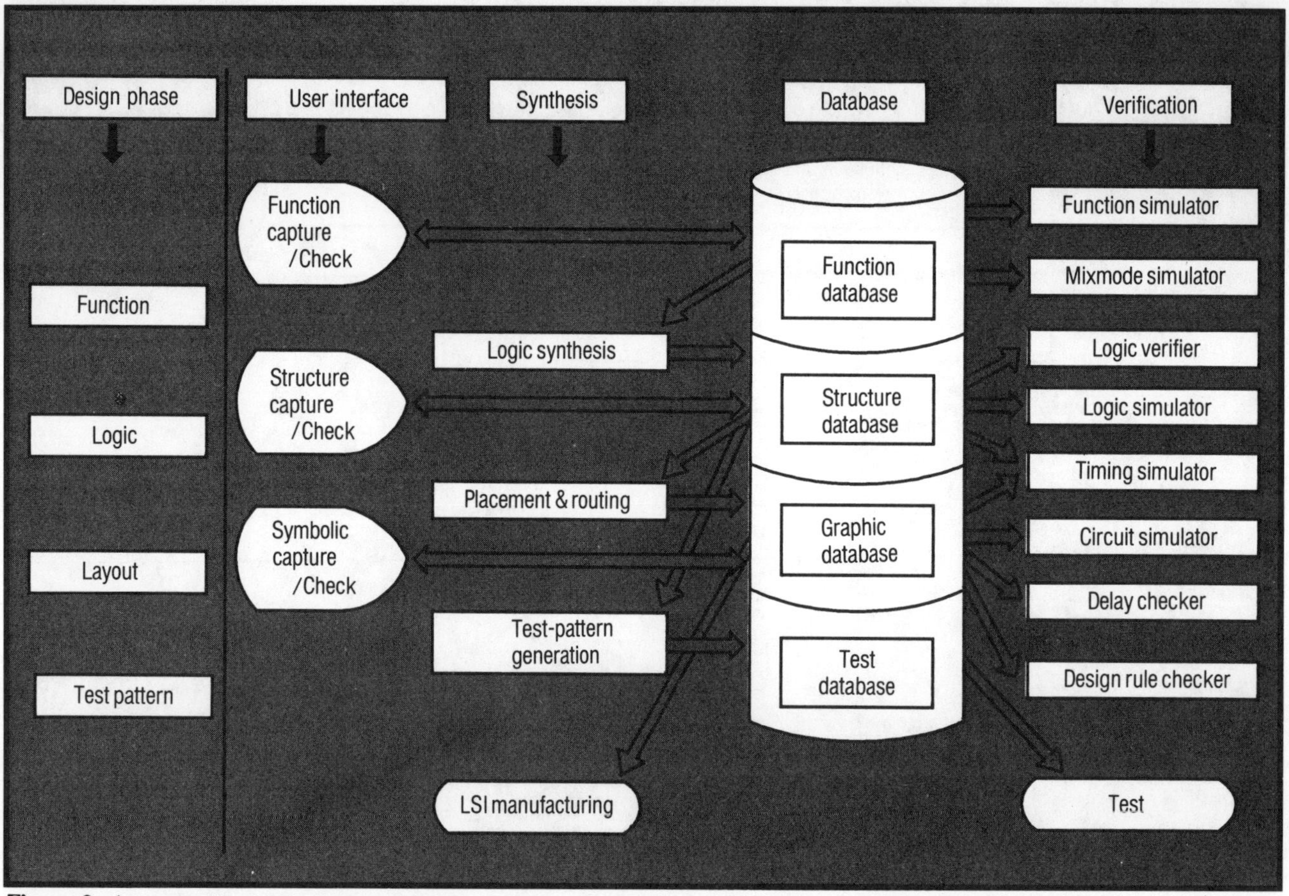

Figure 2. An overview of the integrated DA system.

scriptions, has been developed. HSL-FX is a combination of HSL and FDL, and was developed to give us better automatic logic synthesis results and to broaden the coverage of the DA system. HSL-FX features:

- Support capability from system-level to circuit-level design. As design objectives expand from the device level to the system level, a design language must support behavioral-, firmware-, and structural-level design. HSL-FX can describe VLSI circuit specifications in terms of behavioral, firmware, functional, logical and electronic circuit levels.
- Accommodation of hierarchical design methodology. To cope with large design data, design languages must possess hierarchical description capability; HSL-FX has such a capability. A VLSI system specification is usually coded as a set of modules which are units of descriptions made up of hierarchically partitioned blocks. The module specifications can be on either the behavioral or structural level, or a combination of the two. Top-down and bottom-up design methodologies can be used with the HSL-FX language.
- Support for the common input of data from all DA programs. Design integrity must be maintained until the layout design is finished. Therefore, design specifications have to be the common input of all DA programs to prevent unexpected errors through manual recoding.
- A bridge between the logical and physical structure. Major differences between the logical and physical structure of a network are found in bidirectional pins, dotting functions and power lines. These differences are overcome by the HSL-FX language's description capabilities.
- Suitable for logic synthesis. As the design of a VLSI circuit becomes more complex, an automatic logic synthesis technology can decrease design effort. However, conventional design languages are used mainly for verification purposes. HSL-FX, on the other hand, has description capabilities that are suitable for logic synthesis.

Function-level design

At the first stage in the design, the VLSI function is specified at the behavioral level by HSL-FX and is checked using a mixed-mode simulator, Elsa-FX. An HSL-FX coding example is shown in Figures 3a and 3b. HSL-FX has the capability to hierarchically describe the VLSI function, matching the top-down and bottom-up design methodologies and enabling designers to share tasks by providing a clear interface between functional blocks. HSL-FX makes accurate specification of a VLSI function possible and helps check for design errors in the early design stage. These capabilities lead to a reduction in logic design effort. Moreover, HSL-FX allows microcode to be defined mnemonically, so firmware development and debugging can be executed in parallel with hardware design, as shown in Figure 4.

Automatic logic synthesis. To overcome design limits in function and logic, an automatic logic synthesizer, Angel,[2,6] which synthesizes logic circuits from a behavioral description, has been developed. The design flow incorporating logic synthesis is shown in Figure 5. The logic synthesis process flow is shown in Figure 6.

The steps in logic synthesis are:

(1) Compile: The HSL-FX compiler compiles HSL-FX source code into an object format—a set of operation statements.

(2) Boolean equation simplification: The operation statements are then converted into a set of very simple assignment operation statements

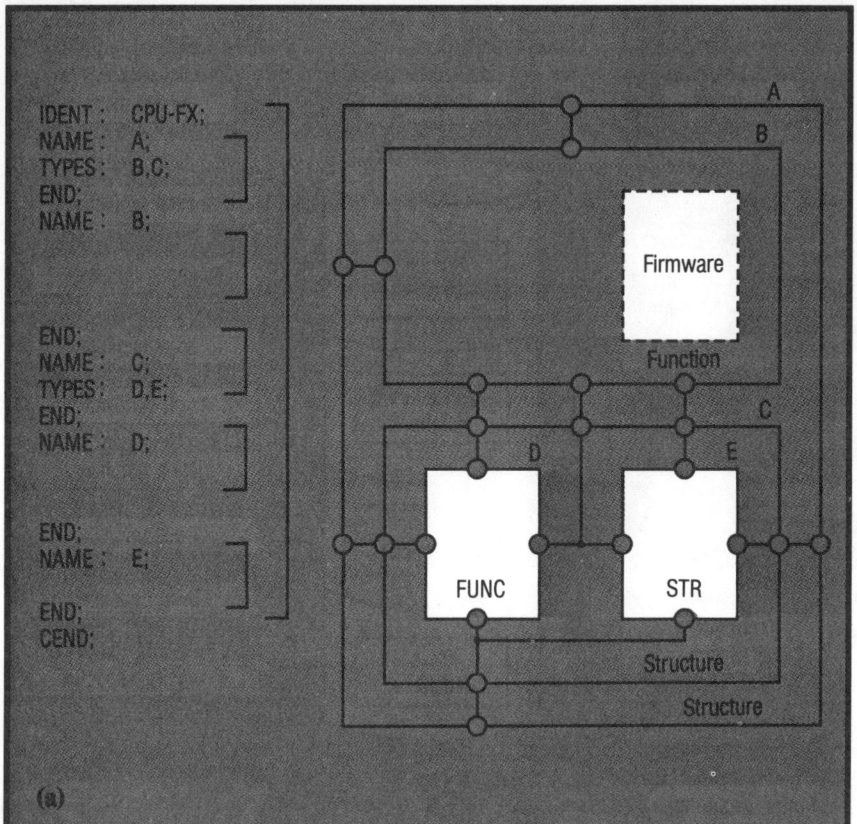


```
IDENT    : EX01;
VERSION  : V00.00;
NAME     : EX01 ;
PURPOSE  : FUNCSIM, LOGSIM ;
LEVEL    : CHIP ;
EXT      : CLK,RSET,START;
INPUTS   : .RSET,.START;
CLOCK    : .CLK;
BEHAVIOR-SECTION;
REGISTER :   PC<0:7>, A<0:7>,MEMD<0:7>,OP1<0:7>,OP2<0:7>,C;
RAM      :   MEM<0:255, 0:7>;
TERMINAL:    MEMOUT<0:7> ;
BEGIN
  IF .START THEN
       AT .CLK DO-->IFET**IFT;
       RESET(PC, OP2)
       END-DO END-IF;
  CASE OP1 OF
      #8H01 CLR      #8H82 LDA #8H83 STA #8H84 BRC
      #8H85 ADD      #8H86 JMP
  END-CASE;
END;


AUTOMATON :IFET(IFT):.RSET:.CLK;
FETCH        :BEGIN MEMOUT:=MEM<PC>;PC:=INC(PC);
              OP1 := MEMOUT;
              IF MEMOUT<0> THEN->OP2FET
                                ELSE-->EXEC**EXT;->FETCH END-IF;
             END;

OP2FET       :BEGIN OP2:=MEM<PC>;
              PC:=INC(PC);
              -->EXEC**EXT;->FETCH
              END;
END-AUTO ;


AUTOMATON:EXEC(EXT):.RSET:.CLK;
FIRST        : BEGIN
  CASEOF
              #ADD  MEMD:=MEM<OP2>
              #CLR  A:=0
              #LDA  A:=MEM<OP2>
              #STA  MEM<OP2>:=A
              #JMP  PC:=OP2
              #BRC  IF C=1B1 THEN PC:=OP2 END-IF
  END-CASEOF;
  CASEOF   #CLR, LDA, STA, JMP,BRC ->FIRST;      -->IFET**IFT
     *     #ADD                    ->SECOND;     - >>IFET**IFT
  END-CASEOF;
END;


SECOND       : BEGIN
  C!!A:=A+MEMD;
  ->FIRST; FINISH(EXT);
END;
END-AUTO;
END-SECTION;
END;
CEND;
```

(b)

Figure 3. Coding scheme of HSL-FX. (a) Hierarchical coding feature; (b) behavioral description example.

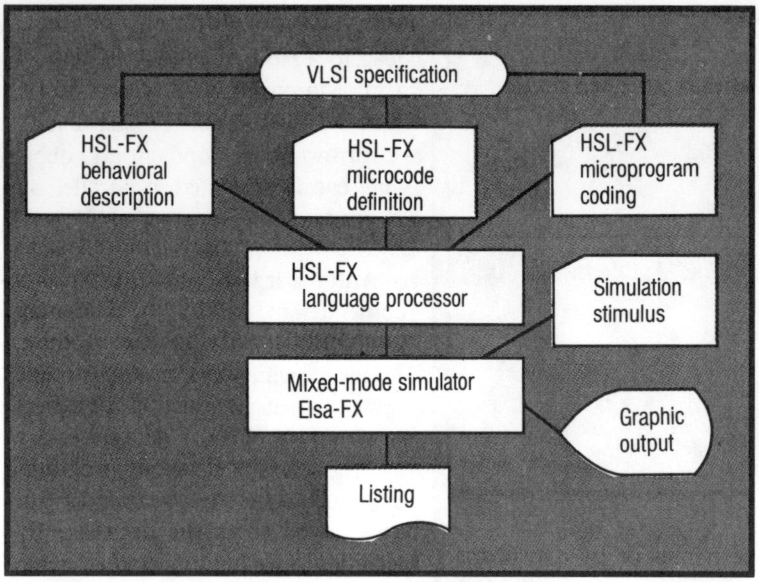

Figure 4. The function design environment.

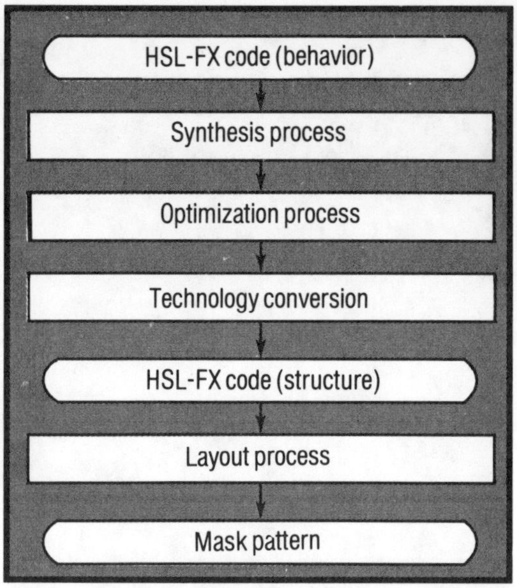

Figure 5. The design flow, incorporating logic synthesis.

which express almost the same structure as the network to be generated.

(3) Network generation: The network is generated from the simplified operation statements.

(4) Bit-wise separation: The multiple bit-size facility is transformed into a bit-wise structure.

(5) Optimization process: Several optimization routines reduce gate count. This is done by deleting a gate such as an AND gate which has a constant 0 input, and so on; deleting unreferenced gates which may have been produced in previous gate deletions; deleting single-input gates with the same input-output polarity: eliminating redundant common subexpressions and common pass switches; and minimizing and factoring a circuit in cubic notation.

Once logic synthesis is accomplished, the technology converter, LConv,[7] converts one circuit technology to another. Target technologies include TTL, ECL, IIL, and CMOS. Some examples are shown in Figure 7.

Table 1 shows a comparison between a manually designed circuit (A) and an automatically designed circuit (B). Gate counts generated for automatically designed circuits vary from 0.5 to 1.7 times that of manually designed circuits.

Structural design system. Structural design data can be handled mainly by

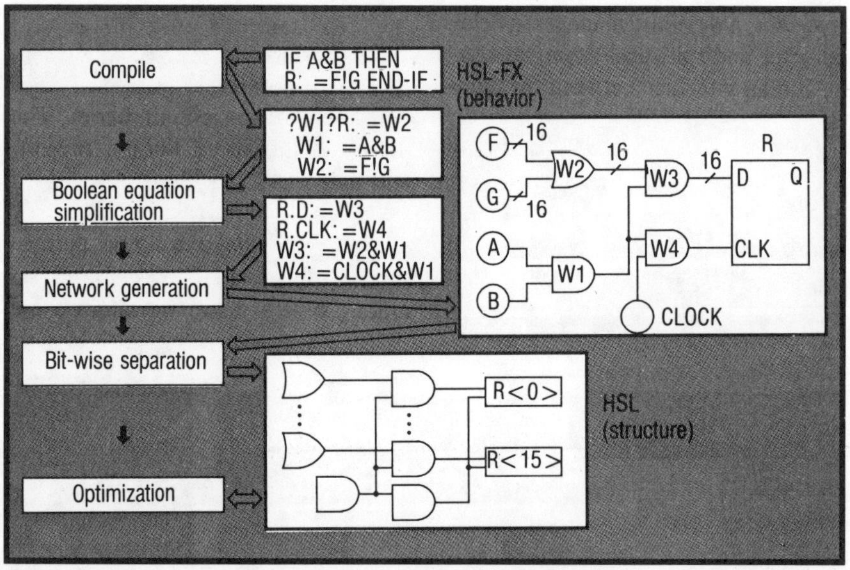

Figure 6. The flow of the logic synthesis process.

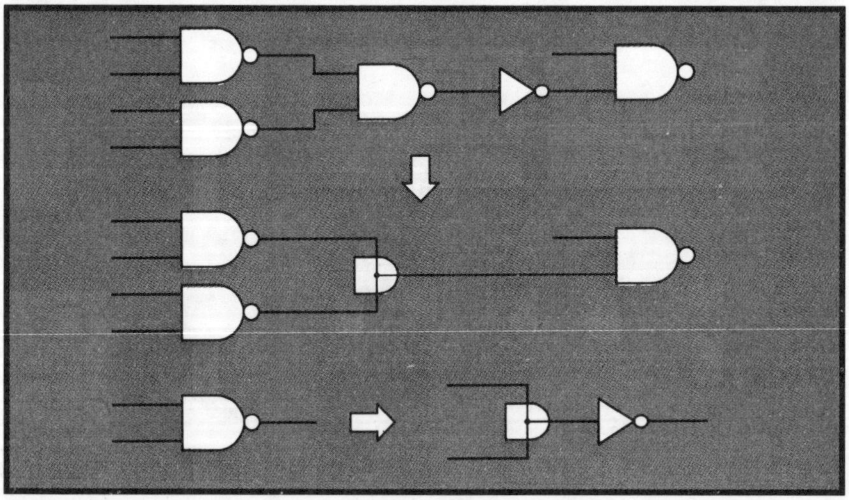

Figure 7. An example of technology conversion for dotting function.

Table 1.
Gate counts for manually designed vs. automatically designed circuits

Circuit name	Manual design* (a)	Logic-synthesizer output without optimization	Logic-synthesizer output with optimization	Technology converter output(B)*	Ratio B/A
A	78	23	8	37	0.47
B	80	147	72	101	1.26
C	341	368	204	317	0.92
D	614	752	739	925	1.50
E	1889	2623	2345	3239	1.71

*Target technology is LCML with multiemitter output max. FI = 3, max. FO = 6, max. out = 6, max. dot = 6

the Hidemap[8] system, which has an HSL compiler and decompiler, a macro expander, a block divider (Scissor), a wired function and bidirectional bus processor, a database manager, a delay extractor and statistical timing analyzer, and an interface between the database and application programs.

Compiler. The HSL compiler checks the input data described by HSL for syntax errors, and then converts it to the object file. The object file is a collection of modules directly accessible by module names or by data items. The file can be registered directly into the database, and converted back to the HSL source file or list.

Database manager. In the database, modules are managed by module name and other management items. The database consists of library, project, and user files. The library file, which contains cell libraries, is accessible by any user who has the access permission. Thus, design data can be shared among several users working on the same LSI development project. Database control commands handle the collection of modules and easily retrieve statistical information.

Macroexpander. The macro-expander program transfers higher-level modules to lower-level modules, as shown in Figure 8. Macro expansion is also controlled by the Hidemap commands specifying top module, and stop-level names or stop-module names. Using this function, designers can set up the network data necessary for the hierarchical layout program. Input data to the macro expander can be retrieved from the database by specifying only the top module name and management item priority.

Block divider. The block divider (Scissor) is also a useful tool for hierarchical VLSI circuit design. In a hierarchically oriented system, design is done in a top-down manner. Often, some blocks are too large and others too small to be processed in a single run of the layout program. Therefore, the block structure needs to be reorganized into a proper size for the routing process. The block divider

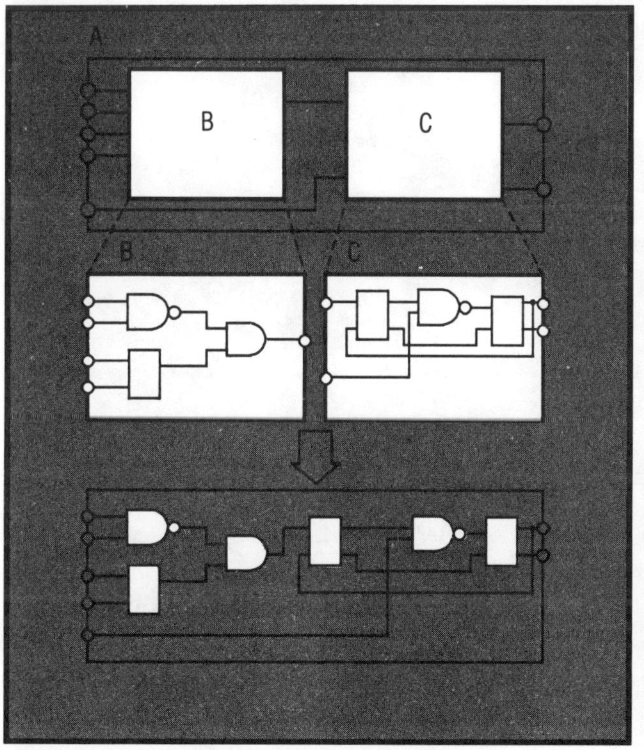

Figure 8. A schematic of macro expansion.

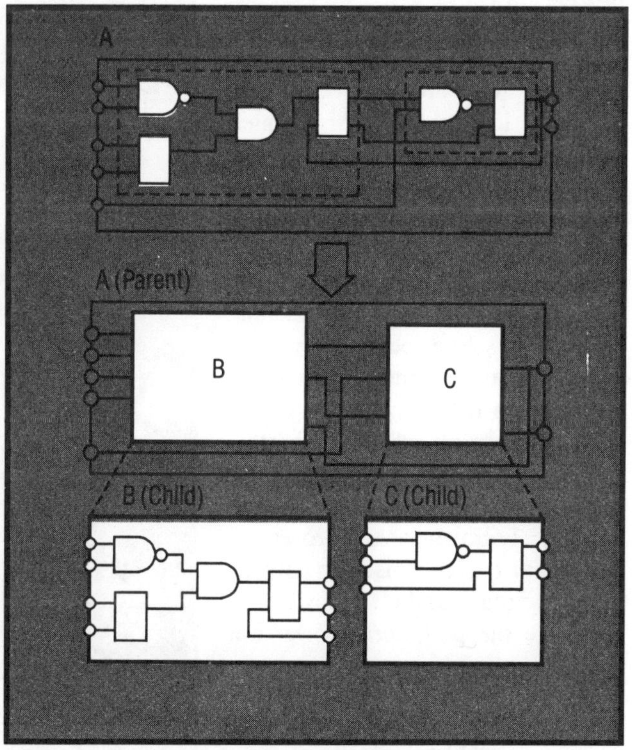

Figure 9. A schematic of block divider.

provides such a function – automatically changing structures with simple commands, as shown in Figure 9.

Layout-level design

In the present hierarchical VLSI chip structure, a chip is constructed with standard-cell and macrocell blocks such as RAMs, ROMs or PLAs, as shown in Figure 10. Each block has a rectangular shape. The number of cell rows for blocks and block locations is adjusted to form the common channel regions when the blocks are merged into a chip. The macrocell heights are measured by a pseudocell row unit. The pin locations of the macrocell are restricted to the upper and/or lower edge of each macrocell. The signal connections are made by the first and second metal layers running horizontally and vertically, respectively. The second layer can be used to pass over the standard cells when the layer does not contain intracell connections. Polysilicon is also used for vertical connections having short paths. The power bus runs perpendicular to the cell rows.

Layout system configurations. The layout system configuration is illustrated in Figure 11. The logical network data described in HSL are processed by the Hidemap system and stored in the database. Using subprograms in Alpha, the block interconnection data and the block area estimation parameters are extracted from the logic network data and the cell library (where the physical pattern data of standard cells and macrocells are stored) and entered into Champ. There, block-level placement is performed semi-automatically, and the block width and the number of cell rows are determined for each standard-cell block. Alpha then automatically performs placement and routing. Following this process, a peripheral router is used to connect internal circuits and I/O pads, and to complete the peripheral power bus connections.

Chip floor plan. In a chip floor-plan placement problem, blocks of arbi-

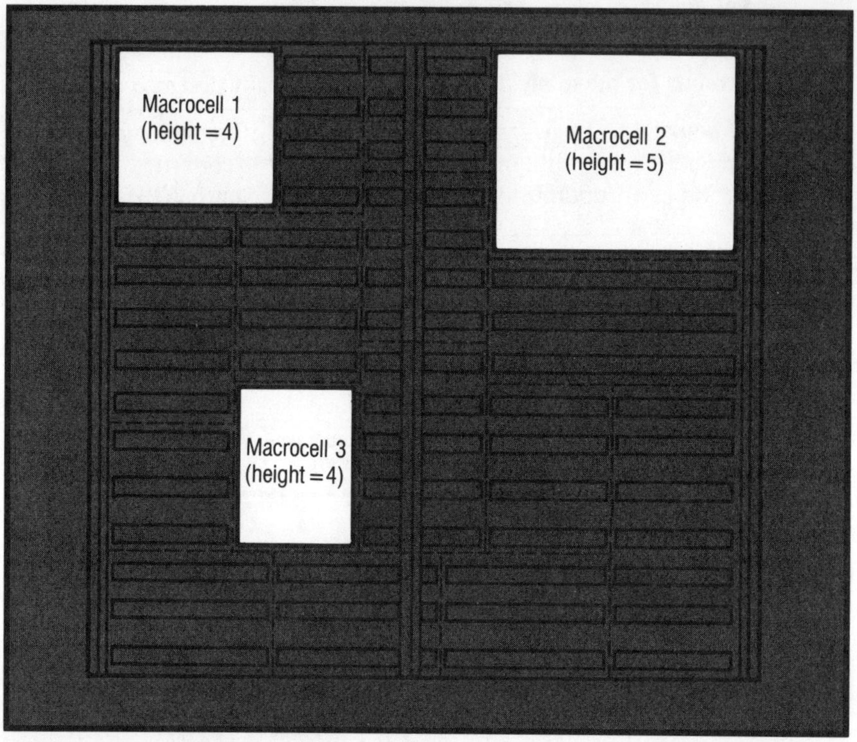

Figure 10. Hierarchical VLSI chip structure.

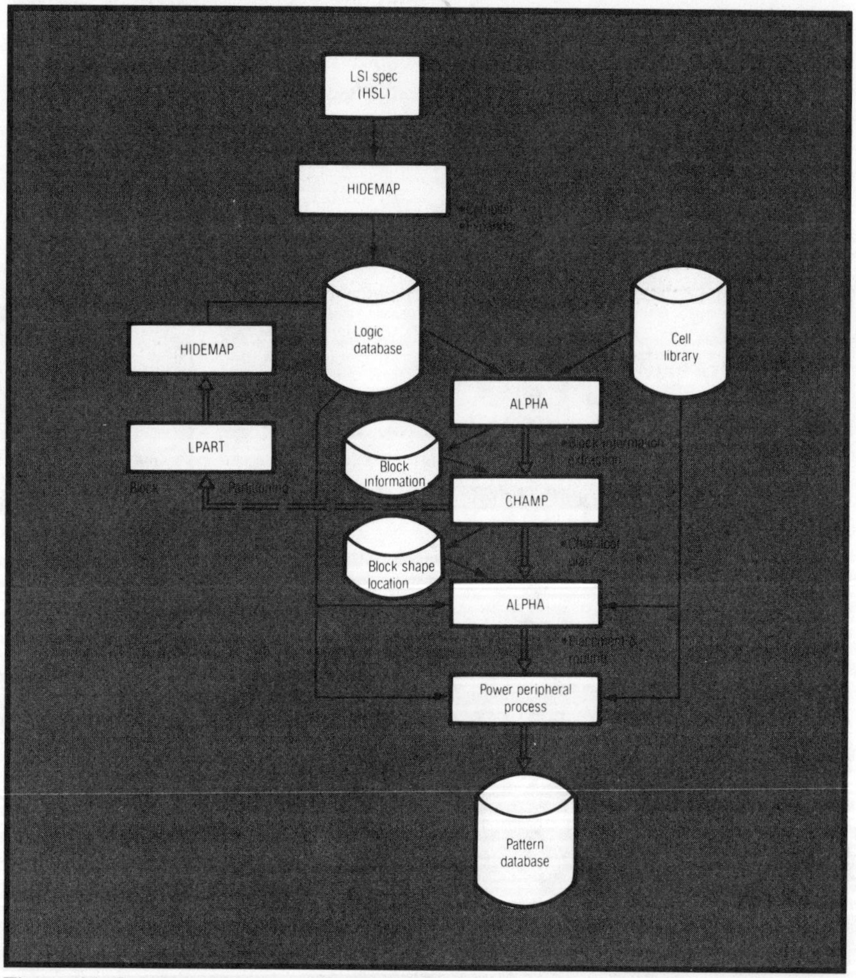

Figure 11. Layout system configuration.

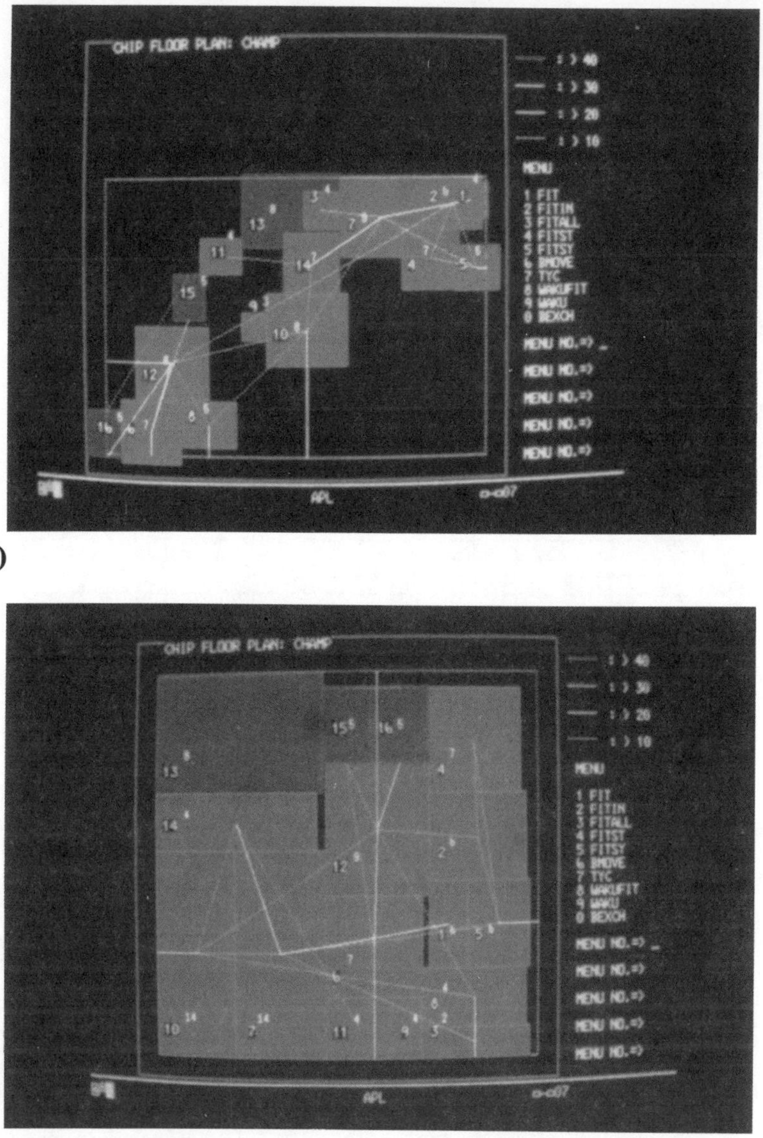

(a)

(b)

Figure 12. The block-packing process executed by CHAMP. (a) Intermediate packing scheme; (b) final floor plan.

trary sizes and shapes must be handled. This unit irregularity is one of the most important factors to be considered in the floor plan and distinguishes this problem from more standard placement problems such as printed-circuit boards or gate-array LSI circuits.

Our method solves this problem by making extensive use of variable block (standard-cell block) flexibility in the block-packing process. An initial placement of blocks is automatically or manually obtained, then the automatic, interactive block-packing process moves or reshapes the blocks as the chip boundary is shrunk.

The initial block placement can be obtained manually or automatically using an attractive- and repulsive-force method, which offers a good starting point for minimization of total-wire-length in inter-block connections. After an initial placement is obtained, each block is given a rectangular shape and an appropriate chip boundary that encloses all of the blocks is formed. In the block-packing process, overlaps between blocks or between blocks and the chip boundary are eliminated by moving and/or reshaping blocks as the chip boundary is shrunk. The moving distance and/or reshaping ratio are set to low values in order to prevent abrupt changes.

Chip area estimation is performed using empirically obtained equations that total interblock, intrablock, and dead-space areas. A number of interactive commands are provided to facilitate manual handling during the block-packing process. The block-packing process is shown in Figures 12a and 12b.

Hierarchical cell placement and routing. Following the design of the floor plan, Alpha executes cell placement and routing. In the present approach, the hierarchical layout is top-down. This globally optimizes cell placement and interconnection net path over the entire chip area. In order to attain this optimization, macroscopic (block-level) and microscopic (cell-level) global routing procedures are used. For routing, the extended hierarchical placement and non-hierarchical routing method is used – a method applicable for chips which contain macrocells and standard cells. This method also reduces the number of re-entries needed to determine a floor plan. Alpha proceeds through six major steps:

(1) Block-level global routing. In this step, the optimum terminal directions for each block and chip are determined. The process is carried out net by net for all interblock nets, and the terminal directions are sorted for every standard-cell block and chip.

(2) Interblock cell placement. The two-dimensional standard-cell placement is performed block by block for every standard-cell block. This process is done in two steps. First, the initial placement is constructed by linear placement and folding. Second, a pairwise exchange and shift insertion of cell groups to reduce the total wire length is done to improve placement.

(3) Block merge. In this step, all standard-cell blocks and macrocells are merged together into a chip and, as a result, the block boundary is removed. Another function of this step is power bus insertion.

(4) Placement improvement. Following the block merging process, final placement improvement is carried out beyond the block boundary for global optimization of placement. However, target cell groups are dissected only within the cell row, to save computation time.

(5) Cell-level global routing. After all the standard-cell and macrocell locations are fixed, the global net routes are determined. The factors considered when prescribing optimum routes are which terminals are electrically equivalent, and the feedthrough position for every net connected over the cell rows. The cell-level global routing step is divided into two phases. The first is the initial global route selection, and the second is the optimization of the global routes.

(6) Modified channel routing. The channel routing algorithms used in this step are based on the well-known left-edge trunk assignment and trunk division algorithms. New strategies have been introduced for the purposes of routing of the chip including macrocells, improvement of electrical performance, and reduction of chip area.

Application results

NTT has widely used the Integrated Design Automation System to develop of a number of VLSI circuits.

Figures 13a and 13b show a layout pattern of a 16-bit CPU, which includes an ALU and a parallel multiplier, that was designed using this system. The specification of the CPU is described with about 580 lines in the HSL-FX (behavioral) source codes. The synthesized CPU has about 11.7K gates and the HSL (structural) source code are about 16K lines.

The layout was done with the Alpha/Champ system. The area is 8.8 mm × 8.1 mm. The cell library consists of an inverter, a buffer, a 2-input NOR and a 3-input NOR NTL gate with 1μm design rule SST technology. The CPU design effort was two man-months. Most of the effort has been consumed for behavioral design coding in the HSL-FX and the function simulation.

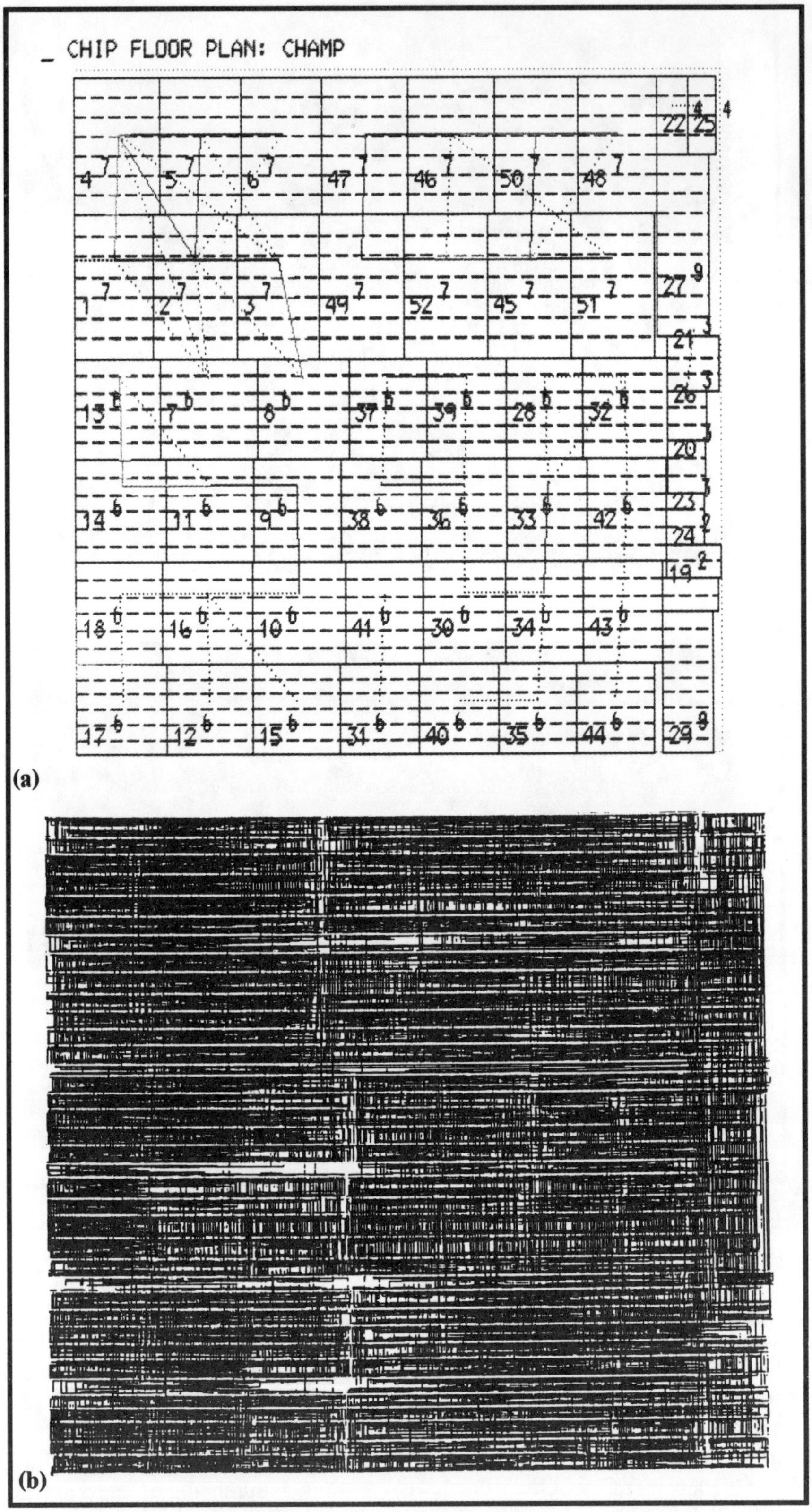

Figure 13. A 16-bit CPU layout. (a) Chip floor plan; (b) layout pattern.

The Integrated Design Automation System has been described. The automatic logic synthesis program and hierarchical layout system have been implemented and the total design effort reduction has been achieved in magnitude. The design system is configured from four tightly coupled major components, which are the database, the design processing tools, the design verification tools and the user interfaces.

Using this system, the 10K-gate CPU has been designed from the function-level description to the layout pattern in two man-months. The use of this methodology has enabled an error-free VLSI design in a much shorter period. In the future, development of new technologies, such as an expert system and DA-oriented processors, is very important to perform much denser chip layout design, to decrease design effort still more and to broaden design coverage of the DA system.

References

1. T. Sudo, T. Otsuki, and S. Goto, "CAD System for VLSI in Japan," *Proc. IEEE*, Vol. 71, No. 1, Jan. 1983, pp. 129-143
2. T. Hoshino, M. Endo, and O. Karatsu, "An Automatic Logic Synthesizer for Integrated VLSI Design System," *Proc. Custom IC Conf.*, Rochester, N.Y., May 1984, pp. 356-360.
3. K. Ueda, H. Kitazawa, and I. Harada, "CHAMP: Chip Floor Plan for Hierarchical VLSI Layout Design," *IEEE Trans. Computer Aided Design*, Vol. CAD-4, No. 1, Jan. 1985, pp. 12-22.
4. T. Adachi et al., "Hierarchical Top-down Layout Design Method for VLSI Chip," *Proc. 19th Design Automation Conf.*, Las Vegas, Nev., June 1982, pp. 785-791.
5. T. Hoshino, O. Karatsu, and T. Nakashima, "HSL-FX: A Unified Language for VLSI Design," *Proc. 7th Int'l Symp. CHDL*, Tokyo, Japan, 1985, pp. 321-336.
6. O. Karatsu et al., "An Automatic VLSI Synthesizer," *Proc. 1985 ISCAS*, Kyoto, Japan, June, 1985, pp. 403-406.
7. K. Matsuhiro, "Logic Conversion Program," *1982 All-Japan Electronic Communication Soc. Conf.* (in Japanese).
8. T. Hoshino, O. Karatsu, and T. Sudo, "VLSI Hierarchical Specification Language and Design System," *Monograph of Technical Group on Computer Architecture of Information Process Soc. Japan*, Vol. 45, No. 4, June 1982, pp. 1-6.

Author Index

A

Adachi, T., 150
Allen, J., 41

B

Beyers, J., 131
Bose, A. K., 111
Brzezinski, D., 131

C

Chawla, B. R., 111

D

Director, S., 86, 97
Dutton, R., 50

E

Endo, M., 150

F

Fichtner, W., 117

G

Geiger, D. J., 117
Goto, S., 135
Gray, J. P., 25
Gummel, H. K., 111

H

Hoshino, T., 150

I

Infante, B., 131

K

Karatsu, O., 150
Kitazawa, H., 150
Kowalski, T. J., 117

L

Lang, C. R., 25

M

Maly, W., 97
McCalla, W., 131

N

Newton, A. R., 3, 62, 76

O

Ohtsuki, T., 135

P

Parker, A. C., 86
Pedersen, D. O., 62, 76
Penfield, P., Jr., 41

R

Rhyne, T., 127
Rowson, J. A., 25
Rutenbar, R. A., 97

S

Sangiovanni-Vincentelli, A.L., 3, 62, 76
Sequin, C. H., 62
Shen, J. P., 97
Siewiorek, D. P., 86, 97
Strojwas, A. J., 97
Sudo, S., 135

T

Thomas, D. E., Jr., 86, 97
Trimberger, S., 25

U

Ueda, K., 150

W

Wolf, W. H., 117

Editor's Biography

Alberto L. Sangiovanni-Vincentelli (M'74–SM'81–F'83) received the Dr.Eng. degree (summa cum laude) from the Politecnico di Milano, Italy, in 1971.

From 1971 to 1977, he was with the Istituto di Elettrotecnica ed Elettronica, Politecnico di Milano, Italy, where he held the position of Research Associate, Assistant, and Associate Professor. In 1976 he joined the Department of Electrical Engineering and Computer Sciences of the University of California at Berkeley, where he is presently Professor and Vice Chairman. He is a Consultant in the area of computer-aided design to several industries. His research interests are in various aspects of computer-aided design of integrated circuits, with particular emphasis on VLSI simulation and optimization. He was Associate Editor of the IEEE Transactions on Circuits and Systems, and is Associate Editor of the IEEE Transactions on Computer-Aided Design of Integrated Circuits and a member of the Large-Scale Systems Committee of the IEEE Circuits and Systems Society and of the Computer-Aided Network Design (CANDE) Committee. He was the Guest Editor of a special issue of the IEEE Transactions on Circuits and Systems on CAD for VLSI. He has been elected Executive Vice-President of the IEEE Circuits and Systems Society in 1983. In 1981 he received the Distinguished Teaching Award of the University of California. At the 1982 IEEE-ACM Design Automation Conference he was given a Best Paper and a Best Presentation Award. In 1983 he received the Guillemin–Cauer Award for the best paper published in the Transactions on CAS and CAD during 1981–1982. At the 1983 Design Automation Conference he received a Best Paper Award.

Dr. Sangiovanni-Vincentelli is a member of ACM and Eta Kappa Nu.